PYROLYSIS

Edited by **Mohamed Samer**

Pyrolysis
http://dx.doi.org/10.5772/65193
Edited by Mohamed Samer

Contributors

Raymond Taziwa, Sharifah Shahnaz Syed Bakar, Sibel Başakçılardan Kabakcı, Şeyma Hacıbektaşoğlu, Oleksandr Ivanovich Malik, Francisco Javier De La Hidalga-Wade, Roger Ruan, Yaning Zhang, Shiyu Liu, Liangliang Fan, Nan Zhou, Min Min, Yanling Cheng, Erik Anderson, Yunpu Wang, Yiqin Wan, Yuhuan Liu, Paul Chen, Bingxi Li, Peng Peng, Miguel Dominguez, José Alberto Luna López, Francisco Flores, Jun Mu, Zongyuan Lai, Qiumin Zhang, Fan Nie, Tao Meng, Anatoliy Levin, Alexander Nikolayevich Kozlov, Vitaly Shamansky, Alexandre Keiko, Denis Svishchev, Andrzej Białowiec, Paweł Sępień, Jakub Pulka, Andrew Odeh, Elina Bastos Caramao, Gabriela P. da Silva Maciel, Juliana Macedo Da Silva, Laiza Canielas Krause, Rosângela Assis Jacques, Claudia Alcaraz Zini, Zaira Chowdhury, Kaushik Pal, Miriam Arabiourrutia Gallastegui, Gorka Elordi, Martin Olazar, Javier Bilbao

Published by InTech
Janeza Trdine 9, 51000 Rijeka, Croatia

Publishing Process Manager Marijana Francetic
Technical Editor SPi Global
Cover InTech Design team

First published June, 2017
Printed in Croatia
Legal deposit, Croatia: National and University Library in Zagreb

Additional hard copies can be obtained from orders@intechopen.com

Pyrolysis, Edited by Mohamed Samer
p. cm.
Print ISBN 978-953-51-3311-7
Online ISBN 978-953-51-3312-4

For EU product safety concerns:
IN TECH d.o.o., Prolaz Marije Krucifikse Kozulić 3, 51000 Rijeka, Croatia,
info@intechopen.com or visit our website at intechopen.com.

Meet the editor

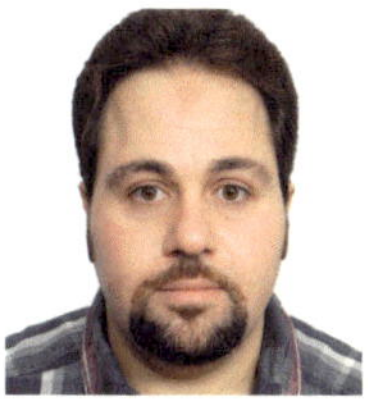

Dr. Mohamed Samer holds the position of associate professor at the Cairo University. He held the position of research scientist at Leibniz Institute for Agricultural Engineering Potsdam-Bornim (ATB), Germany. He was conferred with a PhD degree by the University of Hohenheim, Germany. He speaks English, German, French, and Arabic. He has more than 90 publications. He attended more than 25 conferences. He led four research projects as the principal investigator. He participated in several other projects. He is a member of 11 scientific societies. He serves as a reviewer for more than 20 periodicals. He reviewed more than 100 articles. He taught 15 undergraduate and graduate courses. He supervised several PhD, MSc, and BSc students. His main research and teaching interests are engineering in biosystems, environment, and energy.

Contents

Preface

This book provides useful information about pyrolysis, which includes the pyrolysis of biomass and pyrolysis of fossil fuels and petrochemicals. Additionally, this book elucidates and illustrates further innovative pyrolysis processes such as catalytic pyrolysis, spray pyrolysis, and microwave-assisted pyrolysis. This book discusses the production of semiconductors and nanomaterials through the pyrolysis process.

Each chapter of the book provides precious and up-to-date knowledge from basics to apex, allowing readers to understand more deeply. This book will be very helpful for academics, scientists, scholars, researchers, and undergraduate and graduate students worldwide who are specialized in energy engineering, environmental engineering, and bioresource engineering. Additionally, it will be very helpful for NGOs, universities, and research institutes and centers.

Dr. sc. agr. Mohamed Samer
Associate Professor
Department of Agricultural Engineering
Faculty of Agriculture, Cairo University
Giza, Egypt

Section 1

Pyrolysis of Biomass

Chapter 1

Pyrolysis: A Sustainable Way to Generate Energy from Waste

Chowdhury Zaira Zaman, Kaushik Pal, Wageeh A. Yehye, Suresh Sagadevan, Syed Tawab Shah, Ganiyu Abimbola Adebisi, Emy Marliana, Rahman Faijur Rafique and Rafie Bin Johan

Additional information is available at the end of the chapter

http://dx.doi.org/10.5772/intechopen.69036

Abstract

Lignocellulosic biomass is a potentially more valuable renewable resource that can be utilized effusively as a chief source of heat for cooking and can correspondingly subsidize the production of electricity, heat, biofuels and chemicals including solid fuel like char or carbon. Lignocellulosic residues are mixed and burnt with coal to generate electricity. Presently, crude oil is replaced by bioethanol and biodiesel produced from biomass substrate. Some special class of chemicals can be derived from biomass that can subsequently replace the usage of non-renewable resources of oil and coal. Pyrolysis of woody biomass to obtain pyroliginous acid was started hundreds of years ago, which has versatile applications. The range of products that can be derived from biomass is huge, prompting extent of research using different types of thermal conversion technologies, including pyrolysis, gasification, torrefaction, anaerobic digestion and hydrothermal processing. This chapter provides insights about the stages of reaction during pyrolysis and the outcome of reaction conditions on the products. Technical development and adjustment of process condition can offer a suitable environmentally benign scheme to increase the energy density of the lignocellulosic residues.

Keywords: biomass, cellulose, hemicellulose, lignin, pyrolysis

1. Introduction

Lignocellulosic biomass is considered as a promising environmentally friendly substitute resource for carbon-based fuels and chemicals. Existing global supply of energy depends on non-renewable fuels such as oil, gas and coal formed naturally beneath the earth crust. However, the amount of fossil fuel is limited now. Due to the growing population of world, the consumption of energy per capita is increasing. Thus the inevitability for continuing alternative to generate the possible sources of energy is evident. Utilization of biomass to produce value-added products is receiving great attention by researchers. Furthermore, the inorganic constituent of biomass is negligible and it contains minor quantity of nitrogen, sulphur and ash. Therefore, combustion of biopmass is advantageous as it produces less toxic gas such as nitrogen oxides (NOx), sulphur dioxide (SO_2) and smoke compared to other conventional fuels. Even the emission of carbon dioxide (CO_2) can be controlled by recycling it by photosynthesis [1]. Though many theoretical methods were undertaken for the conversion in the short run; what is required are practical phase application and demonstration with appropriate calculation of material and energy balance. Industrial-scale thermochemical production of liquids, bio-oils, by fast or flash pyrolysis has been established but it has so far not been implemented for commercialization of the overall practice.

Different types of thermochemical even biological processes have been adopted to convert biomass into value-added products. Among those processes, pyrolysis is more convenient since it has several advantages of storing, transportation and flexibility in solicitation such as turbines, combustion appliances, boilers, engines, etc. In some cases, solid biomass and waste are precisely challenging to process for pyrolysis research. It is until now at a preliminary stage in terms of expansion and yet requires resolving numerous practical obstacles to contend with conventional fossil fuel–centred procedures [2, 3]. The preparation of liquid biofuels including other products like solid char and gas by pyrolysis of various lignocellulosic residues has been comprehensively explored earlier. Some of these biomass species are beechwood [4], bagasse [5] woody biomass [6, 7], straws [8], seedcakes [9] and municipal solid waste (MSW) [10, 11]. **Figure 1** illustrates different types of existing biomass conversion process with their respective output.

Pyrolysis is defined as the thermal decomposition of lignocellulosic derivatives under inert condition in oxygen-deficient environment. The word is resulting from two Greek words: 'pyro', which means fire, and 'lysis', which means disintegration into integral parts. Pyrolysis technology is very old and earlier it was first used for preparation of charcoal in Middle East and Southern Europe before 5500 years ago [12]. Egyptian people used this technique to produce tar for sealing boats [13]. Subsequently then, practice of pyrolysis processes have been growing and are extensively carried out for charcoal and coke fabrication. Burning of charcoal can produce intensively high temperature to melt tin with copper to obtain bronze. Consequently, pyrolysis has been getting further consideration as an effective technique for transforming biomass into bio-oil throughout the modern eras [14]. The eventual objective of pyrolysis is to yield high-value energy products for contending with and gradually supplanting non-renewable fossil fuels. Nevertheless, the expansion of progressive know-hows is the ensuing challenge for the investigators to accomplish the objectives. It is required to transform biomass into bio fuels for uninterrupted usage in vehicles, trains, ships and aero-planes to substitute diesel and petrol [15, 16]. Additional improvement of pyrolysis technology is enduring to produce solid fuel like char or carbonaceous materials, syngas, etc. Typically a pyrolysis system unit contains the equipment for lignocellulosic residues pre-processing, the pyrolysis reactor, and subsequent unit for downstream processing. Mainly it can be classified as units that produces only heat and biochar (using slow pyrolysis) or units that produce biochar and bio-oils (using fast pyrolysis). **Figure 2** shows simple layout of pyrolysis units with its main products.

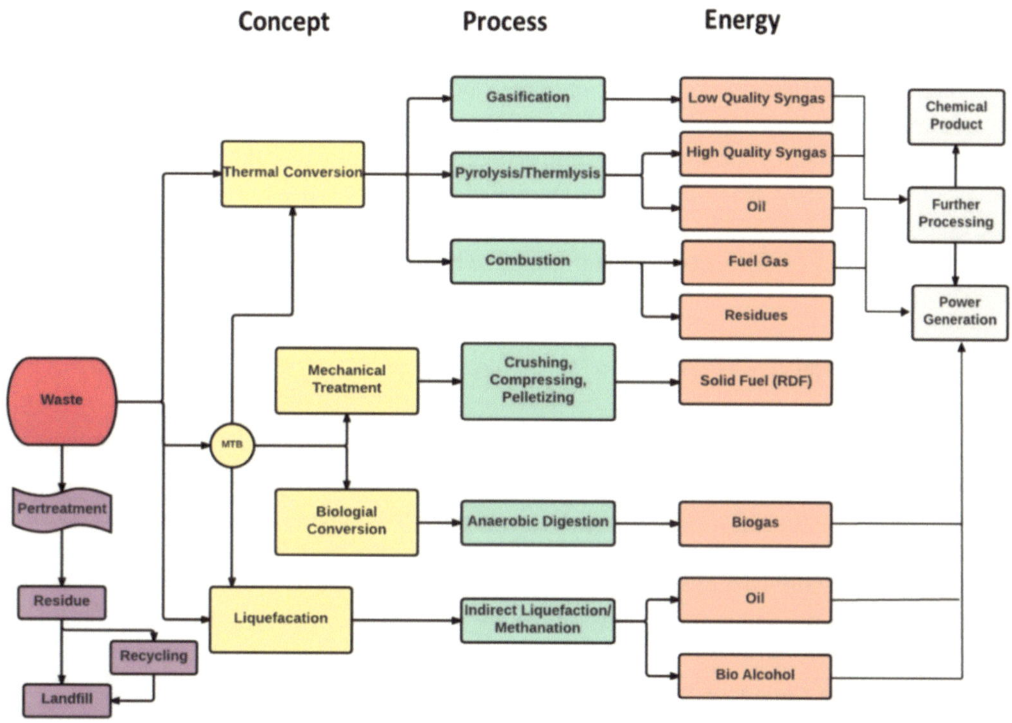

Figure 1. Biomass conversion process to obtain value-added products.

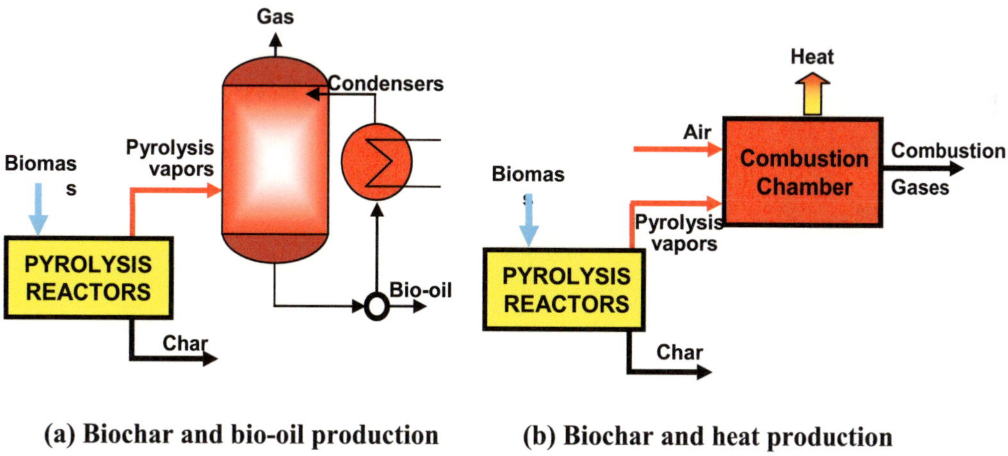

Figure 2. Simplified flow diagram for typical pyrolysis unit. (a) Biochar and bio-oil production. (b) Biochar and heat production.

Extensive amount of research has been conducted in recent years for thermochemical conversion of biomass into biofuels (bio-oil, biochar and biogas) using pyrolysis technology. Compared to other thermochemical conversion technologies, pyrolysis process has lot of advantages based on process parameter optimization. However, this technology still needs to be updated with respect to its commercial applications. In this chapter, emphasis has been given to discuss the current status of pyrolysis technology and its prospective for commercial applications for biofuel, syngas and biochar production. Aspects of pyrolysis technology such as types of pyrolysis, pyrolysis principles, biomass compositions and characteristics, pyrolysis reactor design, pyrolysis products and their physiochemical properties and economics of biofuel production are presented. We have pointed out some of the inherent properties of bio-oil that cause complications for the end use of the products. Finally, we take a brief look at some processes including catalytic pyrolysis process that aim to valorize bio-oil by conversion to higher value liquid fuel products.

2. Basic principles of pyrolysis

The thermal decomposition process of pyrolysis using lignocellulosic biomass takes place in the absence of oxygen under inert atmosphere. As an inert atmosphere argon or nitrogen gas flow is usually needed. The fundamental chemical reaction is very complex and consists of several steps. The end products of biomass pyrolysis consist of biochar, bio-oil and gases. Pyrolysis process emits mainly methane, hydrogen, carbon monoxide and carbon dioxide. The organic materials present in the biomass substrate starts to decompose around 350–550°C and it can proceed until 700–800°C without the presence of air/oxygen [17, 18]. Biomass is mainly composed of long polymeric chain of cellulose, lignin, hemicellulose, pectin and others. The larger molecules of organic materials start to decompose to yield smaller molecules, which are released from the process stream as gases, condensable vapours (tars and oils) and solid char during pyrolysis process. The proportion of each end product depends on the temperature, time, heating rate, and pressure, types of precursors and reactor design and configuration. **Figure 3** illustrates the decomposition process of main lignocellulosic residues at different temperature. The moisture content of biomass also plays a vital role in pyrolysis processes. The moisture content of the feedstock should be around 10% during fast pyrolysis process [18].

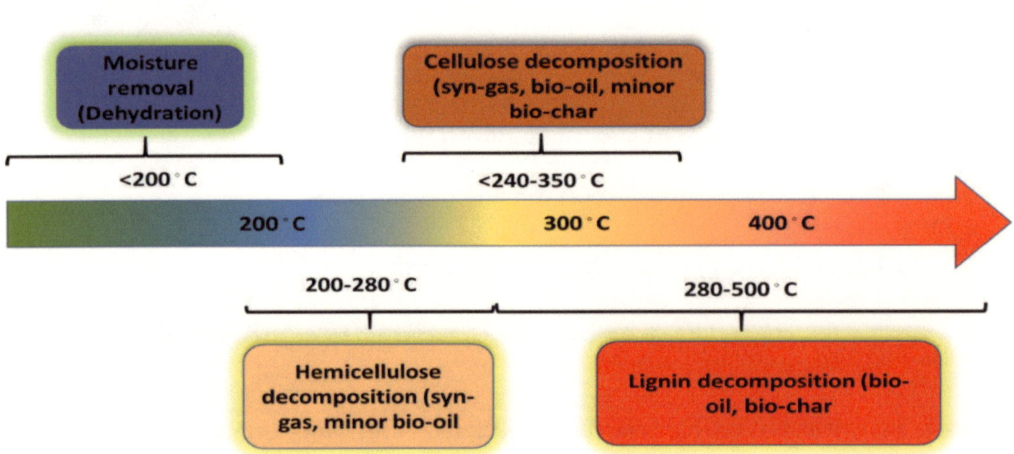

Figure 3. Decomposition behaviour of biomass constituents at different temperature [19].

Due to high moisture content, major products become liquids and if there is low level of water, there is high risk that the process produces huge amount of dust instead of oil. Thus sludge derived from waste stream and meat-processing wastes require drying before exposing them finally to pyrolysis environment. Less than 450°C when the heating rate is slow, the main yield is biochar. However at higher temperature that is more than 800°C when the heating rate is high then larger fraction of ash and gaseous products are produced. Bio-oil can be produced applying intermediate temperature using relatively high heating rates. During the beginning of the process around temperature 250–300°C, volatile materials are released at almost 10 times quicker than the subsequent step [20].

Woody biomass was initially used to produce charcoal. The charcoal based on wood during heating produces negligible amount of smoke. Earlier it was extensively used for melting of ore to extract iron. However, the process had drawbacks of less yield percentages, less energy and excessive air pollution. After that, modern technology was developed to extract maximum possible energy from biomass using combustion (exothermic), gasification (exothermic) and pyrolysis (endothermic) [21]. Combustion deals with the burning of biomass in presence of oxygen to produce heat. The competence of this practice is not satisfactory [22, 23]. Gasification also takes place under oxygenated atmosphere which will yield gaseous fuels. Nevertheless, pyrolysis is the leading phase for both gasification and combustion processes [24, 25]. Consequently pyrolysis can be considered as part of gasification and combustion [26]. The decomposition products yield of biomass during pyrolysis is provided by following **Figure 4** [27].

Table 1 summarizes the list of main pyrolysis reactions at different temperature.

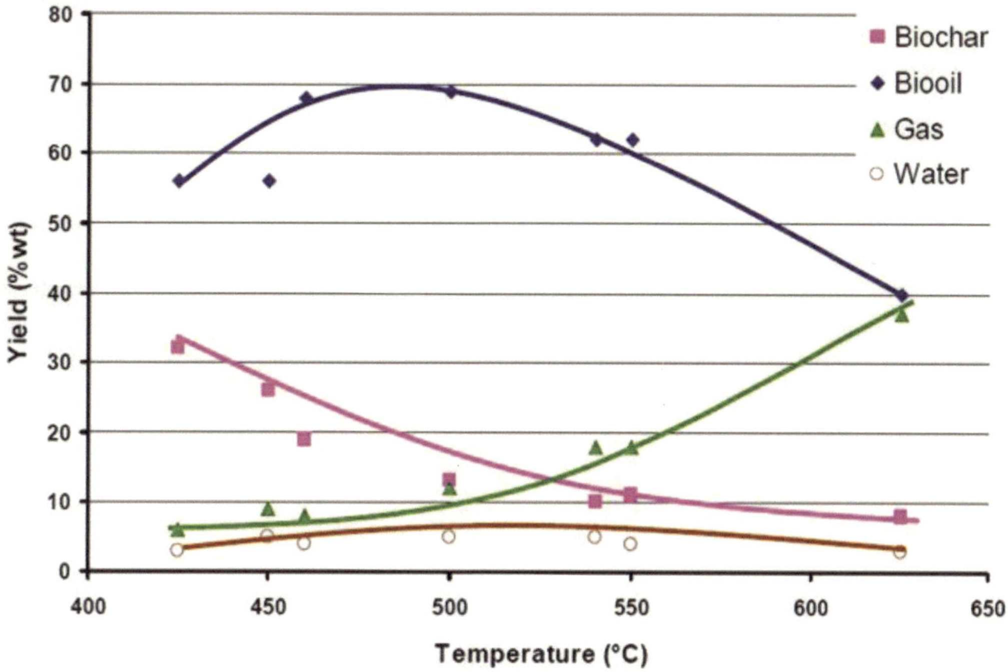

Figure 4. The decomposition products of pyrolysis of biomass [27].

Temperature	Type of reaction	End products
Less than 350°C	Moisture loss, depolymerization, free radical generation	carbonyl and carboxyl group production, CO and CO_2 gas liberation, biochar formation
Between 350°C and 450°C	Substitution for breaking of glycoside chain of polysaccharide	Tar production containing levoglucosan, anhydrides and oligosaccharides
Above 450°C	Dehydration, rearrangement and fission of sugar units	acetaldehyde, glyoxalin and acrolein production
Above 500°C	A mixture of all above processes	A mixture of all above products
Condensation	Unsaturated products condense and cleave to the char	A highly reactive char residue containing trapped free radicals

Table 1. Pyrolysis reactions at different temperature [28].

3. Lignocellulosic biomass feedstock

3.1. Type and composition of biomass feedstock

The structure of biomass is complex and usually composed of three main natural biomacromolecules: Cellulose, hemicellulose and lignin. Besides that it also has extractives and some minerals. The proportion and these basic constituents vary from biomass to biomass [11, 29, 30]. During pyrolysis, cellulose and hemicellulose yield condensable vapours or liquids and gas. Lignin decomposes to give liquid, gas and solid char. Extractives also produce liquid and gas due to simple volatilization or decomposition. The ash fraction inside the char matrix contains minerals. This distribution of components into products is shown schematically in **Figure 5**.

The vapours produced from initial decomposition of biomass undergoes for secondary reactions to yield soot which also varies due to slow and fast pyrolysis process. Alkali metals act like catalyst by enhancing the char yield. The presence of minerals affects the ignition properties of biochar matrix [11]. It was observed that bio-oil mainly derived from cellulosic substrate around 500°C [31] whereas biochar may be extracted from lignin. Thus the biomass substrate which contains greater proportion of lignin derivatives can yield more bio-oils yield. **Table 2** shows a list of selected biomass containing different proportion of cellulose, hemicellulose and lignin substrate [11, 32–37].

3.2. Physiochemical properties of biomass

Based on process parameters and design of the reactor, presence of moisture can have significant effect on products yield [11]. The charcoal-making process proceeds through two distinct steps: drying and pyrolysis steps. During the initial phase of drying, combined water in the pores represented as free water is expelled around 110°C. As much as water is present, it takes more energy to evaporate. After that between temperatures 150 and 200°C combined water present inside the cellulosic chain of wood will be reduced. In the early stage of carbonization, water evaporates as white smoke from charcoal kiln. Fast pyrolysis process

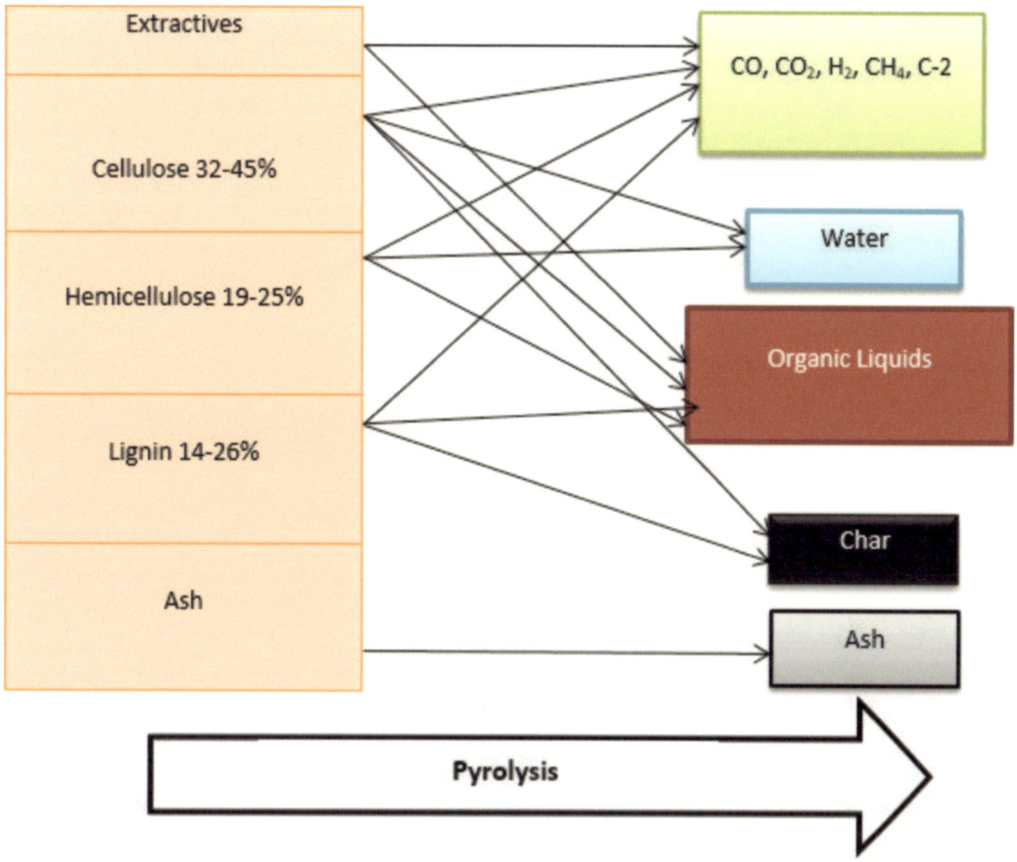

Figure 5. Product distribution during pyrolysis [29].

Feedstock	Lignin (%)	Cellulose (%)	Hemicellulose (%)
Wood	25–30	35–50	20–30
Wheat straw	15–20	33–40	20–25
Switch grass	5–20	30–50	10–40
Sugarcane bagasse	23–32	19–24	32–48
Miscanthus	17	24	44
Corn stover	16–21	28	35
Hazelnut shell	42.9	28.8	30.4
Olive husk	48.4	24	23.6
Corncob	15	50.5	31
Tea waste	40	30.20	19.9
Walnut shell	52.3	25.6	22.7
Almond shell	20.4	50.7	28.9

Feedstock	Lignin (%)	Cellulose (%)	Hemicellulose (%)
Sunflower shell	17	48.4	34.6
Nut Shell	30–40	25–30	25–30
Paper	0–15	85–99	0
Rice straw	18	32.1	24
Stored refuse	20	60	20
Leaves	0	15–20	80–85
Cotton seed hairs	0	80–95	5–20
Barley straw	14–15	31–34	24–29
Oat straw	16–19	31–37	24–29
Bamboo	21–31	26–43	15–26
Rye straw	16–19	33–35	27–30
Coastal Bermuda grass	6.4	25	35.7
Jute fibre	21–26	45–53	18–21
Banana waste	14	13.2	14.8

Table 2. Chemical constituent of selected biomass.

is effective to dry the feed, so that the rate of temperature rise is not restricted by evaporation of water [38]. Typically 15–20% moisture is present in wood [11]. During producing the activated carbon also, moisture can significantly affect the properties of final carbon sample [39]. The particle size of the biomass matrix will have greater influence between char and liquid yield. Larger proportion of char is formed when particle sizes are big. Larger particles restrict the rate of disintegration, resulting in the increased scope of secondary char forming reaction [11]. Thus larger particle size is good to get more carbon yield whereas smaller particles are required to maximize liquid fractions during fast pyrolysis process. Higher proportion of lignin and fixed carbon also can contribute in better yield of biochar substrate if pyrolysis is carried out at medium temperature of 500°C whereas higher percentages of volatile materials can generate higher yield of bio-oil and syngas (**Table 3**) [28]. Therefore, the precursors like hazel nut shell, olive stone, walnut shell is better to produce good quality biochar due to their lignin content (**Table 2**). The biomass like cereal straw, grasses, energy crops like woody biomass that die to their low mineral and nitrogen content are suitable for bio-oil and syngas production (**Table 4**) [40].

Based on composition, physiochemical properties as well as transformation mechanism, lignocellulosic residues can yield different value-added products as illustrated by **Figure 6**.

3.3. Controlling temperature profile and heating values of biomass

To optimize the product yield, controlling the temperature profile is the most important factor as it can partially influence the pressure, heating rate, peak temperature and contact time between

Feedstock	Density (Kg/m^3)	Moisture content (%)	Ash content (%)	Volatile matter (%)	Fixed carbon (%)
Wood	1186	20	0.4–1	82	17
Bituminous coal		11	8–11	35	45
Hybrid polar	150	45	0.5–2	–	–
Switchgrass	108	13–15	4.5–5.8	–	–
Miscanthus	70–100	11.5	1.5–4.5	66.8	15.9
Sugarcane baggage	1198		3.2–5.5	–	–
Barley straw	210	30	6	46	18
Wheat straw	1233	16	4	59	21
Danish pine		8	1.6	71.6	19
Rice straw	200		6 4.3	79	10.7
Fire wood	–	7.74	1.98	80.86	17.16
Grateloupia filicina	–	4.93	22.37	55.93	17.01
Birch	125	18.9	0.004	–	20
Pine	124	17	0.03	–	16
Polar	120	16.8	0.007	–	–

Table 3. Physical properties of selected biomass feedstock [41–44].

Feedstock	Carbon (%)	Hydrogen (%)	Oxygen (%)	Nitrogen (%)	Ash (%)
Wood	51.6	6.3	41.5	0.1	1
Cypress	55	6.5	38.1	–	0.4
Olive baggage	66.9	9.2	21.9	–	2
Wheat straw	48.5	5.5	3.9	0.3	4
Barley straw	45.7	6.1	38.3	0.4	6
Scots	56.4	6.3	–	0.1	0.09
Birch	44	6.9	49	0.1	0.004
Pine	45.7	7	47	0.1	0.03
Polar	48.1	5.30	46.10	0.14	0.007
Willow	47.78	5.90	46.10	0.31	1.30
Switchgrass	44.77	5.79	49.13	0.31	4.30
Reed canary grass	45.36	5.81	48.49	0.34	5.10
Dactylis lomarata	42.96	5.70	49.44	1.90	7.50
Festuca arundinacea	42.22	5.64	50.65	1.50	7.30
Lolium perenne	43.12	5.80	49.80	1.28	6.20

Table 4. Chemical characteristics of some selected biomass materials [35, 41, 44].

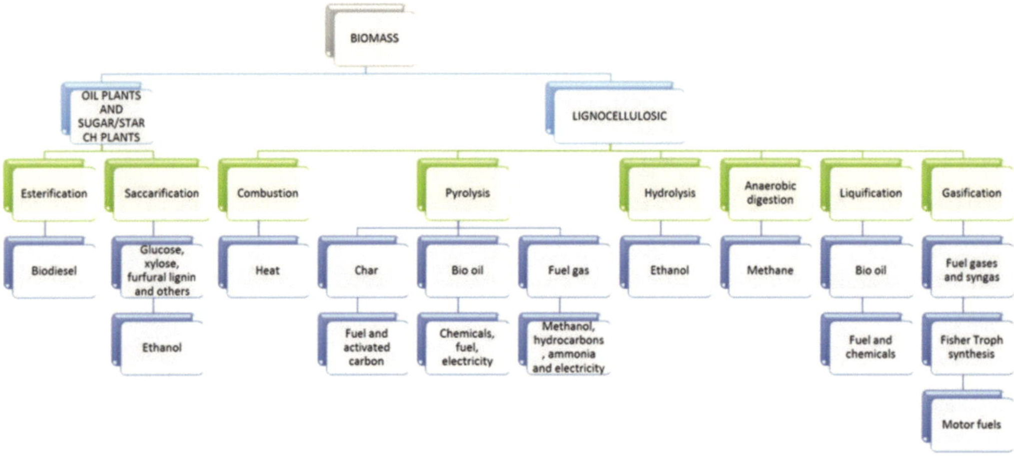

Figure 6. Biomass transformation to value-added products.

solid and gaseous phases. Rapid heating and cooling rate is required for minimizing the extent of secondary reactions during fast pyrolysis. This will reduce liquid yield but product quality will be less. Even it will give a more complex mixture having higher viscosity [38]. On the other hand, the slow pyrolysis process uses slow heating rates which lead to higher char yields, but this is not consistent [11]. Higher temperature can ensure release of more volatile fractions to increase the carbon content of the char. However, longer residence time at higher temperature will significantly drop the product yield. The effect of temperature on liquid and gaseous fraction is far more complex. When the pyrolysis temperature reaches up to 400–550°C, liquid yields are higher. Above this temperature, secondary reactions take place by decomposing the condensable vapour, which finally gives lesser liquid fractions. For fast pyrolysis, maximum liquid is obtained around 500°C [18]. It was reported also that the liquid yields was 28–41% at temperatures between 377 and 577°C, depending on feedstock during the slow pyrolysis process [13]. Around 42–45% liquid fractions were obtained around temperature of 385–400°C using different straw feeds [45].

3.4. Effect of gas flow rates

Gas flow rate during the pyrolysis process affects the degree of secondary char formation. Lower flow rate is favourable for char formation during slow pyrolysis process whereas higher gas flows are provided during fast pyrolysis process to effectively strip off the vapours as soon as they are formed. Higher pressure intensifies the activity of vapours within the reactor and at the surfaces of char particles to increase the secondary char formation. Conversely, pyrolysis under vacuum gives little char and gives more liquid fractions. For pyrolysis under pressure, moisture in the vapour phase can systematically upsurge the production of carbon. Because, in that case water is acting as catalyst by reducing the activation energy for pyrolysis reactions [46].The gas flow rate significantly influences the thermodynamics of the process. At higher pressure, the reaction is more exothermic using lower gas flow rates. Higher char yields can be ensured when pyrolysis process is exothermic and such conditions will favour the overall energy balance of the processes pursuing the carbon or char as main product. Thus it can be concluded that, any factor of pyrolysis conditions that increases the contact between primary vapours and hot char surface, including high pressure, lower flow rate of gas, larger particles size or slow heating is expected to favour the char formation with lower liquid yield.

4. Existing pyrolysis process

4.1. Fast pyrolysis

4.1.1. Garret pyrolysis

Pyrolysis, especially pyrolysis of coal, is an age-long activity but biomass pyrolysis is a completely new entrant. The process is aimed to produce biofuel. In the garret process, solid waste (Biomass) is allowed to mix with hot char and hot recycle gas in a specially designed chamber. This is then followed by pyrolysis at high temperature, usually above 800°C, and at a holding time of about 10 s. After pyrolysis, the char is the removed while the liquid portion is collected. The resulting formed tar is then separated and further processed to produce the process heat as well as the char feed for further pyrolysis. The process is generally expected to lead to production of at least 40% liquid yield but has been found to produce more of gas at the carbonization temperature and time, thereby making the process uneconomical [47]. The whole process can be summarized into three main steps: The formation of turbulent gaseous stream by intermixing the carrier gas, the solid biomass and the hot char using a designed mixing zone, passing the gaseous steam into the pyrolysis chamber and allowing to go through pyrolysis at temperature of about 800°C for about 10 s, and finally removing the pyrolyzed gaseous stream from the pyrolysis chamber [47].

4.1.2. Georgia Tech entrained bed process

In this process, the main feed (biomass) is crushed and sieved into about 1 mm particle size. The precursor is then dried to about 10% moisture content and fed into the reactor where it is pyrolyzed using a preheated inert gas. At pyrolysis temperature of about 500°C, a maximum yield of about 50% liquid and 30% gas is expected. The holding carbonization time is usually calculated based on the reactor height and the gas flow rates but usually made of several seconds [48]. One major problem of this technique is the low heat produced by the entraining gas which usually leads to low liquid yield since biomass requires high heat for a high liquid yield. Another problem is that fresh solid waste exerts a catalytic effect on bio-oil cracking leading to production of more char and gas [49].

4.2. Fluidized bed pyrolysis processes

The fluidized bed pyrolysis process possesses highly excellent mass transfer characteristics as it offers an effective and highly positive means of heating of finely chopped biomass in a rapid manner to achieve the pyrolysis temperature to the desired level. It is a well-established pyrolysis technique that can be used on a large-scale pyrolysis process as it is capable of processing hundred tons of biomass per day. The advantages of this technique over conventional pyrolysis technique include improved performance of the system as well as lower viscosity coupled with higher energy content of the produced bio-oil.

4.2.1. *Waterloo flash pyrolysis process (WFPP)*

The Waterloo flash pyrolysis process involves the production of organic liquids from biomass materials using continuous atmospheric pressure in the absence of oxygen. Generally, it is a carefully controlled process that produces a high liquid yield. The process has been widely demonstrated using hardwood solid waste to produce organic liquid yield as high as 70% of the feed material [50]. It is a process in which the pyrolysis reactor works on a very unique principle in which the char is not allowed to accumulate in the bed while the treatment of the sand may not be necessary. One big advantage of the process is that the liquid product obtained is usually acid and pours easily with relative stability [51].

4.2.2. *RTI process*

In order to satisfy certain criteria which have been left unsatisfied by other fast pyrolysis techniques, the RTI process was developed. This involves the use of deep fluid bed using very low temperatures with moderate heating rates and relatively long holding pyrolysis time. Satisfactory results have been obtained from pyrolysis of most biomass using this technology which effectively leads to high liquid yields at temperature range of between 400 and 450°C with volatile holding time of about 0.8 s [52]. Fine sand is usually employed as the bed heat carrier. This allows a very low gas flow rate and this coupled with indirect heat applied results in a highly effective thermal efficiency. This is a very big advantage from economic point of view when capital and operating costs are considered [52].

4.2.3. *Dynamotive process*

This technology, incorporated in 1991 aimed at producing value-added products from bio-oil especially to produce biolime. The pyrolyzer operating heat usually comes from the gas or char, that is, the by-products of the pyrolysis, while the fluidizing gas comes off from the pyrolysis gas [53]. The liquid product obtained is then utilized in the production of materials such as biolime, slaked lime and in the control of SO_x and NO_x during combustion of coal [53].

4.2.4. *Ensyn process*

This process involves the utilization of wood and other lignocellulosic materials for the production of fermentable sugar. The process involves the following notable steps: subjecting the biomass material to dilute acid treatment (usually dilute sulphuric acid is used to dissolve the hemicellulose while cellulose content is unaffected), separation of the solid residue containing the cellulose, pyrolyzing the separated solid residue at a controlled temperature (400–600°C), atmospheric pressure with short vapour holding time in a fluidized bed reactor, formation of aqueous phase through controlled adjustment of the content of the crude product and finally the separation of the aqueous phase follows [53].

5. Types of pyrolysis

Overall the pyrolysis process can be classified as slow and fast depending on the heating rate. In slow pyrolysis process, the time of heating the biomass substrate to pyrolysis temperature is longer than the time of retention of the substrate at characteristic pyrolysis reaction temperature. However in fast pyrolysis, the initial heating time of the precursors is smaller than the final retention time at pyrolysis peak temperature. Based on medium, pyrolysis can be of another two types namely hydrous pyrolysis and hydro-pyrolysis. Slow and fast pyrolysis is usually carried out in inert atmosphere whereas hydrous pyrolysis is carried out in presence of water and hydro-pyrolysis is carried out in presence of hydrogen. The residence time of vapour in the pyrolysis medium is longer for slow pyrolysis process. This process is mainly used to produce char production. It can be further classified as Carbonization and Conventional. On the contrary, the vapour residence time is only for seconds or milliseconds. This type of pyrolysis, used primarily for the production of bio-oil and gas, is of two main types: (1) flash and (2) ultra-rapid. **Table 5** summarizes some basic characteristics of different types of pyrolysis process.

Pyrolysis types	Retention time	Rate of heating	Final temperature (°C)	Products
Fast	<2 s	Very high	500	Bio-oil
Flash	<1 s	High	<650	Bio-oil, chemicals and gas
Ultra-rapid	<0.5 s	Very high	1000	Chemical and gas
Vacuum	2–30 s	Medium	400	Bio-oil
Hydro-pyrolysis	<10 s	High	<500	Bio-oil
Carbonization	days	Very low	400	Charcoal
Conventional	5–30 min	Low	600	Char, bio-oil and gas

Table 5. Different types of pyrolysis process.

5.1. Fast pyrolysis

During the fast pyrolysis process, biomass residues are heated in absence of oxygen at high temperature using higher heating rate. Based on the initial weight of the biomass, fast pyrolysis can provide 60–75% of liquid biofuels with 15–25% of biochar residues [54]. It can also yield 10–20% of gaseous phase depending on the biomass used [54]. The process is characterized by small vapour retention time. However, quick chilling of vapours and aerosol can ensure higher bio-oil yield [54]. It can provide liquid biofuel for turbine, boiler, engine, power supplies for industrial applications. Fast pyrolysis technology is getting implausible acceptance for producing liquid fuels due to certain technical advantages [55–57]:

1. It can ensure preliminary disintegration of the simple oligomer and lignin portions from lignocellulosic biomass with successive upgrading.
2. The scaling up of this process is economically feasible.

3. It can utilize second generation bio-oil feed stocks such as forest residues, municipal and industrial wastes.
4. It provides easy storability and transportability of liquid fuels.
5. It can ensure secondary transformation of motor fuels, additives or special chemicals.

5.2. Flash pyrolysis

The flash pyrolysis process of biomass can give solid, liquid and gaseous products. The bio-oil production can go up to 75% using flash pyrolysis [58]. This procedure is carried out by speedy devolatilization under inert atmosphere using higher heating rate with high pyrolysis temperatures around 450 and 1000°C. In this process, the gas residence time (less than 1 s) is too little [59]. Nevertheless, this process has poor thermal stability. Due to catalytic effect of the char, the oil becomes viscous and sometimes it contains some solid residues also [60].

5.3. Slow pyrolysis

Slow pyrolysis can yield good quality charcoal using low temperature and low heating rates. The vapour residence time can be around 5–30 min in this process. The volatile organic fractions present in vapour phase continue to react with each other to yield char and some liquid fractions [61]. The quality of bio-oil produce in this process is very low. Longer residence time initiates further cracking to reduce the yield of bio-oil. The process suffers from low heat transfer values with longer retention time leading to enhance the expenditure by higher input of energy [62, 63]. The stoichiometric equation for production of charcoal is shown by [11].

$$C_6H_{10}O_5 \rightarrow 3.74C + 2.65H_2O + 1.17CO_2 + 1.08\,CH_4 \tag{1}$$

Table 6 below gives the theoretical equilibrium yield of cellulose at different temperatures using slow pyrolysis [11].

% of products	Temperature (°C)				
	200	300	400	500	600
C	32	28	27	27	25.2
H_2O	36.5	32.5	27	27	22.5
CH_4	8.5	10	10	10	9
CO_2	23.9	28	35	35	36
CO	0	0	1.2	1.2	4.5

Table 6. Equilibrium concentration of gaseous products at different temperature.

6. Catalytic pyrolysis

A mixture of hydrocarbon was produced earlier from methanol over zeolites like ZSM-5 [64]. Another patent suggested passing the vapours from pyrolyzer over the bed of zeolite ZSM-5 to produce short chain hydrocarbon [65]. It was reported that the catalyst of ZSM-5 can convert bio-oils generated from the pyrolyzer to alkylated benzene [66]. The disadvantage of using ZSM-5 as catalyst was coke formation [66]. The researchers concluded that low H/C ratio of bio-oil caused rapid catalyst deactivation resulting in considerable amount of coke as waste materials [66]. However these disadvantages can be overcome by using circulating fluid bed technology where the fluidized bed can be prepared using different types of catalyst instead of sand [53]. A small-scale pilot plant level (0.1 to 0.35 kg/h) has also been developed by RTI international [53]. This plant can successfully carry out the pyrolysis of locally available pine biomass to produce bio-oil. This plant is also trying to carry out the catalytic pyrolysis of wooden chips where by 1 ton of biomass residues can yield 60 gallons of pyrolytic oil per day [53]. Recently other attempts have been made to produce aromatic compound specially benzene, xylene and toluene from biomass substrate [53]. Recently *KiOR Inc.* in Texas, USA, [53], has stated progress in case of scaling up of this kind of technology [53].

Recently attempt has been taken to develop catalyst from renewable sources. The ash generated from gasified biomass contains 70–87% of silica in amorphous form which has been used by researchers to produce ZSM-5 and ZSM-48 catalyst for bio-oil upgrading [67]. The biochar derived from duckweed showed excellent catalytic activity for reforming of CH_4-CO_2 around 800°C [68]. Catalysts can be mixed with the lignocellulosic substrate earlier to pyrolysis process or separately with the gaseous reactants to obtain desired products. It was revealed that parting of the catalyst and biomass was more operative for the transformation of the required products [69]. This research used chromite ($FeCr_2O_4$) as catalyst and it demonstrated favourable outcomes in terms of restricted water production. **Table 7** provides summary of zeolite-based catalyst used for upgrading the lignocellulosic residues until now.

Catalyst	Temp. (°C)	Feedstock	Catalyst effects	Refs.
HZSM-5 with varying Si/Al_2O_3 ratios	500–764	Kraft lignin	Decreasing the SiO_2/Al_2O_3 ratio from 200/1 to 25/1 and increasing the catalyst-to-lignin ratio from 1:1 to 20:1 decreased the oxygenates and increased the aromatics. Aromatics yield increased from 500 to 650°C and then decreased at higher temperatures. Under optimal reaction conditions, the aromatic yields were 2.0% (EHI 0.08) and 5.2% (EHI 0.35).	[70]
HZSM-5, Na/ZSM5, HBeta, and H-USY	650	Alkaline lignin	H-USY had the largest pore size and lowest Si/Al ratio (7) and had the best liquid yield of 75% and aromatic yield of 40%.	[71]

Catalyst	Temp. (°C)	Feedstock	Catalyst effects	Refs.
ZSM-5, Al/MCM-41, Al-MSU-F, ZnO, ZrO_2, CeO_2, $Cu_2Cr_2O_5$, Criterion-534, alumina-stabilized ceria MI-575, slate, char and ashes derived from char and biomass	500	Cassava rhizome	ZSM-5, Al/MCM-41, Al-MSU-F type, Criterion-534, alumina-stabilized ceria-MI-575, $Cu_2Cr_2O_5$, and biomass-derived ash were selective to the reduction of most oxygenated lignin derivatives. ZSM-5, Criterion-534, and Al-MSU-F catalysts enhanced the formation of aromatic hydrocarbons and phenols. No single catalyst was found to reduce all carbonyl products but ZSM-5, Criterion-534 and MI-575 could reduce most of the carbonyl products that contained hydroxyl groups. ZSM-5, Criterion-534, Al/MCM-41, Al-MSU-F, copper chromite, char and ashes increased acetic, formic, and lactic acid. MI-575 did not increase acids.	[72]
Dolomite	500–800	Waste olive husks	Dolomite increased cracking and gas production.	[73]
HZSM-5, Al/MCM-41, Al-MSU-F, and alumina-stabilized ceria MI-575, pore sizes 5.5, 31, 15, and NA, respectively	500	Cassava rhizome	HZSM-5 was the most effective catalyst for the production of aromatic hydrocarbons, phenols, and acetic acid and the reduction of oxygenated lignin-derived compounds and carbonyls containing side chain hydroxyl groups. Only MI-575 showed a decrease in acetic acid yields. MI-575 also showed the most increase in methanol with HZSM-5 a close second.	[74]

Table 7. Summary of zeolite-based catalysts used for biomass upgrading.

7. Catalytic hydro-pyrolysis

Catalytic hydro-pyrolysis is a kind of catalytic pyrolysis where the pyrolysis is carried out using fluidized bed reactor under the flow of hydrogen. In this process the fluidized bed is replaced by a transition metal catalyst. It was reported that the replacement inert sand with Ni-based catalyst under atmospheric pressure can convert the bio-oil into low molecular weight hydrocarbons within short contact time [75]. *Recently Gas Technology Institute*, Illinois, USA, reported a new process where the overall process is carried out under 7–34 bar pressure [76]. Due to high pressure C1–C3 gases are evolved which after reforming produce large amount of hydrogen. However, the system is also very complex as it is combination of hydro-pyrolysis and reforming. There are some technical challenges like feeding biomass solids into the pressurized pyrolyzer under hydrogen must be overcome. Overall the establishment of this process is costly also.

8. Types of reactor

The importance of appropriate reactor in any process involving pyrolysis cannot be underestimated. Reactors have been designed in such a way as to satisfy specific conditions giving considerations to parameters such as heating temperature, vapour product residence time and required pressure, for a high bio-oil yield. In considerations of the above, many types of reactors have been developed by researchers for specific assignments. These reactors include the following:

8.1. Fixed bed reactor

This is a very simple technology that gives priority to the production of bio-oils, which are relatively uniform in size with low fines content [42]. It is made up of two basic components, that is, the gas cooling compartment and the cleaning system by filtering through cyclone, wet scrubbers and dry filters. During reaction, the solid sample is allowed to pass through a vertical shaft where it encounters an upwardly moving counter current gas stream product. This reactor can be made using either steel, firebricks or concrete and composed of the feeding unit (fuel), a unit for removing the ash and the gas escape unit [77]. The reactor, which has its priority for applications involving small-scale heat and power, has high ability to conserve carbon and can operate for long time for solid residence, low gas velocity and of course with a low ash carry-over. It has its limitation in the problem usually encountered during tar removal [78].

8.2. Fluidized bed reactor

This reactor consists of a mixture of two phases, the solid and the liquid and usually accomplished by passing a pressurized fluid through the solid material. It is very popular for fast pyrolysis as it has the following advantages [79]:

1. The provision of heat transfer is rapid.
2. It has a good grip of pyrolysis reaction and vapour holding time control.
3. It has a sufficiently high surface area for contact between the two phases in the mixture.
4. The heat transfer in the system is exemplary, and
5. The relative velocity between the phases is very high.

There are different types of fluidized bed reactors, which include bubbling fluidized, circulating fluidized, ablative reactor, vortex reactor, rotating-disk reactor, vacuum pyrolysis reactor, rotating cone reactor, PyRos reactor, auger reactor, plasma reactor, microwave reactor and solar reactor with each designed with different operating systems and for specific applications.

8.2.1. Bubbling fluidized bed reactor

The construction and operation of this reactor is very simple and is illustrated by **Figure 7** [80]. The high presence of solid density in the bed ensures a better temperature control, smooth contact between gas and solid, good transfer of heat and excellent storage capacity. The biomass is heated in an environment devoid of oxygen and decomposed into gas, vapour, aerosols and char and these components are finally collected from the reactor. While the charcoal is collected using the cyclone separator and stored, the vapour is cooled rapidly and condensed into high-quality bio-oil and stored with about 70% yield of biomass weight (dry weight) [80].

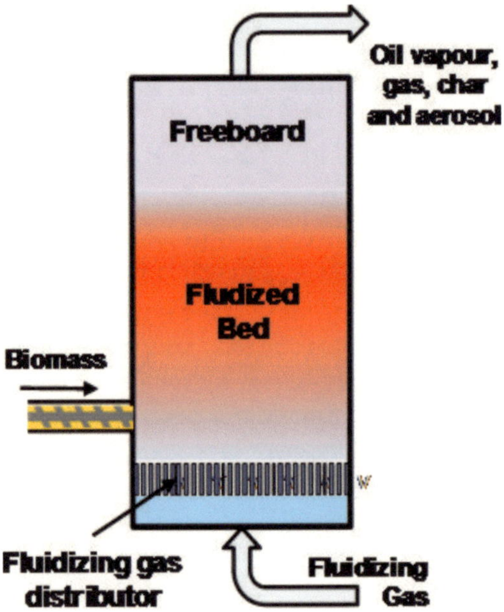

Figure 7. Bubbling fluidized bed reactor [80].

8.2.2. Circulating fluidized bed reactors

The features of this reactor is similar to that of a bubbling fluidized bed reactor described above except the fact that the residence time for the vapours and char is shorter. This makes the gas velocity and the content of char in the bio-oil to be higher. However, it has a large throughputs advantage. Single and double types of this reactor are available [81]. The basic layout of this reactor is shown by **Figure 8** [80].

8.3. Ablative reactor

In this reactor, heating is done through a molten layer on the surface of the hot reactor and in the absence of fluidizing gas. Biomass melting is done by pressing mechanically the biomass against the wall of a heated reactor and as the melted sample is moved, the pyrolysis vapours evaporate as oil. While this reactor allows for a large biomass particle size (up to 20 mm), the materials does not require excessive grinding [82]. However, the configuration of the reactor is a bit complex owing to the process nature which is mechanical. The reactor does not benefit from the same scale of economy as other reactors due to the fact that scaling functions linearly as heat transfer since it is surface area-controlled. Two types of this reactor, that is, ablative vortex and ablative rotating disk are used commonly [82].

8.4. Vacuum pyrolysis reactor

This is a slow pyrolysis reactor with heat transfer rate very low. This results in a lower bio-oil yield usually in the range of 35–50 wt% [83]. The design is highly complicated and requirement

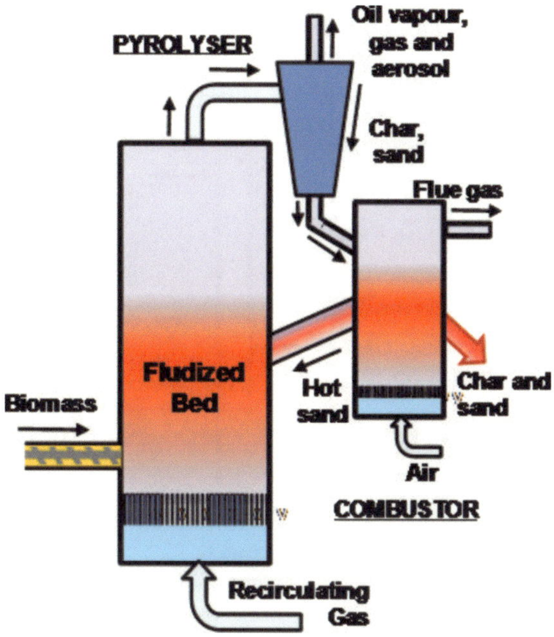

Figure 8. Recirculating fluidized bed reactor [80].

for investment and maintenance is always high thereby making the technology uneconomically suitable. The biomass is conveyed into the vacuum chamber with a high temperature with the aid of a conveyor metal belt with periodical stirring of the biomass by mechanical agitation [83]. The heat carrier is usually made of a burner while the biomass is melted by heating inductively using molten salts. It has the ability to process larger particle size biomass but requires special solids feeds special discharging devices in order to have an effective seal all the time [83]. The basic lay out of this type of reactor is shown by **Figure 9** [83].

8.5. Rotating cone reactor

Unlike the fluidized bed reactor, the rotating cone reactor requires the mixing of biomass and hot sand is done mechanically and does not require the use of inert gas. The operating feature is shown by **Figure 10** [74]. The feed and the hot sand are fed in from the bottom of the cone while they are transported to the lip of the cone during spinning using centrifugal force and as they get to the tip, the vapour generated is condensed by the condenser [74]. The char and the sand are combusted with the sand being heated up again and reintroduced to remix with fresh feedstock at the bottom of the cone. Though the design of this reactor might be complex, its high bio-oil yield makes it desirable.

8.6. PyRos reactor

The aim of this reactor is to produce a bio-oil that will not contain any particle. It uses a reactor that is cyclone in nature integrated with hot gas filter. Both the biomass feedstock and the inert

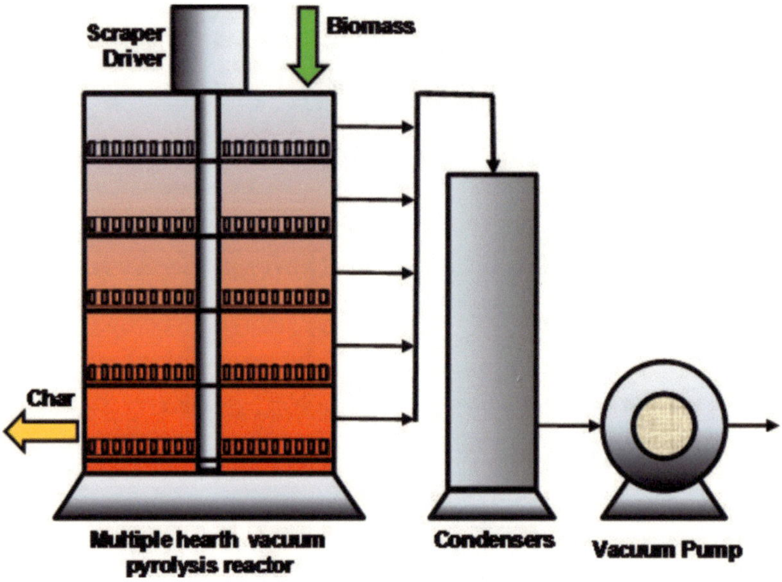

Figure 9. Vacuum reactor [83].

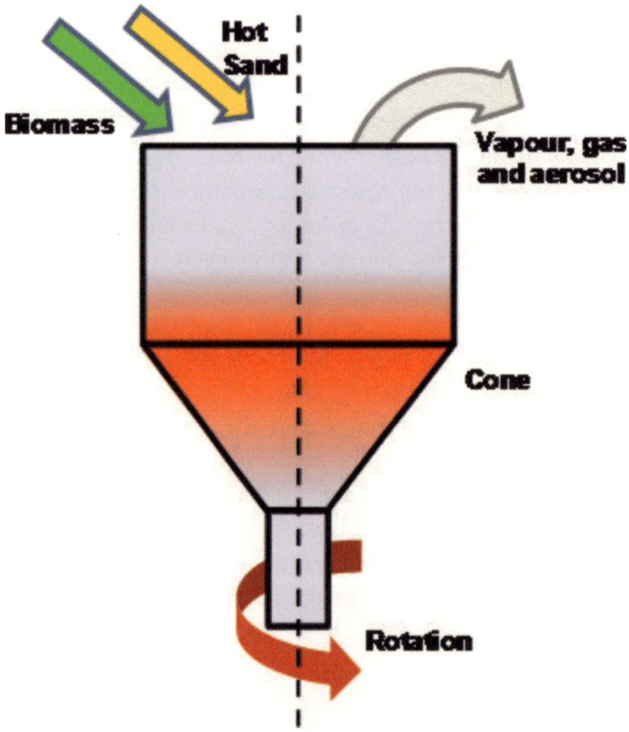

Figure 10. Rotating cone reactor [74].

heat are fed into the cyclone as particles while the vapours that are recycled are made to transport the solids during the process. The particles move down to the bottom of the cyclone through a centrifugal force, during which time drying, heating up and devolatilization take place simultaneously. The heating temperature is usually 450–550°C with residence time of 0.5–1 s. The reactor is highly economical in terms of investment and bio-oil yield [84].

8.7. Auger reactor

This reactor makes use of auger to move the sample feed through a cylindrical tube that is heated up and devoid of oxygen. During this process, the feedstock is pyrolyzed, devolatilized and gasified at a temperature between 400 and 800°C, leading to production of char and condensation of gases into bio-oil [30].

8.8. Plasma reactor

This reactor is made up of a quartz tube that is cylindrical and fitted with two electrodes made of copper. Feeding of the feedstock is done at the middle using screw with variable speed screw at the top of the tube [85]. The gas flows in the tube is powered by thermal energy produced by the electrodes connected to electrical power source. Inert gas is used to remove oxygen from the compartment as well as producing plasma. Apart from its high consumption of energy, it exhibits the ability to guide against the generation of tar as could be witnessed in slow pyrolysis [86].

8.9. Microwave reactor

This is one of the latest developments in pyrolysis. Here, transfer of energy occurs as a result of interaction between the molecules and atoms using microwave. The whole process of drying and pyrolysis are carried out in a microwave oven chamber connected to electricity source. The carrier gas is inert and is also used to create oxygen-free chamber. The reactor has proven to be highly effective in chemical recovery from biomass [87]. Among its advantages include effectiveness in heat transfer, ability to effectively control the heating process as well as ability to guide against the formation of undesirable by-products. It can be used effectively on industrial basis [87].

8.10. Solar reactor

With this technology, provision has been made for storage of solar energy as chemical energy. It is made up of quartz tube with external wall that is opaque, usually exposed to high concentration of solar radiation, capable of high temperature (>700°C) generation in the reactor [88, 89]. Pollution is reduced with this reactor as the feedstock is never tampered with during heating process unlike the slow pyrolysis where the process heat is generated by a part of the feedstock. Start up and shut down time is also very fast.

Table 8 illustrates the advantages and disadvantages of different types of reactors.

Reactor type	Advantages	Disadvantages
Fixed bed	Simplicity in design Reliable results Biomass size independent	High carbon conservation Long solid residence time Low ash carry over Difficult to remove char
Bubbling fluidized bed	Simple design Easy operational procedures	Good temperature control Suitable for large-scale application Small particle sizes are needed
Circulating fluidized bed	Well-understood technology Better Thermal control Larger particle sizes can be processed	Large-scale production difficult Complex hydrodynamics Char is too finer
Rotating cone	Centrifugal force circulates hot sand and biomass substrate No carrier gas required	Difficult operational process Smaller particle sizes needed Large-scale application is difficult
Vacuum	The oil is clean Can process larger particles of 3–5 cm No carrier gas required Lower temperature required Condensation of liquid product is easier	Slow process Solid residence time is too high Require large-scale equipment Poor heat and mass transfer rate Generates more water
Ablative	Inert gas is not required Larger particles can be processed	System is more intensive Moderate temperature required Reactor is costly Lower reaction rate
Augercompact	No carrier gas required	Lower process temperature Moving parts in hot zone Heat transfer in larger scale is not appropriate
PyRos	Compact and low cost Efficient heat transfer Short gas residence time	Complex design Solids in the oil Alkali dissolved in the oil High temperature required
Plasma	High energy density High heat transfer Effective Process control High electrical power consumption	High operating costs Small particle sizes required
Microwave	Efficient heat transfer Exponential control Compact structure Higher heating rate Large-size biomass can be processed Uniform temperature distribution	High temperature High electrical power consumption High operating costs
Solar	Use renewable energy Higher heating rate	High temperature High costs Weather dependent

Table 8. Advantages and disadvantages of different types of reactor [52, 90, 91].

9. Pyrolysis products

9.1. Biochar

Biochar is solid amorphous carbonaceous materials obtained from thermal degradation of lignin and hemicellulose polymer during the pyrolysis process. The physiochemical properties of biocar matrix greatly vary with reactor type and design, biomass composition, particle size and extent of drying, chemical activation, heating rate, reaction time, pressure, flow rate of inert gas, *etc.* [29, 92–96]. If higher heating rate up to 105–500°C/s is used for less retention time and finer particle size, finer biochar is produced during fast pyrolysis process, whereas larger particle-sized feedstock during slow pyrolysis results in a coarser biochar. Usually woody biomass results in coarser biochar whereas crop residues and manures yield more fragile-structured biochar [97]. Earlier investigation demonstrated that biochar yield varies for different temperature regions in a fluidized bed pyrolysis reactor [98]. The results showed that at a low temperature of around 450–500°C, the yield of biochar was high as the rate of devolatilization was low. At a temperature around 550–650°C, the yield of biochar was decreased. At that temperature maximum yield achieved was about 8–10% [98]. However at higher temperature, around 650°C, biochar yield was very low. Biochar predominantly contains larger portion of fixed carbon along with moisture, volatile materials, hydrogen and various other constituents in two structures: stacked crystalline graphene sheets and randomly ordered amorphous aromatic structures [99]. The aromatic portion of biochar contains H, O, N, P and S. These inorganic species have a pronounced impact on the physical and chemical properties of a biochar [100]. The percentages of these constituents depend on the type of biomass and the process of the pyrolysis process [101–103]. Biochar can be utilized as solid fuel in boilers. After catalytic pre-treatment, it can be used to produce activated carbon, carbon nanotubes and gaseous fractions, etc.

9.2. Syngas

Based on biomass composition and process parameters of pyrolysis, the composition of syngas varies. Usually gaseous products obtained after pyrolysis mostly comprises of H_2 and CO. It also contains negligible fraction of CO_2, N_2, H_2O, mixture of alkanes, alkenes and alkynes, such as CH_4, C_2H_4, C_2H_6, tar, ash, etc. [104]. Higher pyrolysis temperature leads to endothermic reaction. With the increase of pyrolysis, the vaporization of moisture from the biomass takes place initially. After that, thermal degradation and devolatilization take place. At this stage, tar is produced and volatile species are released. A series of secondary reactions such as decarboxylation, decarbonylation, dehydrogenetaion, deoxygenation and cracking takes place to produce mixture of syngas. Therefore, higher temperature initiates the tar decomposition, which results in the production of syngas with decreased yield of oil and char. For a given temperature, dry biomass yields the highest amount of gas at the early stage of pyrolysis, whereas with wet biomass the production of the maximum quantity happens later in the process. This is evident and expected as increase in humidity results in increase in drying time. The cracking of hydrocarbon produces hydrogen at higher temperature. Due to presence of oxygen in biomass, CO and CO_2 are produced. The presence of oxygenated polymer that is cellulose determines the evolution of carbonated oxides produced [105].

The lighter hydrocarbons such as CH_4, C_2H_4, C_2H_6, etc. is formed due to reforming and cracking of heavier hydrocarbons and tar in the vapour phase [106]. Plasma reactor using radio frequency can produce up to 76.64% syngas [85]. The advantages of using syngas are that it produces a considerably minor quantity of unburnt hydrocarbon (HC) and carbon monoxide (CO) with higher emissions of nitrogen oxides (NO*x*). It is reported that, CO and H_2 in syngas have comparatively elevated flame speed and temperature which produce greater temperatures in engines to increase the speed of production of $CO_2$and NO*x* [97–109]. Slow pyrolysis processes give about 10–35% of biogas. At higher temperature flash pyrolysis gives more syngas [110]. Calcined dolomite was used at 750–900°C as a catalyst using fixed bed reactor to produce syngas [110].

9.3. Bio-oil

The oil extracted after pyrolysis is a mixture of about 300–400 compounds [111]. The oil obtained after pyrolysis has tendency to become viscous due to ageing as numerous physical and chemical changes with subsequent loss of volatile matters take place. However, the ageing process can be slowed down by storing them in cool places [35]. Previously it was found that energy crops can yield oil with high ash/metal content and water [112]. The presence of water will lower the heating values as well as it will make the phase separation difficult [112]. Thus for commercial application, presence of ash and lignin inside the biomass substrate should be carefully monitored. Earlier thermal efficiency of the pyrolysis oils was compared with diesel, but they demonstrated unwarranted delay in ignition [113]. On the other hand, the quantity, quality and constancy of pyrolysis oil can also be improved by method variables such as heating rate, temperature and retention times [114]. Type of different reactors (ablative and fixed), particle size and char accretion can disturb the amount and feature of the pyrolysis oil. Till currently, there is no inclusive research to diminish these things. Therefore, additional research is obligatory in order to achieve a complete representation of thermochemical transformation processes to yield superior quality pyrolysis oil. Bio-oil which should be used commercially should preserve its chemical and physical properties such as constancy and viscosity. If the oil contains low molecular weight compound, it is possible. The oil contains high molecular weight compounds if the starting biomass contains larger proportion of lignin.

10. Pyrolysis technology: current status

The utilization of food crops such as soybean, maize and sugarcane for producing ethanol and biodiesel may not endure for long since these crops are primarily cultivated for consumption. The need therefore arise for a more sustainable means of generating these materials from other sources such as biomass materials in addition to others already being researched into. However, none of these has proven to be feasible economically yet, but there is great hope on utilizing lignocellulosic biomass for this purpose through pyrolysis process even though it is still faced by some teething challenges. Some tangible efforts have been made by Ensyn and Dyna Motive companies to commercialize the utilization of biomass materials and other agricultural wastes in the generation of biofuels through fast pyrolysis process. These materials are readily available at little or no cost thereby making their utilization highly economical. While Dyna Motive concentrates on how to make energy systems from the fuels produced more environmentally friendly, Ensyn on the other hand is looking towards utilizing the chemicals that may be produced from the system as co-products for other usage such as food smoking. These efforts have since led to production of biofuels from biomass materials. Other notable pyrolysis companies that have been involved in these activities are Pyrovac and Renewable Oil International which use vacuum pyrolysis technique in addition to other smaller pyrolysis plants which are available worldwide. From available data, it clear that fluidized bed reactor are mostly in use for production of bio-oil using biomass while this is followed by other technologies.

11. Conclusions

The perusal of the literature showed that the transformation of biomass to value-added products still needs to resolve some trials such as determining the relation between the starting precursors or feedstock and the overall operation of the pyrolysis plant, upgrading the consistency of the pyrolysis reactions in terms of complete energy and material balances to become sustainable for profitable applications. This chapter elaborately described about the principle of pyrolysis technology including the choice of effective parameters for pyrolysis, types of reactor, *etc.*, depending on the preferred output (bio-oil, biochar or syngas) from the process. However, a comprehensive understanding of the typical process will permit to get maximum output. The chapter highlighted the resulting conclusions and recommendations for additional studies:

1. The major challenge of pyrolysis process is to improve the process by enhancing the product quality and quantity as well as lessening the costs and reduce hazardous environmental impact.

2. Appropriate selection of biomass is a crucial factor to obtain high bio-oil yields. Biomass containing high cellulose content can be selected, as bio-oils are mostly derived from it whereas lignin-based biomass can be used for biochar production. Furthermore, biomass with low moisture content is appropriate to decrease the drying costs and enhance the quality of the extracted oil.

3. The kinetics of pyrolysis of biomass can proceed by several parallel paths. However, application of low temperature would produce lower activation energy to yield mainly char and gas. On the contrary, an enhanced temperature will lead higher activation energy to produce mainly condensable vapours, oils and liquid aerosols. In order to obtain maximum liquid fuels, it is required to heat the biomass speedily at a suitable elevated temperature. Nevertheless, rapid heating of biomass needs smaller particle sizes of the precursors which can initiate constant particle heating. In this regard, fluidized beds are frequently used as efficient reactor type. The disadvantages of fluidized reactor have been overcome by using ablative pyrolysis and auger pyrolysis methods. These types of novel approaches can endure a wider range of variable sizes of the particles.

4. The separation process of the biochar should be effective and fast to reduce contamination of the bio-oil.

5. Amendments should be done for upgrading the engine, turbine and boiler combustion systems for proper utilization of pyrolysis bio-oil while the effect of physiochemical properties of the oil, emission of small particles, combustion efficiency and slag and carbon deposition during the burning process should be taken under considerations.

6. Until recently, sustainable industrial-scale catalytic pyrolysis systems are not identified. Only some transition metal catalysts have been developed and studied for lab-scale approach to improve gas production.

7. Though a lot of studies have been done on pyrolysis economy but most of those were restricted for small- or pilot-scale production. Detail calculation for industrial-scale pyrolysis plant is essential to inaugurate this technology up to a larger extent for practical phase applications.

8. The usage of bio-oil as a renewable liquid fuel is hindered due to its underprivileged physicochemical properties. Presently, commercial projections for liquid fuel uses are dependent on its successful alteration to gasoline, diesel or kerosene, or chemicals such as olefins or aromatics. However, these techniques are still in emerging stage.

Author details

Chowdhury Zaira Zaman[1*], Kaushik Pal[2], Wageeh A. Yehye[1], Suresh Sagadevan[3], Syed Tawab Shah[1], Ganiyu Abimbola Adebisi[1], Emy Marliana[1], Rahman Faijur Rafique[4] and Rafie Bin Johan[1]

*Address all correspondence to: zaira.chowdhury76@gmail.com

1 Nanotechnology and Catalysis Research Center (NANOCAT), University Malaya, Kuala Lumpur, Malaysia

2 Wuhan University, Wuchang, PR China

3 Department of Physics, AMET University, Kanathur, Chennai, India

4 Kumoh National Institute of Technology (KIT), Gumi, South Korea

References

[1] Bridgwater AV. Biomass fast pyrolysis. Journal of Thermal Science. 2004;**8**:21-49

[2] Downie A. BEST pyrolysis technology: A solution for the greenhouse challenge. BEST energies, Australia. Thermal Net Newsletter. 2007;**5**:5

[3] Demirbas A. Pyrolysis of ground beech wood in irregular heating rate conditions. Journal of Analytical and Applied Pyrolysis. 2005;**73**:39-43

[4] Asadullah M., Rahman M.A., Ali M.M., Motin M.A., Sultan M.B. Alam M.R. Production of bio-oil from fixed bed pyrolysis of bagasse. Fuel. 2007; 86: 2514-2520.

[5] Demiral I, Sensoz S. The effects of different catalysts on the pyrolysis of industrial wastes (olive and hazelnut bagasse). Bioresource Technology. 2008;**99**:8002-8007

[6] Mohan D, Pittman CU, Bricka M, Smith F, Yancey B, Mohammad J. Sorption of arsenic, cadmium, and lead by chars produced from fast pyrolysis of wood and bark during bio-oil production. Journal of Colloid and Interface Science. 2007;**310**:57-73

[7] Aho A, Kumar N, Eranen K, Salmi T, Hupa M, Murzin DY. Catalytic pyrolysis of woody biomass in a fluidized bed reactor: Influence of the zeolite structure. Fuel. 2008;**87**:2493-2501

[8] Karaosmanoglu F, Tetik E. Fuel properties of pyrolysis oil of the straw and stalk of rape plant. Renewable Energy. 1999;**16**:1090-1093

[9] Jensen PA, Sander B, Dam-Johansen K. Pretreatment of straw for power production by pyrolysis and char wash. Biomass Bioenergy. 2001;20:431-446

[10] Putun E, Uzun BB, Putun AE. Fixed-bed catalytic pyrolysis of cotton-seed cake: Effects of pyrolysis temperature, natural zeolite content and sweeping gas flow rate. Bioresource Technology. 2006;**97**:701-710

[11] Antal MJ, Grönli M. The art, science, and technology of charcoal production. Industrial & Engineering Chemistry Research. 2003;**42**:1619-1640

[12] Mohan D, Pittman CU, Steele PH. Pyrolysis of wood/biomass for bio-oil: A critical review. Energy Fuels. 2006;**20**:848-889

[13] Demirbas A. Partly chemical analysis of liquid fraction of flash pyrolysis products from biomass in the presence of sodium carbonate. Energy Conversion and Management. 2002;**43**:1801-1809

[14] Demirbas A. The influence of temperature on the yields of compounds existing in bio-oils obtained from biomass samples *via* pyrolysis. Fuel Processing Technology. 2007;**88**:591-597

[15] Demirbas A. Combustion characteristics of different biomass fuels. Progress in Energy and Combustion Science. 2004;**30**:219-230

[16] Muradov NZ, Veziroglu TN. 'Green' path from fossil-based to hydrogen economy: An overview of carbon-neutral technologies. International Journal of Hydrogen Energy. 2008; **33**:6804-6839

[17] Bridgwater AV, Peacocke GVC. Fast pyrolysis processes for biomass. Renewable and Sustainable Energy Reviews. 2000;**4**:1-73

[18] Bridgwater AV, Meier D, Radlein D. An overview of fast pyrolysis of biomass. Organic Geochemistry. 1999;**30**:1479-1493

[19] Available from: http://documentslide.com/documents/biomass-pyrolysis-training-on-technologies-for-converting-waste-agricultural.html

[20] Lanzetta M, Blasi DC. Pyrolysis kinetics of wheat and corn straw. Journal of Analytical and Applied Pyrolysis. 1998;**44**:181-192

[21] Frassoldati A, Migliavacca G, Crippa T, Velata F, Faravelli T, Ranzi E. Detailed kinetic modeling of thermal degradation of biomasses. In: Proceeding of the 29th Meeting on Combustion; September 2006; Napoli, Italia. Available from: ci.irc.na.cnr.it/download/proc%2006/documenti/Papers/09-02-frassoldati-039.pdf (Accessed: 20 November 2012)

[22] Pei-dong Z, Guomei J, Gang W. Contribution to emission reduction of CO2 and SO2 byhousehold biogas construction in rural China. Renewable & Sustainable Energy Reviews. 2007;**11**:1903-1912

[23] Thornley P, Upham P, Huang Y, Rezvani S, Brammer J, Rogers J. Integrated assessment of bioelectricity technology options. Energy Policy. 2009;**37**:890-903

[24] Somerville C. Energy from Biomass. Workshop Presentation for the Inter Academy Council Study Report; Lighting the Way: Towards Sustainable Energy Future. Amsterdam, The Netherlands: IAC; 2005

[25] Fisher T, Hajaligol M, Waymack B, Kellogg D. Pyrolysis behavior and kinetics of biomass derived materials. Journal of Analytical and Applied Pyrolysis. 2002;**62**:331-349

[26] Grønli MG, Varhegyi G, Blassi CD. Thermogravimetric analysis and devolatilization kinetics of wood. Industrial & Engineering Chemistry Research. 2002;**41**:4201-4208

[27] International Energy Agency. Annual Report, 2006: IEA Bioenergy: Task 34, Pyrolysis of Biomass. Paris, France: International Energy Agency; 2006

[28] Jahirul ML, Rasul MG, Chowdhury AA, Ashwath N. Biofuel production through biomass pyrolysis- a technological review. Energies. 2012;**5**:4952-5001

[29] Brown R. Biochar production technology. In: Lehmann J, Joseph S, editors. Biochar for Environmental Management. London: Earthscan; 2009. Chapter 8.

[30] Mohan D, Pittman CU, Steele PH. Pyrolysis of wood/biomass for bio-oil: A critical review. Energy and Fuels. 2006;**20**(3):848-889

[31] Yang H, Yan R, Chen H, Lee DH, Liang DT, Zheng C. Pyrolysis of palm oil wastes for enhanced production of hydrogen rich gases. Fuel Processing Technology. 2006;**87**:935-942

[32] Abbasi T, Abbasi SA. Biomass energy and the environmental impacts associated with its production and utilization. Renewable & Sustainable Energy Reviews. 2010;**14**:919-937

[33] Yang H, Yan R, Chen H, Lee DH, Zheng C. Characteristics of hemicellulose, cellulose and lignin pyrolysis. Fuel. 2007;**86**:1781-1788

[34] Wang J, Wang G, Zhang M, Chen M, Li D, Min F, Chen M, Zhang S, Ren Z, Yen Y. A comparative study of thermolysis characteristic and kinetics of seaweeds and fir-wood. Process Biochemistry. 2006;**41**:1883-1886

[35] Fahmi R, Bridgwater AV, Donnison I, Yates N, Jones JM. The effect of lignin and inorganic species in biomass on pyrolysis oil yields, quality and stability. Fuel. 2008;**87**:230-1240

[36] Demirbas A. Current technologies for the thermo-conversion of biomass into fuels and chemicals. Energy Source Part A. 2004;**26**:715-730

[37] Demirbas A. Calculation of higher heating values of biomass fuels. Fuel 1997;**76**:431-434.

[38] Bridgwater AV, Peacocke GVC. Fast pyrolysis processes for biomass. Renewable and Sustainable Energy Reviews. 2000;**4**:1-73

[39] Zanzi R, Bai X, Capdevila P, Bjornbom E. Pyrolysis of biomass in the presence of steam for preparation of activated carbon, liquid and gaseous products. 6th World Congress of Chemical Engineering; 23-27 September 2001; Melbourne, Australia. 2001.

[40] Friedl A, Padouvas E, Rotter H, Varmuza K. Prediction of heating values of biomass fuel from elemental composition. Analytica Chimica Acta. 2005;**544**:191-198

[41] Wang X. Biomass fast pyrolysis in a fluidized bed [Thesis]. The Netherlands: University of Twente, Enscheda; 2006

[42] A Review of fixed bed gasification systems for biomass. Agricultural Engineering International. 2007;**5**:1-23

[43] McKendry P. Energy production from biomass (part 1): Overview of biomass. Bioresource Technology. 2002;**83**:37-46

[44] Yaman S. Pyrolysis of biomass to produce fuels and chemical feedstocks. Energy Conversion and Management. 2004;**45**:651-671

[45] Hornung A, Bockhorn H, Appenzeller K, Roggero CM, Tumiatti V. Plant for the thermal treatment of material and operation process thereof. US Patent Application No.: 10/451018. 2004

[46] Brownsort PA. Review of scope, control and variability. 2009; UK Biochar Research Center, 1-39. Available from:www. Biochar.org.uk

[47] Garrett DE, Mallan GM. Pyrolysis process for solid wastes. US Patent. 1979;**4**:153, 514

[48] Kovac RJ, Gorton CW, Knight JA, Newman CJ, O'Neil DJ. Research on the pyrolysis of hardwood in an entrained bed process development unit. 1991. DOI: 10.2172/5086913

[49] Maniatis K, Baeyens J, Peeters H, Roggeman G. The Egemin flash pyrolysis process: Commissioning and results. In: Bridgwater AV, editor. Advances in Thermochemical Biomass Conversion. Blackie; 1993. pp. 1257-1264

[50] Scott DS, Piskorz J. The flash pyrolysis of Aspen-Poplar wood. Canadian Journal Chemical Engineering. 1982;**60**: 666-674

[51] Scott DS, Piskorz J. The continuous flash pyrolysis of biomass. Canadian Journal Chemical Engineering. 1984;**62**:404-412

[52] Scott DS, Majerski P, Piskorz J, Radlein D. A second look at fast pyrolysis of biomass — the RTI process. Journal Analytical Applied Pyrolysis. 1999;**51**:23-37

[53] Radlein D, Quignard A. A short historical rteview of fast pyrolysis of biomass, oil & gas science and technology. Rev. IFP Energy's Novellas. 2013;**68**(4):765-783

[54] Demibas A, Arin G. An overview of biomass pyrolysis. Energy Source Part A. 2002; **24**:471-482

[55] Venderbosch RH, Prins W. Review: Fast pyrolysis technology development. Biofuel. 2010;**4**:178-208

[56] Chiaramonti D, Oasmaa A, Solantausta Y. Power generation using fast pyrolysis liquids from biomass. Renewable & Sustainable Energy Reviews. 2007;**11**:1056-1086

[57] Bridgwater AV. Fast Pyrolysis of Biomass: A Handbook. Newbury, UK: CRL Press; 2005

[58] Demirbas A. Recent advances in biomass conversion technologies. Energy Education Science and Technology. 2000;**6**:77-83

[59] Aguado R, Olazar M, Gaisan B, Prieto R, Bilbao J. Kinetic study of polyolefin pyrolysis in a conical spouted bed reactor. Industrial & Engineering Chemistry Research. 2002;**41**:4559-4566

[60] Cornelissen T, Yperman Y, Reggers G, Schreurs S, Carleer R. Flash co-pyrolysis of biomass with polylactic acid. Part 1: Influence on bio-oil yield and heating value. Fuel. 2008;**87**:1031-1041

[61] Bridgwater AV, Czernik S, Piskorz J. An overview of fast pyrolysis. Progress in Thermochemical Biomass Conversion. 2001;**2**:977-997

[62] Demirbas AH. Yields and heating values of liquids and chars from spruce trunkbark pyrolysis. Energy Source Part A. 2005;**27**:1367-1373

[63] Tippayawong N, Kinorn J, Thavornun S. Yields and gaseous composition from slow pyrolysis of refuse-derived fuels. Energy Source Part A. 2008;**30**:1572-1578

[64] Weisz PB, Haag WO, Rodewald PG. Catalytic production of high-grade fuel (Gasoline) from biomass compounds by shape-selective catalysis. Science. 1979;**5**:57-58. DOI: 10.1126/science.206.4414.57

[65] Frankiewicz TC. Process for converting oxygenated hydrocarbons into hydrocarbons. US Patent 4308411. 1981

[66] Diebold J, Scahill J. Biomass to gasoline. In: Soltes J, Milne TA, editors. Pyrolysis Oils from Biomass, ACS Symposium Series. Washington, DC: American Chemical Society; 1988. p. 376. DOI: 10.1021/bk-1988-0376.fw001

[67] Kuen-Song L, Wang HP, Chang NB, Jou C, Hsiao M. Synthesis of ZSM-type zeolites from bio-waste gasification ashes. Energy Sources. 2003;**25**:565-576

[68] Muradov N, Fidalgo B, Gujar AC, Garceau N, T-Raissi A. Production and characterization of lemna minor bio-char and its catalytic application for biogas reforming. Biomass Bioenergy. 2012;**42**:123-131

[69] Samolada MC, Papafotica A, Vasalos IA. Catalyst evaluation for catalytic biomass pyrolysis. Energy Fuels. 2000;**14**:1161-1167

[70] Li X, Su L, Wang Y, Yu Y, Wang C, Li X, Wang Z. Catalytic fast pyrolysis of Kraft lignin with HZSM-5 zeolite for producing aromatic hydrocarbons. Frontiers of Environmental Science & Engineering. 2012;**6**:295-303

[71] Ma Z, Troussard E, van Bokhoven JA. Controlling the selectivity to chemicals from lignin via catalytic fast pyrolysis. Applied Catalysis A: General. 2012;**423-424**:130-136

[72] Pattiya A, Titiloye JO, Bridgwater AV. Evaluation of catalytic pyrolysis of cassava rhizome by principal component analysis. Fuel. 2010;**89**:244-253

[73] Encinar JM, Gonzalez JF, Martinez G, Roman S. Catalytic pyrolysis of exhausted olive oil waste. Journal of Analytical and Applied Pyrolysis. 2009;**85**:197-203

[74] Pattiya A, Titiloye JO, Bridgwater AV. Fast pyrolysis of cassava rhizome in the presence of catalysts. Journal of Analytical and Applied Pyrolysis. 2008;**81**:72-79

[75] Radlein DSAG, Mason SL, Piskorz J, Scott DS. Hydrocarbons from the catalytic pyrolysis of Biomass. Energy Fuels. 1991;**5**:760-763

[76] Marker TL, Felix LG, Linck MB, Roberts MJ. [Tcbiomass2011] integrated hydropyrolysis and hydroconversion (IH2) for the direct production of gasoline and Diesel fuels or blending components from biomass, Part 1: Proof of principle testing. Environmental Progress Sustainable Energy. 2012;**31**:191-199. DOI: 10.1002/ep.10629

[77] Altafini CR, Wander PR, Barreto RM. Prediction of the working parameters of a wood waste gasifier through an equilibrium model. Energy Conversion and Management. 2003;**44**:2763-2777

[78] Rao MS, Singha SP, Sodhaa MS, Dubey AK, Shyam M. Stoichiometric, mass, energy and exergy balance analysis of countercurrent fixed-bed gasification of post-consumer residues. Biomass Bioenergy. 2004;**27**:155-171

[79] Lv PM, Xiong ZH, Chang J, Wu CZ, Chen Y, Zhu JX. An experimental study on biomass air–steam gasification in a fluidized bed. Bioresource Technology. 2004;**95**:95-101

[80] Sadaka S, Boateng AA. Pyrolysis and Bio-Oil, Agriculture and Natural Resources (2008); FSA1052. Fayetteville, AK, USA: University of Arkansas. Available from: http://www.uaex.edu/Other_Areas/publications/PDF/FSA-1052.pdf [Accessed: 5 August 2010]

[81] Li XT, Grace R, Lim CJ, Watkinson AP, Chen HP, Kim JR. Biomass gasification in a circulating fluidized bed. Biomass Bioenergy. 2004;**26**:171-193

[82] Jones SB, Holladay JE, Valkenburg C, Stevens DJ, Walton CW, Kinchin C, Elliott DC, Czernik S. Production of Gasoline and Diesel from Biomass via Fast Pyrolysis, Hydrotreating and Hydrocracking: A Design Case; Report No. PNNL-18284. Springfield, VA, USA: U.S. Department of Energy; 2009

[83] Roy C, Blanchette D, Korving L, Yang J, DeCaumia B. Development of a novel vacuum pyrolysis reactor with improved heat transfer potential. In: Bridgewater AV, Boocock DGB, editors. Developments in Thermochemical. Biomass Conversion. London, UK: Blackie Academic and Professional; 1997. pp. 351-367

[84] Bramer EA, Holthuis MR. Clean liquid fuel through flash pyrolysis. In: The Development of the PyRos Process; AFTUR Final Report. Enschede, The Netherlands; University of Twente; 2005

[85] Tang L, Huang H. Plasma pyrolysis of biomass for production of syngas and carbon adsorbent. Energy Fuels. 2005;**19**:1174-1178

[86] Chen G, Andries K, Luo Z, Spliethoff H. Biomass pyrolysis/Gasification for product gas production. The overall investigation of parametric effects. Energy Conversion and Management. 2003;**44**:1873-1884

[87] Fernández Y, Menéndez JA. Influence of feed characteristics on the microwave-assisted pyrolysis used to produce syngas from biomass wastes. Journal of Analytical and Applied Pyrolysis. 2011;**91**:316-322

[88] Boutin O, Ferrer M, Lede J. Flash pyrolysis of cellulose pellets submitted to a concentrated radiation: Experiments and modeling. Chemical Engineering Science. 2002;**57**:15-25

[89] Hofmann L, Antal MJ. Numerical simulations of the performance of solar fired flash pyrolysis reactors. Solar Energy. 1984;**33**:427-440

[90] Menendez JA, Dominguez A, Inguanzo M, Pis JJ. Microwave pyrolysis of sewage sludge: Analysis of the gas fraction. Journal of Analytical and Applied Pyrolysis. 2004;**71**:657-667

[91] Zhao X, Hunang H, Wu C, Li H, Chen Y. Biomass pyrolysis in an Argon/Hydrogen plasma reactor. Chemical Engineering & Technology. 2001;**24**:197-199

[92] González JF, Román S, Encinar JM, Martinéz G. Pyrolysis of various biomass residues and char utilization for the production of activated carbons. Journal of Analytical and Applied Pyrolysis. 2009;**85**:134-141

[93] Downie A, Crosky A, Munroe P. Physical properties of biochar. In: Lehmann J, Joseph S, editors. Biochar for Environmental Management: Science and Technology. London, UK: Earthscan; 2009

[94] Cetin E, Moghtaderi B, Gupta R, Wall TF. Influence of pyrolysis conditions on the structure and gasification reactivity of biomass chars. Fuel. 2004;**83**:2139-2150

[95] Lua AC, Yang T, Guo J. Effects of pyrolysis conditions on the properties of activated carbons prepared from pistachio-nut shells. Journal of Analytical and Applied Pyrolysis. 2004;**72**:279-287

[96] Dawei A, Zhimin W, Shuting Z, Hongxing Y. Low-temperature pyrolysis of municipal solid waste: Influence of pyrolysis temperature on the characteristics of solid fuel. International Journal of Energy Research. 2006;**30**:349-357

[97] Sohi S, Lopez-Capel E, Krull E, Bol R. Biochar, Climate Change and Soil: A Review to Guide Future Research; CSIRO Land and Water Science Report. Canberra, Australia: CSIRO; 2009

[98] Rocha JD, Olivares-Gómez E, Mesa-Pérez JM, Cortez LAB, Seye O, Brossard-González LE. The demonstration fast pyrolysis plant to biomass conversion in Brazil. In: Proceedings of World Renewable Energy Congress VII (WRC 2002); Cologne, Germany. 2002

[99] Verheijen F, Jeffery S, Bastos AC, van der Velde M, Diafas I. Biochar application to soils: A critical scientific review of effects on soil properties, processes and functions. European Commission Report No. EUR 24099 EN; European Communities, Ispara, Italy, 2010. Energies. 2012;**5**:4997

[100] Bourke J, Manley-Harris M, Fushimi C, Dowaki K, Nunoura T, Antal Jr. MJ. Do all carbonised charcoals have the same structure? A model of the chemical structure of carbonized charcoal. Industrial & Engineering Chemistry Research. 2007;**46**:5954-5967

[101] Van Zwieten L, Kimber S, Morris S, Chan KY, Downie A, Rust J, Joseph S, Cowie A. Effects of biochar from slow pyrolysis of papermill waste on agronomic performance and soil fertility. Plant and Soil. 2009;1-12

[102] Amonette JE, Jospeh S. Characteristics of biochar: Microchemical properties. In: Lehmann J, Joseph S, editors. Biochar for Environmental Management Science and Technology. London, UK: Earthscan; 2009 (Second Edition)

[103] Gaskin JW, Steiner C, Harris K, Das KC, Bibens B. Effect of low-temperature pyrolysis conditions on biochar for agricultural use. Transactions of the ASABE. 2008;**51**:2061-2069

[104] Wei L, Thomasson JA, Bricka RM, Batchelor WD, Columbus EP, Wooten JR. Experimental Study of a Downdraft Gratifier; ASABE Meeting Paper No. 066029. St. Joseph, Michigan: American Society of Agricultural and Biological Engineers; 2006

[105] Couher C, Commandre JM, Salvador S. Failure of the component additively rule to predict gas yields of biomass in flash pyrolysis at 950°C. Biomass Bioenergy. 2009;**33**:316-326

[106] He M, Xiao B, Liu S, Hu Z, Guo X, Luo S. Syngas production from pyrolysis of Municipal solid waste (MSW) with dolomite as downstream catalysts. Journal of Analytical and Applied Pyrolysis. 2010;**87**:181-187

[107] Papagiannakis RG, Rakopoulos CD, Hountalas DT, Giakoumis EG. Study of the Performance and exhaust emissions of a spark-ignited engine operating on syngas fuel. International Journal of Alternative Propulsion. 2007;**2**:190-215

[108] McMillian MH, Lawson SA. Experimental and modeling study of hydrogen/syngas production and particulate emissions from a natural gas-fueled partial oxidation engine. International Journal of Hydrogen Energy. 2006;**31**:847-860

[109] Shudo T, Nagano T, Kobayashi M. Combustion characteristics of waste pyrolysis gases in an internal combustion engine. International Journal of Automotive Technology. 2003;**4**:1-8

[110] Kantarelis E., Zabaniotou A. Valorization of cotton stalks by fast pyrolysis and fixed bed air gasification for syngas production as precursor of second generation biofuels and sustainable agriculture. Bioresource Technology. 2009;**100**:942-947

[111] Oasmaa A, Meier D. Norms and standards for fast pyrolysis liquids: 1. Round robin test. Journal of Analytical and Applied Pyrolysis. 2005;**73**:323-334

[112] Shihadeh A, Hochgreb S. Diesel engine combustion of biomass pyrolysis oils. Energy Fuels. 2000;**14**:260-274

[113] Abdullah N. An assessment of pyrolysis for processing empty fruit bunches [Thesis]. Birmingham, UK: Aston University; 2005

[114] Evans R, Milne T. Molecular characterization of the pyrolysis of biomass 1. Energy Fuel. 1987;**1**:123-137

Chapter 2

Organic Waste Torrefaction – A Review: Reactor Systems, and the Biochar Properties

Paweł Stępień, Jakub Pulka and Andrzej Białowiec

Additional information is available at the end of the chapter

http://dx.doi.org/10.5772/67644

Abstract

Torrefaction is a thermochemical process in a narrow temperature ranging from 200 to 300°C, where primarily hemicellulose fibers are depolymerized. This process is carried out under atmospheric pressure and in anaerobic conditions; heating ratio is low (<50°C/min) and the residence time is relatively long, up to 1 h. During the process, a biomass is partially decomposed and forms different condensing and noncondensing gases. The final product is a constant substance rich in carbon, which is called a torrefied biomass—biochar and biocarbon. Currently an increase in energy demand is impacting the environment considerably. For this reason, in this chapter the organic waste torrefaction technology will be presented, including the reactor systems review. Torrefaction process may be conducted in different types of reactors, with diverse technologies. From this variety, two main groups of reactors can be distinguished, with direct and indirect heating. Direct heating group consists of reactors with multiple design, such as Multiple Hearth Furnace, microwave reactor, moving bed, vibrating belt, the reactor belt, and auger. Indirect heating reactors are less common and this group consists of rotating drum and auger reactor. All mentioned reactor types will be presented and discussed.

Keywords: organic waste, torrefaction, thermal treatment, biocarbon, torrefaction reactors

1. Introduction

Torrefaction is a biomass/waste thermal decomposition process that produces a carbon-rich product—Biochar [1]. Biomass partly decomposes during this process, generating both condensable and noncondensable gasses. The resulting product is a solid substance rich in carbon, referred to as biochar, torrefaction biomass, or biocarbon [2]. In industry and literature, the torrefaction process is also referred to as roasting, slow and mild pyrolysis, wood cooking, and high-temperature drying [3].

Temperature and retention time are two main parameters that influence torrefaction process efficiency [4]. Torrefaction is usually conducted at temperatures between 200 and 300°C and the designated temperature is maintained for 15–60 min [5]. Choosing specific value of those two key parameters for different types of biomass is essential for cost-effective biomass treatment.

Torrefaction is a biomass treatment method for future utilization in cofiring in gasification process [3]. The process is commonly applied for lignocelluloses biomass treatment [6]. Lignocelluloses are built of three polymers: hemicelluloses, lignin, and cellulose. Hemicelluloses are the most reactive form of those three polymers, and their carbonization and devolatilization occur at temperatures below 250°C [6]. Vegetable biomass is used most commonly as the stock in torrefaction process. This biomass can be divided into two groups—green waste and energetic forestry products. Plants with the highest lignocelluloses percentage compared to sugars and fats have best energy potential [6]. Torrefaction feedstock used commercially or in research is mainly lignocellulosic (wood pellets or chips, crop residue, or tree bark) although organic nonlignocellulosic waste (bagasse from sugarcane industry, olive mill waste, poultry waste and litter, paper sludge, dairy cattle manure, or distillers grain) are being used more often [7].

Furthermore, all the considered biomass types are not just lignocellulosic by nature. Some waste biomass types, such as sewage sludge, digestate from biogas plants and agricultural animal waste, food waste, and spacecraft solid wastes (chemical composition of fecal simulant), consist of fats, proteins, and other organic matter, with very low lignocellulose content [7–9].

Due to the above studies, torrefied biomass/waste with very low lignocellulose wide-scale urbanization, production of such waste has increased substantially and the torrefaction process may help utilize this large volume of nonlignocellulosic biomass, including refuse derived fuel (RDF). The current absence of direct research in this particular area renders torrefaction decidedly underutilized.

Due to that, this technology is being found perspective, but the relation between process parameters, and biomass, and biocarbon properties should be still optimized. One of the methods is design and application of efficient torrefaction reactor.

Torrefaction process may be conducted in different types of reactors, with diverse technologies. From this variety, two main groups of reactors can be distinguished, with direct and indirect heating. The review of torrefaction reactor types will be presented and discussed.

2. Reactors classification

Torrefaction reactor can be divided into two main groups, based on the substrate heating—reactors with indirect and direct heating. Two subgroups can be distinguished in indirect heating reactors group: auger and rotary type. Direct heating group may be divided because of the oxygen content in the heating medium into several subgroups: (1) the reactors in which the heating medium does not contain oxygen and (2) reactors wherein the heating medium contains a small amount of oxygen and other types (**Figure 1**).

The specific types of torrefaction reactors are described further.

2.1. Auger reactors

Auger-type reactor is constructed of one or more screw conveyors (auger). Its location relative to the ground may be vertical, horizontal, or at an angle. Biomass is fed to the reactor and then transported by a screw conveyor. During transport, the biomass is indirectly heated by a heating medium or directly by the heating elements located in the reactor wall. In both cases, there is a problem with uneven heating of biomass and excessive charring of the product. This phenomenon is linked to insufficient mixing of the substrate and local heating of the material [11, 12]. The residence time in the reactor depends on the length and speed of the conveyor.

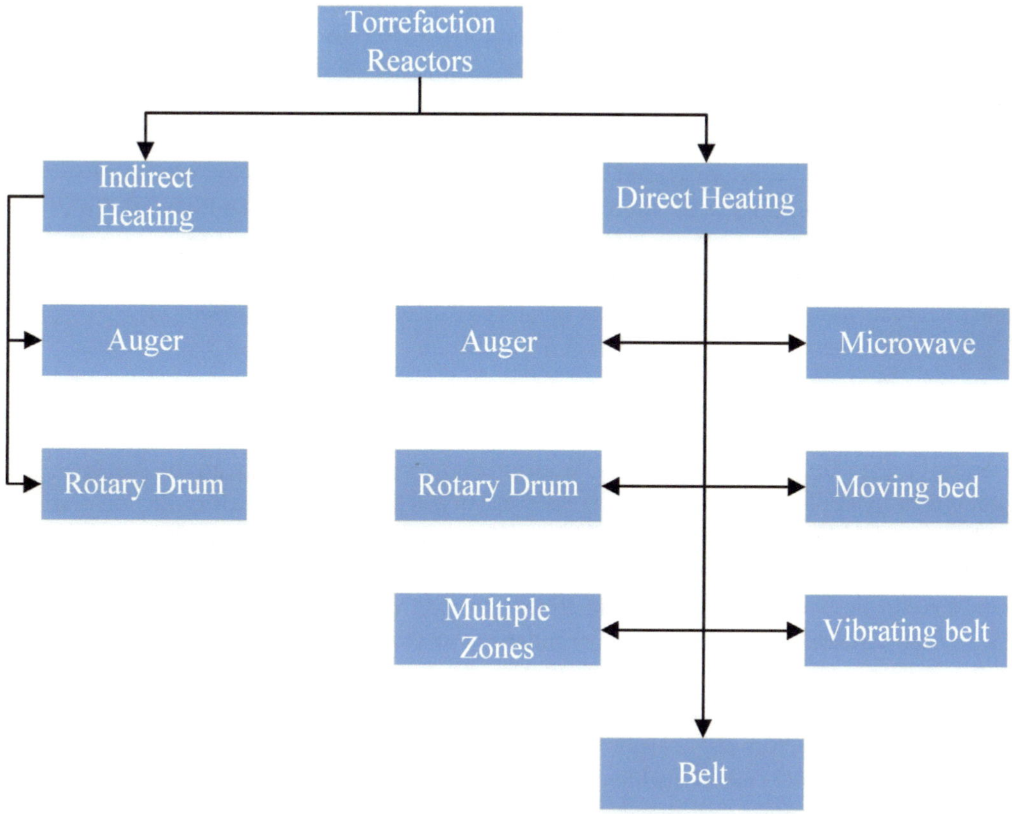

Figure 1. Torrefaction reactors division based on Ref. [10].

The advantage of auger-type reactors is their relatively low price, simplicity of adaptation to a large industrial scale, and low inert gas demand. The disadvantages include limited production capacity [11]. One example of the auger type reactor with indirect heating was described in patent published 15 October 2015 titled *Torrefaction plant, its operation and maintenance*. This design consists of five parts: feeder transporting a substrate, drying reactor, torrefaction reactor, ventilation and heating system, and auger. This reactor scheme can be seen in Ref. [13].

The material supplied for processing is fed to the feeding screw from where it is transported to the first reactor, where drying process is conducted. After passing through the first reactor, material falls by gravity into a second reactor, wherein torrefaction is carried out. At the end of the torrefaction reactor in the lower part, there is an opening that allows the material to fall on the conveyor transporting the product to the storage. Lower part of the reactor is equipped with rods that can be replaced when wear down caused by friction of the material during transportation. Auger transporting material inside the reactor for drying and torrefaction are powered by two independent electric motors (which can rotate in both directions). Conveyors drive shafts have been secured by a special latch (couplings), allowing quick removal of the tray in case of failure by inspection hatches located on the side of the engine.

The ventilation system for gases produced during the process is divided into two parts: an exhaust for gases discharge from the drying and torrefaction reactors. Gasses produced in the first process are not used due to the high moisture that affects its low calorific value. Torgas formed in the second reactor is purged of dust and partially of the condensate in cyclone, and is combusted to provide heat for the process (in the case of excessive production gas may be stored). Heating is indirect, provided by heat exchangers inside the reactor. Literature review shows that auger reactors with direct heating system does not differ significantly from the patent described above, due to that fact, their description is omitted. It is worth mentioning that torrefaction technology based on the chamber equipped with several screw conveyors that can create autonomous chambers or one large compartment can be seen [14].

2.2. Rotating drum reactor

Rotating reactor is a technology that allows for continuous operation without stoppage for loading or unloading. Process heat can be supplied directly or indirectly. In the first type, the heat is usually applied by the medium in the form of gas produced in the torrefaction process, which is recycled to the reactor and heated by the heat generated in the combustion of overabundance torgas. Direct heating is performed by the drum walls. Drum torrefaction reactors can be controlled by rotating speed adjustment or length and angle of drum inclination. The construction of the reactor ensures good substrate mixing, resulting in the uniform heating. This technology is simple and easy to scale. The disadvantages of such solutions include the production of a significant amount of fines that is formed by friction between walls and the substrate. Drum reactors also have a lower capacity than fluidized bed reactors, within the range of 1.5–4.5 $mg{\cdot}h^{-1}$ [15]. In the literature and in registered technologies at the patent office, various torrefaction technologies using the drum reactor can be found.

Patented reactor by Teal W. B. and R. J. Gobel is equipped with screw conveyor or other mechanism for biomass feeding. The inlet is equipped with a gutter, followed by a single or double lock. The lock is designed to prevent the oxidant penetration into the processing chamber during substrate feeding [15, 16]. The reactor heating system consists of three parts: a furnace, heat exchanger, and ventilation system. During reactor start-up, the fuel is supplied from the outside. Heat produced during combustion is used to warm up the air drawn from the outside, which is then transferred to a heat exchanger by a fan, positioned in front of the drum reactor. Heat is exchanged between the air heated in the furnace and the gas circulating in a closed circuit between the drum and the heat exchanger. Following a star up, the surplus torgas produced during torrefaction is burned inside the furnace. It is worth mentioning that biomass is heated before entering the drum reactor [15]. This reactor scheme can be seen in Ref. [16].

Main part of the reactor, the drum, rotates around its vertical axis. It is driven by the electric motor, which can be controlled to regulate the amount of drum rotation. Torrefaction chamber interior is equipped with a special blade to move or mix the processed substrate [16].

Behind the drum, there is a separator, which isolates biocarbon and torgas. Particles of biocarbon descend under gravity to the bottom of the separator and are disposed by screw conveyor or other transporting mechanism. In the bottom of the separator, valves are installed to prevent oxygen from getting to the system, which could adversely affect the process. Produced torgas from the separator is sucked by the fan and then directed to the combustion furnace or heat exchanger. Cyclone is installed before the fan to purify the gas from the fine particles and dust [16].

A. D. Livingston and B. J. Thomas registered patent proposing another drum reactor with indirect heating technology. Fuel delivery and its heating (ventilation inlets) differ this technology from the previously discussed reactor [17]. The biomass fed to the reactor first goes to the screw conveyor driven by an electric motor. This mechanism transports the substrate directly into the drum and in contrast to previous technology is not mixed with the heating medium.

The next element is a drum. In this case, it rotates inside a sealed casing and is driven by an electric motor. Shape of the blades responsible for moving and mixing of the material inside the process chamber differs from previous technology [17].

The system of heating the reactor operates in the same manner as in the first case. The difference is the method of heating medium delivery into the reactor chamber. Three air inlets were installed and located in the upper part of the drum casing. This reactor scheme can be seen in Ref. [17].

Output unit behind the drum acts as a gravity separator. The solid fraction falls to the bottom and the volatiles escape through a hole located in the top of the unit. Openings to receive the products are equipped with locks, tasked with preventing oxidant to enter the reactor. In addition, the separator is equipped with inspection doors, allowing reactor review without the demolition of the individual elements [17].

Direct heating reactor was divided into three parts: the substrate input, the drum reactor, and the products output, where the latter element is coupled with the reactor heating system.

Technical line begins with an airlock, which prevents air from entering into the reactor. Behind the latch, mechanical feeder is located for transporting the substrate into the processing chamber. The next element is the drum, which is mounted on bearings, allowing its rotation. Rotation is provided by electrical motor. The drum itself is sloping toward the outlet end allowing material movement in its interior. The authors assumed that the reactor should be tilted by about ½ inch per foot of the drum length. Collection of solid products takes place at the end of the reactor. Biochar falls by gravity to a conveyor installed at the end of the reactor through the rectangular holes.

The resulting exhaust gasses can be drawn through the ventilation system. The hood is positioned in the upper part of the back of the reactor forming a metal casing through which the process gases escape. This process is mechanical, powered with a fan. Produced gasses can be used for the purpose of the process. Ventilation system allows creating small vacuum for technological purposes.

A heating system is located in the rear part of the reactor. It consists of a rotary joint connecting the inlet and outlet of the reactor heating medium with the wires forming a heat exchanger inside the reactor (they are divided into sections and their number depends on the size of the reactor). The principle of the system is very simple. A heating medium which may be water, oil, propylene glycol, or other thermal transfer fluid is heated in a heating system to 315°C. Then, the medium is transported in tubes to the rotary coupling; there, depending on the size of the reactor, it is split into heat exchanger sections. The tubing forms a ring inside the drum and is attached to it in the front part of the reactor prior to the inlet of the substrate. Pipes forming a heat exchanger are equipped with thermal expansion joint in order to prevent damages during operation. The liquid after transferring heat to the reactor is recycled to the rotary coupling and then to the heating system where the cycle begins again [18]. Schematic drawing of inlet, outlet, and reactor heating systems is shown in Ref. [18].

2.3. Multiple Hearth Furnace

Multiple Hearth Furnace technology is used on an industrial scale because it scales easily, and it can be adjusted to the individual preferences of the customer. Also, it provides stable process temperatures, mixing of substrate, and leak free gas flow. The disadvantages of this technology should include slow heat transfer to the substrate, compared to other direct reactors, the limited volume of the converted substrate, which results in larger dimensions of the reactor and requires good seal of the shaft [15]. Multiple hearth torrefaction reactors do not differ significantly from each other. Design differs mainly on configuration of heating and ventilation system, and therefore, one design will be presented to describe the principles of this technology. Multiple hearth reactors are cylindrical, and their interior is divided into multiple levels formed of trays which are fixed to the centrally placed shaft which rotates about an axis of symmetry. It is driven by a motor with a built-in gearbox. The substrate is fed to the reactor from above by a mechanical conveyor, equipped with airlock located at the end of the conveyor, preventing oxidant from entering the reactor. Biomass can be predried in a separate drying system. In this case, the reactor has only a section in which the torrefaction process occurs. If separate drying system is not installed, the reactor is divided into a drying and torrefaction section [19, 20].

The substrate supplied to the first level begins to be heated and distributed evenly using a roller located over the tray. After one full rotation, overabundant biomass is pushed by the roller to the hole where it falls by gravity to the lower level and the process begins again.

The product is collected at the bottom of the reactor and goes to the cooling system. Heating system can be divided into two types depending on whether the reactor has a drying zone. In the first case, the heat is supplied with heated gas into the drying and torrefaction zone independently. Reason behind this design is that during the process of drying, the moisture contained in the biomass evaporates, hence decreases the gas calorific value (the resulting gas is not suitable for energy production). After heating medium passes through the drying zone, excessive gas is released to the atmosphere, and the remaining volume is returned to the heat exchanger for reheating and back to the reactor. Torrefaction zone is heated in the same way as described above with the difference that the overabundant gas formed in the torrefaction process is used as fuel to provide the heat for the process [19, 20]. Reactors with only torrefaction zone are heated by the heating medium consisting of inert gases circulating in a closed loop between the reactor and the heat exchanger—same design as a two-zone reactor. Excessive gas is used for the purposes of the process as a fuel [19].

2.4. Microwave reactor

In this type of reactor, the heat is provided by microwave radiation. This technology is characterized by rapid and uniform heating of the material. The process duration depends on the type, size, and microwave radiation absorption capacity of the processed material and on the reactor power [15]. The main problem with this technology is the high energy consumption required for the production of microwave radiation. Torgas is not used for process purposes and it adversely affects the process efficiency and increases the operation costs [12]. Technology shown in patent titled *Microwave torrefaction of biomass* schematically illustrates a microwave reactor, wherein the authors indicate that besides torrefaction, other processes like pyrolysis and gasification can be performed. Technological line starts from the biomass storage, where material grinding (hammer mills are used most often) and drying occurs. The heat for the drying process is supplied from the heat exchanger located at the end of the process line responsible for the cooling of the product [21]. Behind the hopper, there is a biomass powder compacting device to form pellets or briquettes, which are then collected by a screw conveyor that acts as a process chamber. The front part of the conveyor is also equipped with an inlet of inert gases to ensure anaerobic conditions. Patent authors suggest that the feeding screw should have cylindrical shape with a constant diameter of not more than 50 cm (optimum diameter is in the range 0.5–10 cm) to ensure uniform biomass radiation. The length of the reactor depends on the process parameters, including the diameter of the substrate, feeding rate, the microwave energy, and the numbers of microwave radiation points. The process chamber is equipped with volatile components outlet located at the top and resulting liquid products outlet located at the bottom. Screw conveyor is surrounded by a microwave chamber, which should be equipped with at least one source of microwave radiation. Number of sources depends on the reactor size and process parameters [21]. Screw conveyor is longer than the microwave chamber. Behind the microwave chamber, there is a cooling section of the solid product. Heat is received by a heat exchanger and is used for biomass drying. Diagram of the technological system is shown in Ref. [21, 22].

2.5. Moving bed

The reactor consists of a closed process chamber, where biomass is fed from the top. The reactor has no moving parts, responsible for moving the biomass that falls down freely during the process. The substrate is heated by a heating medium, a gas that has an inlet located at the bottom part of the reactor. Torgas outlet is located at the top of the chamber. Single cycle duration range from 30 to 40 min, and the maximum temperature that can be obtained is 300°C [12]. Simple design, high bed density, and a good heat transfer are main advantages of this design. Difficulty of controlling the temperature and maintaining heating medium pressure are clear disadvantages of this technology [15].

2.6. Vibrating belt

The main part of the reactor is a vibrating belt that is responsible for biomass transporting. Flow rate of the substrate is controlled by intensity of vibration. Biomass is heated indirectly by the gaseous heating medium [12].

In order to standardize the resulting product, reactor has many levels. The advantages of this type of reactor include simplicity of process time adjustment and the possibility of converting the biomass of larger dimensions. Clogging of the apertures with tar and dust generated during the process (cleaning of the reactor is associated with a long maintenance brake, since it must be disassembled) is a main disadvantage. Temperature control of the process is difficult, because it must be correlated with flow of the heating medium and the intensity of the vibration. These reactors require a large space, which also causes problems with their use if space is limited. High risk of corrosion is also associated with this design [12].

2.7. The belt reactor

Presented torrefaction belt reactor consists of four parts: feeder, the reactor chamber with conveyor belts, screw conveyor, and the heating system [23]. This reactor was presented in Ref. [23].

The biomass supplied to the reactor with conveyor goes into the torrefaction chamber, wherein three conveyor belts segments are located. Each of the conveyors is rotating in the opposite direction as the previous one in order to transport the substrate to the bottom of the reactor. Torrefaction chamber is heated directly using a heating medium, produced during the combustion of torgas or the fuel supplied from the outside. The temperature inside the chamber does not exceed 800°C and is controlled by the volume of injected heating medium. Process chamber is equipped with heating ducts, which have been separated from the gas space of the reactor in order to prevent mixing of heating medium and torgas. After the process, biomass goes to the chute located at the bottom of the reactor and then is received by the externally cooled screw conveyor.

Reactor production capacity range from 100 to 500 kg h^{-1}, and the plant can operate in temperature range 220–350°C.

3. Survey of existing installations

Reactors technology described above has been applied on an industrial scale. There are more than 50 companies involved in the implementation of torrefaction technology [12]. **Table 1** shows the characteristics of said technologies.

Developer	Technology	Heating mode	Capacity, mg·h^{-1}	Country
4 Energy	Belt conveyor	Direct	5.5	The Netherlands
Agritech	Screw conveyor	Indirect	8	USA
AIREX	Cyclonic bed reactor	Direct	0.25	Canada
Atmoclear	Belt	Direct	5	UK
Bio Energy Development North AB	Rotary drum	Direct	3.5	Sweden
Biolake	Moving bed	Direct	5	The Netherlands
BTG	Screw conveyor	Indirect	5	The Netherlands
CanBiocoal	Microwave	Direct	12	UK
EBES	Rotary drum	Direct	1.5	Germany
ECN	Moving bed	Direct	5	The Netherlands
Earth Care Products	Rotary drum	Direct	1.5	USA
ETPC	Rotary drum	Indirect	4.3	Spain
Foxcoal	Screw conveyor	Indirect	4.2	The Netherlands
Horizon Bioenergy	Oscillating belt conveyor	Direct	6.5	The Netherlands
IDEMA	Moving bed	Direct	2.5	France
Integro	Multiple hearth	Direct	2	USA
New Biomass Energy	Screw reactor	Indirect	5	USA
New Earth	Oscillating belt conveyor	Direct	2	USA
RFT	Screw conveyor	Indirect	5	USA
Stramproy	Oscillating belt conveyor	Direct	5	The Netherlands
Thermya/LMK Energy	Moving bed	Direct	2.5	France
Topell	Torbed	Direct	8	The Netherlands
Torr-coal	Rotary drum	Indirect	4.5	The Netherlands
West Creek Energy	Rotary drum	Direct	10	USA
WPAC	Unselected	Unknown technology	5	Canada

Table 1. Survey of existing installations based on Refs. [12, 15].

4. Product characteristics

Torrefaction products (biochar and biocarbon) can be characterised by specific properties. Biochar has high energy density, it contains 80–90% of potential energy, while decreasing its mass to 70–80%, hence energy density can be increased by 30% [24]. Biochar does not absorb moisture or its equilibrium moisture contents drop to 1–3%, thus it can be described as hydrophobic [6]. Fixed carbon content increases during the process, depending on process parameters (temperature and duration), values ranged between 25 and 40%, making biochar a potentially attractive reducing agent [24]. Torrefaction reduced oxygen content significantly, thus reducing O/C ratio, this makes biochar attractive substrate for gasification [6]. Mechanical processing (grindability andpalettization) of biochar improves significantly. The output of a pulverizing mill can increase by 3–10 times [25, 26] comparing it to a raw biomass. Torrefied biomass takes less time to ignite due to lower moisture and it burns longer due to larger percentage of fixed carbon compared to raw biomass [27].

Typical lower calorific value (LCV) of biocarbon from lignocellulosic biomass (LB) wood chips—torrefaction ranges between 18 and 23 MJ/kg [27]. Due to low biocarbon moisture (1–6%), the difference between higher calorific value (HCV), and LCV is small [28]. LB biocarbon has relatively low bulk density 180–300 kg/m^3, it is fragile and homogenous [29]. Additional advantage of LB biocarbon is its hydrophobic nature. The absorption of water by torrefied biomass is strongly limited by dehydration processes during thermal decomposition of organic matter. Destruction of OH− groups causes the inhibition of formation of bonds between water and hydrogen. Therefore, biocarbon may be a storage outdoor without risk of biological decay. Torrefaction of LB brings benefits in biocarbon incineration, due to decreasing ignition temperature and shortening the time of ignition [30]. Additionally, many researchers [24, 27, 31] proved that during torrefaction, biocarbon retains potential energy (around 90%), while decreasing substrate mass to 70–80%. All of these properties make biocarbon a desirable fuel for processes like incineration, co-combustion, and gasification.

Another possible pathway is to recycle biocarbon from LB and nonlignocellulosic biomass (NLB) for improving soil properties agent by its application on weak soils (arable and forest) and on the former land after mining of aggregates such as sand or gravel. Soil deposit of biocarbon from lignocellulosic crops (biological coal) according to many research reports has been considered as the method of effective soil improvement and significant element of carbon sequestration in the process of climate change mitigation [32, 33]. It is known that beneficial effect of biocarbon on soil properties is caused by improvement of soil texture, porosity that reflects in modifications of many physical and chemical properties, and soil biology. But simultaneously, the processes of biochar decomposition and impact on soil biology are fragmented and require closer research attention.

Dissolved organic matter is a labile fraction, which can rapidly respond to changes in carbon pools, as they are potentially easy-mineralizable. These labile parts of organic carbon have been suggested as sensitive indicators of soil organic matter changes and important indicators of soil quality [33]. Mineralization of organic carbon compounds promotes the release of carbon dioxide into ambient air as one of greenhouse gases (GHG). Most "active" and susceptible to transformations form of soil organic carbon (SOC) is labile organic carbon. Soil labile organic carbon (SLOC) is composed of amino acids, carbohydrates, microbial biomass, and other simple organic compounds [34]. SLOC is cycling fast in the environment [35]. Circulation of SLOC lasts for not more than several years, while the refractory carbon cycle may last even several thousand years [35]. Soluble carbon and nitrogen are important, as they have a great impact on dissolved organic fraction concentrations in freshwater [36]. Hot water-extractable carbon is the fraction of organic matter, which is naturally labile and its content is correlated with the mass of microorganisms simultaneously being an excellent indicator of qualitative changes in organic matter [36]. This fraction is potentially the most susceptible to oxidation of CO_2 [33], and therefore has the greatest impact on global climate change.

Introducing biocarbon into soil causes decrease of solubility of SLOC and finally decreases the GHG emission. Therefore, the interesting aspect of biocarbon recycling into soil is proposed in this project examination of GHG emissions from soil enriched biocarbon and the degree of pollutants elution form biocarbon including organic compounds and heavy metals.

Biocarbon has heterogeneous highly porous structure and its outer and inner surfaces are very big and have a lot of "niches" of different water properties – hydrophilic and hydrophobic of basic and acid reaction etc. It makes biochar important in water holding capacity, which is especially an important treatment on weak soils. Thanks to stable nature of biocarbon (with half-lives estimated in broad ranges from hundreds to thousands of years); the positive impact of biochar may be prolonged for years. In this context, it is also important to build the knowledge on long term impact of biochar on groundwater. Méndez et al. [37] examined the influence of biocarbon obtained from sewage sludge on plants. The concentration of cupper in biocarbon was about 80% higher than in raw sewage sludge and about 40% in case of other heavy metals, but their bioavailability and mobility were significantly lower. The increase of torrefaction temperature caused the increase of heavy metal content in biocarbon, but their bioavailability and mobility decreased. Authors determined also that within the increase of temperature up to 300°C, the content of nitrogen slightly increased, but levels of P and K were constant. Presented data indicate that also in NLB torrefaction, it is possible to generate biocarbon with valuable properties.

5. Summary

Given torrefaction reactors review showed a variety of technical and technological solutions. Most of the differences are related to material flow through reactor, material heating mechanism, the source of heat for the process, and torrgas treatment. As the torrefaction process is classified between high temperature drying and low temperature pyrolysis, most reactor systems are similar to those commonly used in biomass/waste drying, and/or pyrolysis. Actually, it is difficult to distinguish a specific type or solution of the reactor, which would be a characteristic only for torrefaction. Therefore, it seems that application of torrefaction of some biomass may be easily implemented just by adaptation of pyrolysis reactors. The problem may be related to torrefaction energy balance due to relatively low calorific value of torrgas or problem with mechanical movement of the feedstock trough the reactor caused by friction and/or melting of such materials like plastics.

Each presented reactor type has its advantages and disadvantages. Some are cheap, easy to construct, and operate. Some have problems with material mass flow, heat flow. Some are good for laboratory test, but some may have potential for industrial purposes. At this stage of the torrefaction technology development, it is hard to specify which type of reactor should be recommended. The torrefaction may be dedicated for different types of biomass and waste. The choice of torrefaction reactor should be based on the biomass/waste type and properties, the components of energy balance, pollution degree of the torrgas, desired biocarbon properties, energy demand, economy, and the current situation of the biomass/waste, and biocarbon utilization market. The torrefaction technology is relatively new and it is perspective. Not all problems have been solved, yet. Many new ideas arise each day at this field. Therefore, there is a room for innovations and inventions, which may move the torrefaction technology at the higher level of development. Intensive research and development activity in this field is then required and justified.

Acknowledgements

Presented work was done under financial support of the project titled "Innovative organic waste conversion technological line into innovative, high-quality solid fuels", run in frame of Action. 1.1. – "Research, and development projects for enterprises", Sub-action 1.1.1. – "Industrial research and development works executed by enterprises", Intelligent Development Operational Program for years 2014–2020, call 1/1.1./2015), cofounded by European Regional Development Found, and The National Centre of Research and Development. Project No. UDA-POIR.01.01.01-00-0334/15.

Author details

Paweł Stępień[1], Jakub Pulka[1] and Andrzej Białowiec[1,2*]

*Address all correspondence to: andrzej.bialowiec@upwr.edu.pl

1 Faculty of Life Sciences and Technology, Institute of Agricultural Engineering, Wroclaw University of Environmental and Life Sciences, Wrocław, Poland

2 Ekopartner-Recykling Sp. z o.o., Lubin, Poland

References

[1] Mimmoa T, Panzacchib P, Baratieri M, Davies CA, Tonon G. Effect of pyrolysis temperature on miscanthus (Miscanthus x giganteus) biochar physical, chemical and functional properties. Biomass and Bioenergy. 2014;**62**:149-157. DOI: 10.1016/j.biombioe.2014.01.004

[2] Lehmann J, Rillig MC, Thies J, Masiello CA, Hockaday WC, Crowley D. Biochar effects on soil biota – A review. Soil Biology and Biochemistry. 2011;**43**(9):1812-1836. DOI: 10.1016/j.soilbio.2011.04.022

[3] Bergman PCA, Kiel JHA. Torrefaction for biomass upgrading. In: 14th European Biomass Conference & Exhibition; 17-21.10.2005, Paris, France; 2005.

[4] Wannapeera J, Fingtammasan B, Worasuwannarak N. Effects of temperature and holding time during torrefaction on the pyrolysis behaviors of wood biomass. Journal of Analytical and Applied Pyrolysis. 2011;**92**(11):99-105. DOI: 10.1016/j.jaap.2011.04.010

[5] Verhoeff F, Arnuelos AA, Boersma AR, Pels JR, Lensselink J, Kiel JHA, Schukken H, editors. Torrefaction technology for the production of solid bioenergy carriers from biomass and waste. Eindhoven: ECN; 2011. 82 p.

[6] Lipinsky ES, Arcate JR, Reed TB. Enhanced wood fuels via torrefaction. Fuel Chemistry Division Preprints. 2002;**74**(1):408-410.

[7] Poudel J, Ohm T, Lee SH, Oh SC. A study on torrefaction of sewage sludge to enhance solid fuel qualities. Waste Management. 2015;**40**:112-118. DOI: 10.1016/j.wasman.2015.03.012

[8] Atienza-Martinez M, Fonts I, Ábregoa J, Ceamanosa J, Geaa G. Sewage sludge torrefaction in a fluidized bed reactor. Chemical Engineering Journal. 2013;**222**:534-545. DOI: 10.1016/j.cej.2013.02.075

[9] Wiśniewski D, Gołaszewski J. Thermal treatment of dewatered digestate for energy use. In: International Anaerobic Digestion Symposium at Biogas World; 23-25.04.2013, Berlin, Germany; 2013.

[10] Dhungana A, Basu P, Dutta A. Effects of reactor design on the torrefaction of biomass. Journal Energy Resource Technology. 2012;**134**(4) 1-11. DOI: 10.1115/1.4007484

[11] Klimiuk E, Pawłowski M, Pokój T. BIOFUELS Technologies for sustainable development. Warsaw: PWN SA; 2012. 326 p.

[12] Koppejan J, Sokhansanj S, Malin S, Madrali S. Status overview of torrefaction technologies. Enschede: IEA Bioenergy Task 32 raport; 2012. 54 p.

[13] Unn E. Torrefaction plant, its operation and maintenance. Patent number EP 2789677 A1; 2014.

[14] Hopkons ChB, Burnette RP. Autothermal and mobile torrefaction devices. Patent number 0250331 A1; 2009.

[15] Nhuchhen DR, Basu P, Acharya B. A comprehensive review on biomass torrefaction. International Journal of Renewable Energy & Biofuels. 2014; **2014** 1-56. DOI: 10.5171/2014.506376

[16] Teal WB, Gobel RJ, Johnson A. Biomass torrefaction system and method. Patent number 0085023 A1; 2012.

[17] Libingson AD, Thomas BJ. High energy efficiency biomass conversion process. Patent number 8388813 B1; 2013.

[18] Thorn M, Bennett A, Griend SV. Rotary torrefaction reactor. Patent number 0265373 A1; 2011.

[19] Causer TP. System and method for preparation of solid biomass by torrefaction. Patent number 0233914 A1; 2012.

[20] Weisselberg E, Bevacqua J, Borre R. System and method for drying and torrefaction. Patent number 0083531 A1; 2010.

[21] Budarin VL, Milkowski KJ, Shuttleworth P, Lanigan B, Clark JH, Macquarrie DJ, Wilson A. Microwave torrefaction of biomass. Patent number 0219679 A1; 2011.

[22] Verhoeff F, Kiel JHA, Zwart RWR. ECNs moving bed torrefaction technology in light of desired product qualities. 2012. Available from: https://www.ecn.nl/docs/library/report/2012/l12079.pdf Accessed: 2017-01-02

[23] Pawlak-Kurczek H, Czerep M. Thermal fuel valorization technology-torrefaction in autotermall reactor. Nowa Energia. 2016;**1**:1-16.

[24] Bergman PCA. Combined torrefaction and pelletisation – The TOP process. Petten: ECN; 2005. 29 p.

[25] Bridgeman TG, Jones JM, Williams A, Woldron DJ. An investigation of the grindability of two torrefied energy crops. Fuel. 2010;**89**(12):3911-3918. DOI: 10.1016/j.fuel.2010.06.043

[26] Phanphanich M, Mani S. Impact of torrefaction on the grindability and fuel characteristics of forest biomass. Bioresource Technology. 2011;**102**:1246-1253. DOI: 10.1016/j.biortech.2010.08.028

[27] Bridgeman TG, Jones JM, Shield I, Williams PT. Torrefaction of reed canary grass, wheat straw and willow to enhance solid fuel qualities and combustion properties. Fuel. 2008;**87**(6):844-856. DOI: 10.1016/j.fuel.2007.05.041

[28] Uslu A, Faaij APC, Bergman PCA. Pre-treatment technologies, and their effect on international bioenergy supply chain logistics. Techno-economic evaluation of torrefaction, fast pyrolysis and pelletisation. Energy. 2008;**33**(8):1206-1223. DOI: 10.1016/j.energy.2008.03.007

[29] Arias B, Pevida C, Femos J, Plaza MG, Rubiera F, Pis JJ. Influence of torrefaction on the grindability and reactivity of woody biomass. Fuel Processing Technology. 2008;**89**(2):169-175. DOI: 10.1016/j.fuproc.2007.09.002

[30] Jakubiak M, Kordylewski W. Biomass torrefaction. Archiwum Spalania. 2010;**10**(1/2): 11-25.

[31] Pulka J, Wiśniewski D, Gołaszewski J, Białowiec A. Is the biochar produced from sewage sludge a good quality solid fuel? Archives of Environmental Protection. 2016;**42**(4):125-134. DOI: 10.1515/aep-2016-0043

[32] Lehmann J. Bio-energy in the black. Frontiers in Ecology and the Environment. 2007;**5**(7):381-387. DOI: 10.1890/1540-9295(2007)5[381:BITB]2.0.CO;2

[33] Creswell ET, Lefroy RDB. The role and function of organic matter in tropical soils. Nutrient Cycling in Agroecosystems. 2001;**61**(1):7-18. DOI: 10.1023/A:1013656024633

[34] Zoua XM, Ruanc HH, Fua Y, Yanga XD, Shaa LQ. Estimating soil labile organic carbon and potential turnover rates using a sequential fumigation–incubation procedure. Soil Biology and Biochemistry. 2005;**37**(10):1923-1928. DOI: 10.1016/j.soilbio.2005.02.028

[35] Parton WJ, Schimel DS, Cole CV, Ojima DS. Analysis of factors controlling soil organic matter levels in great plains grasslands. Soil Science Society of America Journal. 1987;**51**(5):1173-1179. DOI: 10.2136/sssaj1987.03615995005100050015x

[36] Sparling G, Vojvodic-Vukovic M, Schipper LA. Hot-water-soluble C as a simple measure of labile soil organic matter: The relationship with microbial biomass C. Soil Biology & Biochemistry. 1998;**30**(10/11):1469-1472. DOI: 10.1016/S0038-0717(98)00040-6

[37] Méndez A, Gómezb A, Paz-Ferreirob J, Gascó G. Effects of sewage sludge biochar on plant metal availability after application to a Mediterranean soil. Chemosphere. 2012;**89**(11):1354-1359. DOI: 10.1016/j.chemosphere.2012.05.092

Chapter 3

Pyrolysis Characteristics of Wood-Based Panels and its Products

Jun Mu and Zongyuan Lai

Additional information is available at the end of the chapter

http://dx.doi.org/10.5772/67506

Abstract

Wood-based panels containing urea formaldehyde (UF) resin will be focused on to make clear the influence of UF resin on pyrolysis products and transformation mechanism during pyrolysis of waste wood composites. Thermogravimetric analysis (TGA) will be used to study the thermal degradation reaction kinetics according to different heating rate. Thermogravimetric analyzer coupled to a Fourier transform infrared spectrometer analysis (TGA-FTIR), X-ray photoelectron spectroscopy (XPS) and gas chromatography coupled with mass spectrometry (GCMS) will be used to study the gas, solid, liquid products, respectively. Results from TG and differential thermogravimetric analysis (DTG) indicate that UF resin at first accelerated the degradation rate however inhibited the degradation of wooden composites over the whole pyrolysis process. Compared with wood, UF resin had an obvious effect on the release of HNCO and NH_3. Mass loss of hydrogen is significantly inhibited by UF resin and nitrogen is much stable in the char with structure of the pyridine and pyrrole at relative content of 6.65% and 7.45% respectively. Influence of UF resin on pyrolysis liquids of wooden composites is mainly on nitrogen compounds and ketones rather than aldehydes and esters, which is probably due to the chemical reactions of UF resin with lignin constituent in wood.

Keywords: wood-based panels, pyrolysis characteristic, reaction kinetics, pyrolysis products

1. Introduction

Considering energy and resources, major conventional fossil energy resources are at the verge of extinction. Biomass as a renewable resource has been explored as the substitute of conventional energy and chemical resources. As a kind of lignocellulosic bioresource, waste wood-based panels show great potentials to provide raw materials for the production of renewable fuels and chemical feedstocks. Hence, it is of great significance for the efficient utilization of wood-based panels to better understand the thermal decomposition mechanism of wood-based panels and its products distribution.

2. Pyrolysis characteristics of wood-based panels

Nowadays, the main utilization method of waste wood-based panels is through incineration. Although this method can convert waste wood-based panels into energy, owing to the existence of formaldehyde-based adhesives, it is inevitable to release N-containing gases during the combustion process, which could lead to air pollution. Thermal-chemical processes including pyrolysis, gasification, and hydrogenolysis have proven to be a highly efficient process to utilize bioresource including waste wood-based panels. Because of formaldehyde-based adhesives, characteristics and products of pyrolysis of wood-based panels are much different from that of common bioresource and the influence of resins should be investigated.

2.1. Influence of adhesives on conversion of biomass

In order to study the pyrolysis characteristics of wood-based panels, the pyrolysis of main components of wood-based panels has been studied using a thermogravimetric technique. Most of the research results show that the wood, urea formaldehyde (UF) resin, and wood-based composites all show three steps of weight loss, weight loss with elevated temperature, as presented in **Figure 1**. In the first stages, moisture lost from the material and trace weight loss could be found in all samples because samples were predried before thermal analysis. The second stage is the most significant decomposition stage during the whole process, with most of weight loss. During the second stage, the main release of primary volatiles, a charring process that consists in the rearrangement of char skeleton, occurs. The third stage is mainly the slow decomposition of the residual part, without obvious quality change, leading to the formation of char. This behavior is quite common in biomass samples. The samples present a single, broad peak of decomposition. The majority weight loss of UF resin is in an active step over a range of about 180–330°C. Pyrolysis of UF resin produces higher yield of volatile matters compared to wood. This result may be caused by structure difference between them. The UF resin has a network, which mainly consists of groups of imino, carbonyl, and methylene. Pyrolysis of poplar particleboard produces more char and less volatile matter compared to its two components, which means that the thermal behavior of poplar particleboard is not the reflection of the sum of the behavior of poplar and UF resin. UF resin is more likely to accelerate the chemical reactions of wood-based panels at lower temperature, however, inhibited the degradation of solid residue of wood-based panels at higher temperature during the global pyrolysis process. The thermal stability of UF resin is weaker than wood; however, it enhanced the thermal stability of wood-based panels, which confirms former conclusions.

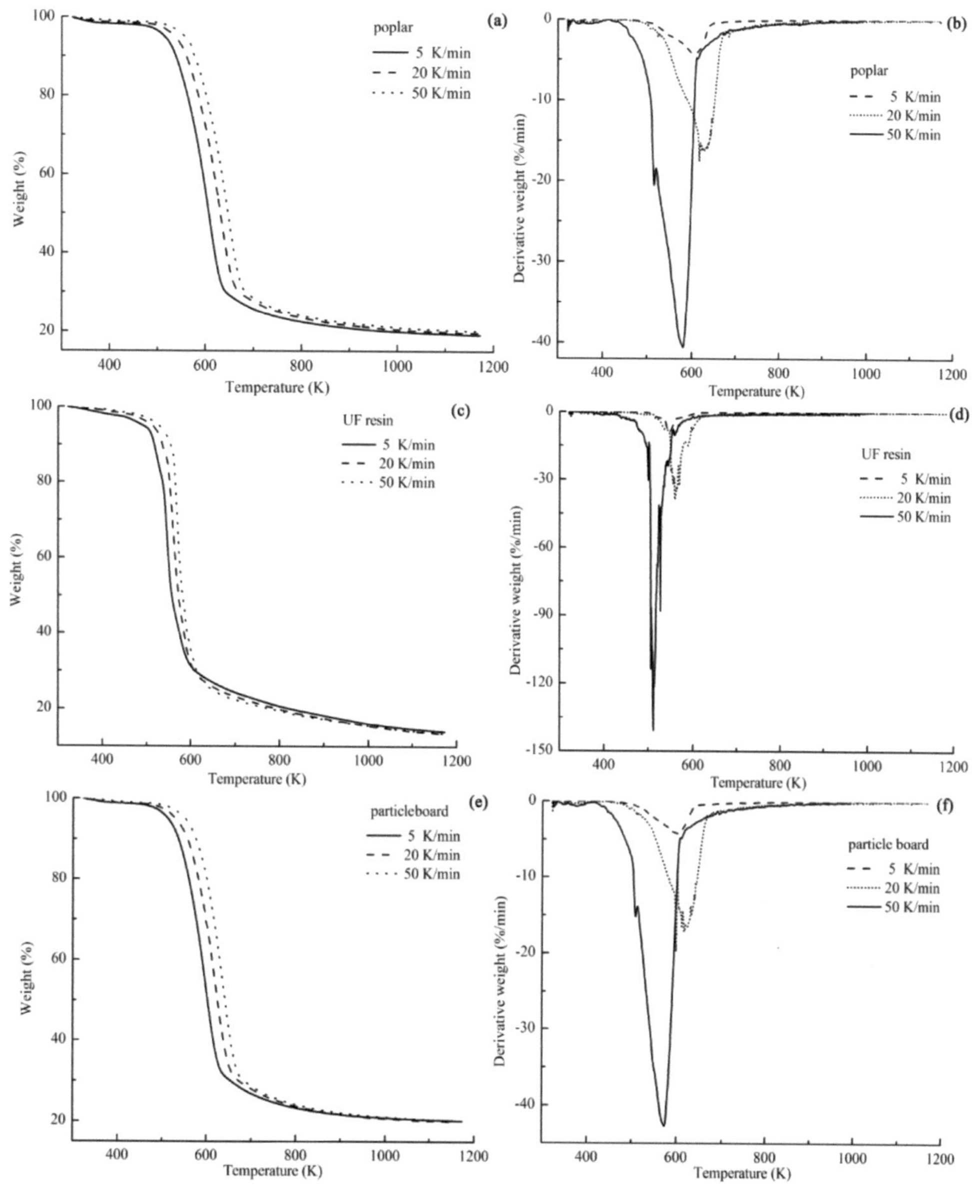

Figure 1. TG curves of poplar (a), UF resin (c) and particleboard (e); and DTG curves of poplar (b), UF resin (d) and particleboard (f) at three different heating rates [1].

2.2. Influence of adhesives on conversion of biomass components

UF resins and modified urea-formaldehyde (MUF) resins have poor thermal stability. A study on pyrolysis of urea formaldehyde resins suggested three reactions resembling wood pyrolysis involved in the pyrolysis, which were initiation reactions, reactions splitting off volatile matters, and reactions forming stabilized structure [2]. Due to the existence of adhesives, characteristics and products distribution of pyrolysis of wood-based panels are different from that of common bioresource and the influence of resins should be investigated.

Cellulose, hemicellulose, and lignin are the main components of biomass, which consist of more than 90% of wood. Given the complexity of wood-based panels, pyrolysis conversion, and the diversity of the obtained products, it may be more convenient to study separately the conversion of each constituent. Based on the research of the main components separately, and the pyrolysis behavior of wood-based panels can be assumed as the sum of the main components weighted summatively.

In this section, the pyrolysis of each of the main biomass components mixed with UF resin is individually discussed. First, the results from the literature concerning thermogravimetric analysis are used to detail the main steps of conversion. The influence of UF resin on the three main biomass constituents is different, as illustrated in **Figure 2**. The results obtained from TG analysis show that three reaction stages involved in the pyrolysis before or after mixed with UF resin. However, the DTG curves representing the mass loss rate of a constituent generally present a different profile with several peaks after addition of UF resin. It is clear that the TG curve of cellulose adding UF resin moved to the higher temperature range after 200°C. Meanwhile, the maximum weight loss rate of cellulose pyrolysis was decreased. So the UF resin is more likely to inhibit the degradation of cellulose after 200°C especially in the suppression of the pyrolysis rate. Former researcher [3] indicated that an acidic environment may accelerate the dehydration reaction of the cellulose. While the alkaline environment caused by UF resin was assumed to be the main reason to inhibit the degradation of cellulose. For xylan (representative of hemicellulose), TG and DTG curves before and after mixed with UF resin generally coincided. So the UF resin had little influence on the pyrolysis behavior of the hemicellulose. For lignin, TG curves added UF resin moved to the low temperature segment after 200°C. With the addition of UF resin, the char yield reduced and the maximum mass loss rate slightly increased. Therefore, the promotion of UF resin in the process of pyrolysis for the lignin after 200°C can be observed. While the promoting effect plays a more evident role in weight loss rather than the degradation rate. It can be concluded that during the process of the pyrolysis of waste wood-based panels, lignin was the one that UF resin mainly impacted among the three main components of wood.

Innovatively, the weight-loss character of the model (made from cellulose, xylan, and lignin, based on the chemical components study of poplar wood), the main components as well as the ones mixed with UF were analyzed by TG-FTIR. According to the results from TG analysis, the residual weight of poplar and model is very close, and the weight loss curve is basically the same. It can be seen that the model has similar pyrolysis characteristics to common biomass. The TG curves show that the UF resin has obvious effects on the pyrolysis of the model. In the TG analysis, it is well known that the two curves coincide basically before 250°C, the differences appear gradually after 250°C. Studies realized that UF resin could promote the pyrolysis of model on the basis of a small diminution in char yield [4].

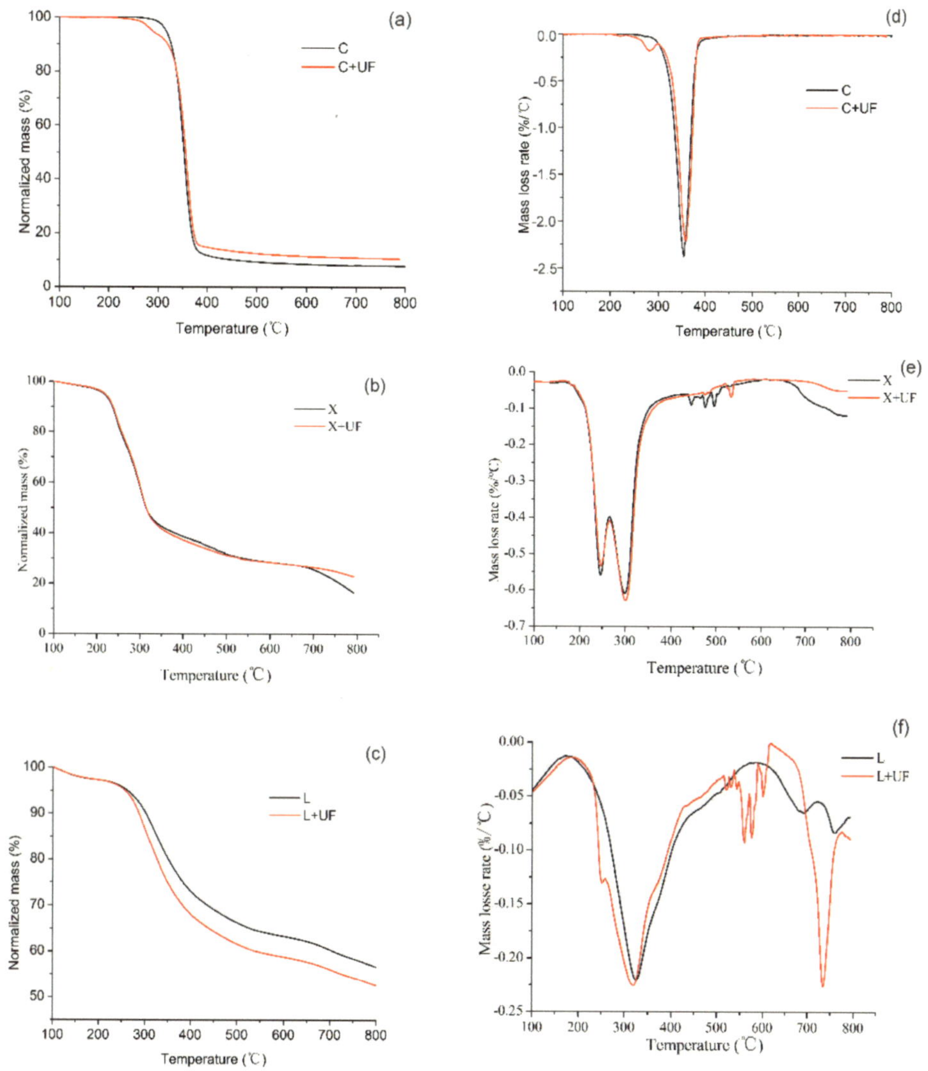

Figure 2. TG and DTG curves of cellulose (C), xylan (representative component of hemicellulose, X) and lignin (L) as well as the ones mixed with UF resin [4].

2.3. Kinetic study

There are variety of models available for analyzing the kinetics of the investigation of biomass thermal decomposition, including first-order [5], discrete activation-energy distributions [6], and sequential models [7] such as models having Gaussian [8], Weibull [9], and Gamma [10] distributions. The distributed activation energy model (DAEM), representing the sequential models, has been employed successfully for analyzing complex reactions in thermal decomposition of biomass. The model assumes that an infinite number of first-order parallel reactions having unique kinetic parameters take place concurrently. The kinetic parameters could be calculated through thermogravimetric analysis (TGA) without considering complex chemical reactions during a thermal decomposition process [11]. The representative equation for the Miura-Maki model [12] is

$$\ln\left(\frac{\beta}{T^2}\right) = ln\left(\frac{k_0 R}{E}\right) + 0.6075 - \frac{E}{RT} \tag{1}$$

To investigate this further, based on the pyrolysis of particleboard, the kinetic parameters were calculated. Zhang et al. [1] reported the plots of $\ln(\beta/T^2)$ versus $1/T$ at selected α for poplar, UF resin and particleboard, as presented in **Figure 3**. For every value of α, the three points corresponding to the three heating rates are positioned in a straight line, as the method predicts. The sum of the pyrolysis behavior of poplar and UF resin cannot embody the thermal behavior of poplar particleboard. The activation energy of poplar and particleboard is 140–200 kJ/mol; however, the activation energy of UF resin is 150–300 kJ/mol, and that the activation energy of particleboard is lower than that of the poplar during the main pyrolysis stage because of the promoting effect of UF resin.

3. Influence of adhesives on pyrolysis products distribution of biomass

Pyrolysis is a capital step of biomass thermochemical conversion as it is the first step of all the process. Biomass by the action of heat in an inert atmosphere can be converted into gas, char, and a liquid composed of a mixture of hundreds of oxygenated organic compounds [13, 14]. More precise information about conversion mechanisms of wood-based panels is provided by coupling TG analysis and on-line analyzer of the evolved products such as FTIR. With this approach, it is possible to study the composition of the evolved products throughout the temperature rise of the sample. Based on the evolution of the structure of the gas products, we can have better understanding of the evolution of the composition of the sample. What is more, the knowledge of formation of the evolved compounds helps to understand the reactions of fragmentation of unstable functions during the main pyrolysis stage and of rearrangement during the charring process [15]. Moreover, the X-ray photoelectron spectroscopy (XPS) technique can be used to probe the chemical changes occurring in the solid residue prepared by pyrolysis, the GC-MS technique can be also used to analyze the composition of the pyrolysis liquid condensed from volatile matters of the pyrolysis of wood-based panels, additional information can be obtained. Based on a review of the study of our team and other authors, the influence of adhesives on the formation of products (char, liquid, gas) can be described.

3.1. Gas

Owing to the complexity of biomass conversion, few authors tried to explain the influence adhesives on the chemical reactions involved at the molecular scale. The study of the volatile matters obtained from the pyrolysis of wood-based panels mainly concentrates on N-containing gases, such as NH_3, HCNO, and HCN. For a better understanding of these mechanisms,

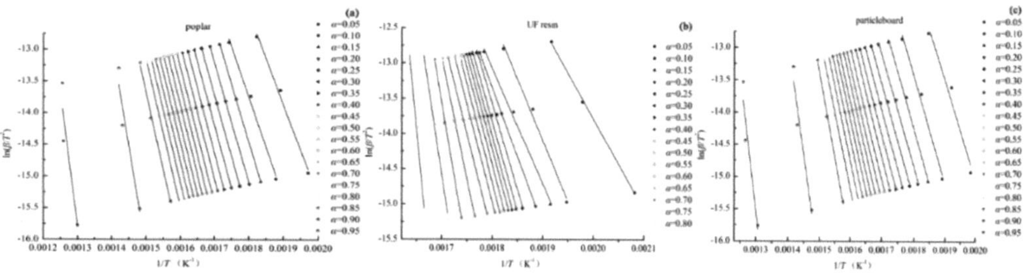

Figure 3. Plots of $\ln(\beta/T^2)$ versus $1/T$ of poplar (a), UF resin (b), and particleboard (c) at different conversion rates [1].

based on a review of the published research works, the influence of the adhesives on the formation of these products is discussed.

The volatile matters obtained from the pyrolysis of wood-based panels and common bioresource contain CH_4, CO, CO_2, NH_3, HCNO, and HCN [16–18]. However, as presented in **Figure 4**, N-containing gases of the pyrolysis of common bioresource derived from the decomposition of extractives in wood, nitrogen gases of the pyrolysis of wood-based panels are significantly influenced with the addition of UF resin. It is concluded that the degradation of urea function in UF resin generates HNCO and a precursor of HCN, which will crack into HCN (the gas is highly toxic, when conducting experiments, operators should be very careful) under 650°C [19]. In regard to NH_3, we could make conclusions that the fragmentation of UF resin unit directly leads to the formation of azyl at the narrow temperature range between 180 and 320°C and then the formation of large quantities of NH_3 with the highest intensity at 210°C [17]. A small amount of CH_4, CO, and CO_2 could also be detected gradually at the low temperature range, which is primarily due to the break of the methoxyl group and single C–C bonds. According to the literature [3, 20], it is assumed that CH_4, CO_2, and CO are formed, respectively, by the cracking of methoxy groups, decarboxylation, and decarbonylation. In the main pyrolysis stage, the break of carboxyl in the lateral chains in furfural acid results in the release of CO_2. The less stable carbonyl in volatiles is more likely to be the precursor of CO. The formation of CO is also largely influenced by second cracking of volatiles during the thermal process [21]. During the third pyrolysis stage in a high temperature range between 400 and 800°C, almost all the gases reach their equilibrium intensity value, which is relatively low. Lignin degradation plays an important role in this stage. Functional groups in the lateral chains of phenyl propane are still cracking and volatiles would undergo further secondary reactions, resulting in the formation of incondensable gas and of small chain organic compounds.

TG-FTIR analysis of each of the main biomass components mixed with UF resin can explicitly understand the influence of UF resin on the formation of gas products. Cellulose is a linear homopolysaccharide of cellobiose monomers, composed of C, H, and O. Hence, the element N of the nitrogen gases obtained from the pyrolysis of UF-cellulose blend only results from the pyrolysis of UF resin. The addition of UF promoted the formation of water and carboxylic acids during cellulose pyrolysis. The study of Shafizadeh and Bradbury [22] showed that the pyrolysis process of cellulose is divided into two stages. Below 300°C, reducing the degree

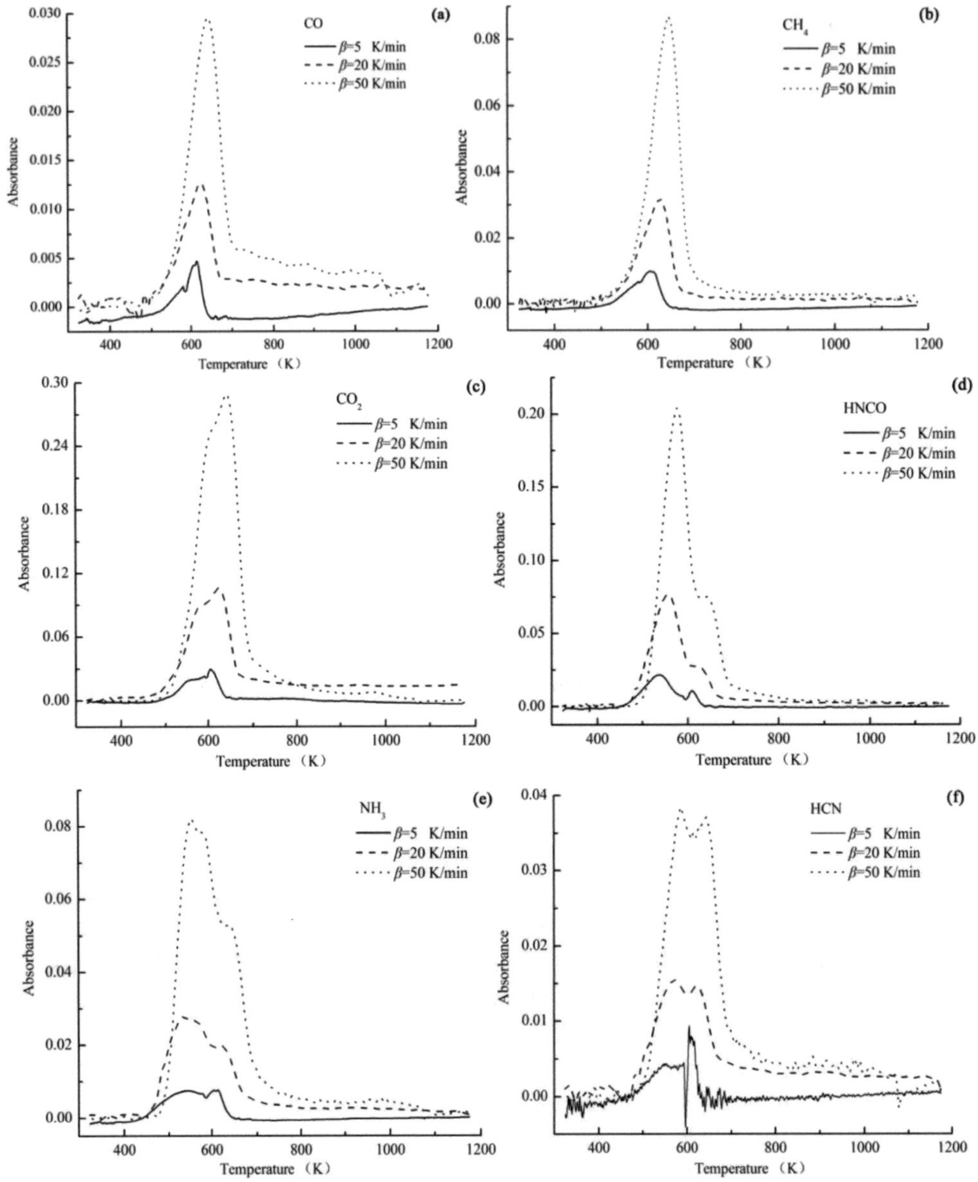

Figure 4. Evolutions of CO (a), CH_4 (b), CO_2 (c), HNCO (d), NH_3 (e), and HCN (f) level in pyrolysis product during pyrolysis of particleboard at three different heating rates [1].

of polymerization caused by bond breaking leads to the formation of free radicals (carbonyl radicals, carboxyl radicals, and hydroperoxyl radicals), the release of H_2O, CO, and CO_2, depolymerization induced by transglycosylation appears at 300°C or above. The influence of UF resin on the pyrolysis of cellulose is mainly in the low temperature range, namely the first stage of cellulose pyrolysis. UF resin promotes the fracture of chemical bonds in cellulose and accelerates the formation of carbonyl and hydroxyl radicals. For lignin, UF resin could promote the release of CO during the pyrolysis process. The interaction between UF resin and lignin leads to the formation of weak stable compounds, then releases large amounts of nitrogen compounds and oxycarbides. The influence of UF resin on wood-based panels could be found mainly in the release of nitric gases. Intensity peaks of nitric gases of wood-based panels are higher and occur earlier than that of wood. Therefore, UF resin is the main factor contributing to the nitric gases of the pyrolysis of waste wood-based panels while the influence on the CO_2 is not obvious.

3.2. Char

Char formation consists in the conversion of biomass in a solid residue, namely char, which presents an aromatic polycyclic structure [23, 24]. Intra- and intermolecular rearrangement reactions occurred in the process of char formation could result in a higher degree of reticulation and in a higher thermal stability of the residue [23, 25]. The significant steps of these reactions are the formation of benzene rings and the combination of these rings in a polycyclic structure. Recombination consists in the combination of volatile compounds lead to the formation of a higher molecular weight molecule, which sometimes is no longer volatile matter under the conditions of temperature of the reactor [26, 27]. When the recombination happens inside the pores of the polymer, this reaction can lead to the information of a secondary char.

The char of wood-based panels obtained from pyrolysis also contains C, H, N, and O. The content of these elements in the char is different from wood-based panels. The content ratio of C in sample is relatively stable both before and after pyrolysis. Although a large amount of element C are lost, C is still the main element in the char, accounting for 70–90%, and the mass loss of these elements could be influenced by UF resin. The influence of UF resin on element C is not significant; however, the mass of element O is the largest one, probably fixed in the form of oxygenated compounds in pyrolysis liquid and CO, CO_2 in gases during process. In contrast of content in original sample, it could be found that UF resin inhibited the loss of H mostly during pyrolysis; however, there was no significant relationship between the loss of element H and the content of UF resin in wood-based panels. The content of N is relatively steady both before and after pyrolysis of particleboard with little changes in each sample, while the content of N in medium density fiberboard (MDF) decreases dramatically. This difference originates probably from the different interactions between UF resin and different lignocellulosic materials. A study on pH value of the char of wood and wood-based panels display as neutral and alkaline; therefore, it is concluded that the element N in UF resin is also retained in the char of wood-based panels.

To investigate this further, the X-ray photoelectron spectroscopy technique was used to research the existing forms of nitrogen compounds in char of wood-based panels, as presented in **Table 1**.

The main form of element N in MDF is primary amine (399.4 eV), and some are amide (399.8 eV). The main form of element N in wood is amide (399.8 eV), which originates from the protein existed in wood extractives. Meanwhile, amide (399.8 eV) is the main form of element N in UF resin, accounting for 32.69%. Both of UF resin and MDF have strong intensities of peaks attributed to the C-N stretching vibration (1690–1590 cm^{-1}). However, the structure of primary and amide is not detected in the samples after pyrolysis. The existing forms of nitrogen compounds in MDF char were pyridine (398.4 eV) and pyrrole (400.2 eV). In the pyrolysis process, the N-containing compounds undergo bond cleavage and rearrangement to form stable cyclic structures of pyridine and pyrrole.

3.3. Liquid

Biomass mainly consists of cellulose, hemicellulose, and lignin, and therefore its pyrolysis behaviors as the integrated performance of these three components [29]. Given the complexity of lignocellulosic material and of biomass pyrolysis conversion, the liquid products obtained from pyrolysis of biomass consist of complex chemical compositions, including acids, alcohols, aldehydes, ketones, and esters [30, 31] and have higher oxygen content and lower heating value than hydrocarbons. Furthermore, its acidity, high water content, and corrosion make it hard to be used as a transportation fuel [32]. In general, the physicochemical properties of pyrolytic products are closely related to future industrial applications. The compositions of pyrolysis liquid of wood-based panels are of obvious difference from those produced from biomass. The pH value of pyrolysis liquid of common bioresource always is shown as acidity, while that of pyrolysis liquid of wood-based panels varied from acidity to alkalescency. These difference occurred in the two types pyrolysis liquid originates from the existence of UF resin in the wood-based panels and conditions for thermal decomposition. And furthermore, the variation of pH value of pyrolysis liquid of wood-based panels may be due to the types of wood-based panels and pyrolysis conditions. The specific gravity of pyrolysis liquid of wood-based panels is relatively stable, varying from 1 to 1.1 [33, 34]. The specific gravity of pyrolysis liquid also is influenced by pyrolysis conditions, such as pyrolysis temperature and the heating rate. The different compositions of pyrolysis liquid of wood-based panels compared to common bioresoure are mainly on the nitrogen compounds. The relative content of nitric compounds in pyrolysis

Samples	**Relative amount (%)**					
	Binding energy (eV)	399.4	399.8	398.4	400.2	403
	Chemical structure	**Primary amine**	**Amide**	**Pyridine**	**Pyrrole**	**Pyrrole oxide**
Original samples	MDF	4.28	1.07	/	/	/
	Wood fiber	/	0.49	/	/	/
	UF	/	32.69	/	/	/
Char	MDF	/	/	6.65	7.45	0.4
	Wood fiber	/	/	0.24	0.71	/
	UF	/	/	1.36	2.29	/

Table 1. Contents of different N constructions of wood fiber, MDF, and UF resin [28].

liquids of wood is very few but different types of amides are detected in pyrolysis liquids of wood-based panels and UF resin. The formation of nitric compounds in pyrolysis liquids originates from the conversion of urea in UF resins as applied in wood-based panels and the interaction between UF resin and wood [1]. The transformation of nitrogen-containing compounds is influenced by the interactions between UF resins and wood constituents. From GCMS analysis of the pyrolysis liquid of UF resin, it can be found that nitrogenous substances are the main components. Feng et al. reported that the two most abundant components in pyrolysis liquids of UF resin are methyl-urea and N, N′-dimethyl-urea [17], while Zhang et al. reported that the most relative content is N-ethylformamide [1]. In addition, other nitrogenous substances, such as 1-Methyl-2,4,5-trioxoimidazolidine, 2,3-Pyridinediamine, Pyrimidine,5-methyl-), urea, formyltrimethyl-, can also be detected in the pyrolysis liquid of UF resin. Nitrogen-containing heterocyclic compounds in the pyrolysis liquid of UF resin, such as pyridine, pyrimidine and piperazine, are alkaline and have irritating smell (see **Table 2**).

The anhydro-saccharides identified correspond to glucose-derived compounds undergone several dehydration reactions. Glycosidic bond cleavage of cellulose and aldol condensation reaction lead to the formation of 1,6;2,3-Dianhydro-4-O-acetyl-beta-D-gulopyranose and 2,3-Anhydro-D-mannosan. Furan derivatives are considerable constituents of pyrolysis liquid of biomass. These furan derivatives are generated by the pyrolysis of holocellulose. These compounds can also be considered as depolymerization products. Various kinds of ketones are detected in pyrolysis liquids, which are probably formed by decarboxylation with active chemical properties. Ketones are much likely to be originated from the large amount of carbonyls in wood. Some pyridone, imidazolone in the form of nitrogen heterocyclic compounds are found in pyrolysis liquids of wood-based panels but little are found in UF resin, suggesting that nitrogen from UF resin is reformed and fixed with chemical groups degraded from wood in the form of ketones in the pyrolysis liquids.

During the pyrolysis process, lignin molecule chains are splitted and fragments are rearranged, and pyrolysis products are mainly phenolic compounds. Phenolic compounds in pyrolysis liquids of particleboard (PB) are mainly originated from lignocelluloses; however, their structures might be influenced by UF resin in terms of nitrogen during the thermal conversion process, resulting in a small amount of phenol (peak area around 1.5%) in PB despite the large amount (peak area over 15%) detected in pyrolysis liquids of wood. The influence of UF resin on other components including aldehydes, alcohols, and esters is not significant since these compounds are not the main components in pyrolysis liquids of PB and not detected in pyrolysis liquids of UF resin.

The influence of UF resin for particleboard is mainly on the nitrogen compounds, as presented in **Table 3**. There is a small amount of nitric compounds in the pyrolysis liquid of wood. However, a large number of nitrogenous compounds in the pyrolysis liquid of wood-based panels are detected, such as pyrimidine,5-methyl-and 1,3,5-triazine-2,4,6(1H,3H,5H)-trione,1,3,5-trimethyl-, just generated by the pyrolysis of UF resin. But others generated by the concerted reaction of wood and UF resin, such as 4-(1H)-pyridinone,2,3-dihydro-1-methyl. Thus, it can be seen that the transformation of nitrogen-containing compounds in the process of wood pyrolysis is influenced by the interactions between UF resin and wood.

As the main constituent in wood, cellulose will go through dehydration reaction during pyrolysis process, forming levoglucose and glycolaldehyde. The glucosyl groups will then crack into smaller structures in high temperature. Research studies show that acids could catalyze the dehydration reactions while UF resin introduces an alkaline atmosphere during the pyrolysis, which inhibits further cracking of glucosyl groups. Compared with cellulose, hemicelluloses are inhomogeneous glycan with weaker thermal stability. The pyrolysis of

Retention time (min)	Designation	Molecular formula	Relative amount (%)
3.30	Acetonitrile	$C_4H_8N_2$	0.54
3.51	N-ethylformamide	C_3H_7ON	54.52
5.12	Pyrimidine,5-methyl-	$C_5H_6N_2$	4.50
6.59	1H-imidazole,1-methyl-	$C_4H_6N_2$	1.18
8.66	Cyclohexanol,3-methyl-	$C_7H_{14}O$	0.30
9.92	Urea, formyltrimethyl-	$C_5H_{10}O_2N_2$	4.30
10.16	Pentanamide, N-(aminocarbonyl)-	$C_6H_{12}O_2N_2$	0.88
10.72	3,4-Pyridinediamine	$C_5H_7N_3$	1.81
11.07	N-N'-dimethyloxamide	$C_4H_8O_2N_2$	1.26
13.45	1H-imidazole[1,2-b] pyrazole,2,3-dihydro-	$C_5H_7N_3$	0.19
13.62	(1H)-Pyrrole-3-carbonitrile,2-methyl-	$C_6H_6N_2$	0.13
13.79	Hydrouracil,1-methyl-	$C_5H_8O_2N_2$	1.40
14.22	2,3-Pyridinediamine	$C_5H_7N_3$	6.79
15.68	Piperazine,1-methyl-	$C_5H_{12}N_2$	0.86
16.01	Hexahydroindole	$C_8H_{13}N$	1.63
17.07	2-Acetyl-3-methyl-3,4,5,6-tetrahydropyridine	$C_8H_{13}ON$	0.50
18.36	1,3,5-Triazine-2,4,6(1H,3H,5H)-trione,1,3,5-trimethyl-	$C_6H_9O_3N_3$	2.38
18.49	S-triazolo[4,3-A] pyridine,8-amino-	$C_6H_6N_4$	2.03
19.81	1-Methyl-2,4,5-trioxoimidazolidine	$C_4H_4O_3N_2$	7.65
21.20	1,2,4-Trizolo[4,3-A] pyridin-8-amine,3-methyl-	$C_7H_8N_4$	2.70
22.94	6-Methyl-7,8-dihydro-2(1H)-pteridinone	$C_7H_8ON_4$	1.75
24.46	2-Pyrrolidinophenol	$C_{10}H_{13}ON$	2.69

Table 2. Main components of pyrolysis liquid of urea formaldehyde resin [1].

http://dx.doi.org/10.5772/67506

Retention time (min)	Designation	Molecular formula	Relative amount (%)
4.55	2-Furanmethanol	$C_5H_6O_2$	14.22
5.29	Acetamide,N,N-dimethyl-	C_4H_9ON	1.84
5.68	2-Cyclopenten-1-one,2-methyl-	C_6H_8O	0.14
6.03	1-Pyrrolidinecarbonitrile	$C_5H_8N_2$	11.24
7.39	2-Cyclopenten-1-one,3-methyl-	C_6H_8O	2.33
7.58	4-Aminopyrimidine	$C_4H_5N_3$	0.50
8.09	Pyrimidine,5-methyl-	$C_5H_6N_2$	3.19
8.98	2-Cyclopenten-1-one,2-hydroxy-3-methyl-	$C_6H_8O_2$	1.45
9.22	2-Sec-butyl-3-methyl-1-pentene	$C_{10}H_2O$	3.19
9.27	2-Cycpenten-1-one,2,3-dimethyl-	$C_7H_{10}O$	0.77
9.51	Dl-citrulline	$C_6H_{13}O_3N_3$	3.25
9.64	Piperidine,1-methyl-	$C_6H_{13}N$	0.02
9.88	Phenol,2-methyl-	C_7H_8O	0.92
10.68	Ethanone,1-(2-methyl-1-cyclopenten-1-yl)-	$C_8H_{12}O$	6.64
10.82	1,3-Cyclopentanedione,4-ethyl-	$C_7H_{10}O_2$	0.71
10.94	2,5-Pyrrolidinedione,1-methyl-	$C_5H_7O_2N$	0.62
11.05	1-H-Pyrazole,1,3,5-trimethyl-	$C_6H_{10}N_2$	0.21
11.37	1,3-Cyclopentanedione,2,2-diethyl-	$C_7H_{10}O_2$	2.12
11.56	Maltol	$C_6H_6O_3$	1.91
11.69	2-Cyclopenten-1-one,3-ethyl-2-hydroxy-	$C_7H_{10}O_2$	1.17
11.85	Bicyclo(2,2,2) octane,2-methyl-	C_9H_{16}	0.10
12.18	4-(1H)-Pyridinone,2,3-dihydro-1-methyl-	C_6H_9ON	2.12
12.99	1H-Imidazole,2,4,5-trimethyl-	$C_6H_{10}N_2$	0.03
13.36	4,5-Octanedione	$C_8H_{14}O_2$	2.04
13.6	Phenol,2-methuxy-4-methyl-	$C_8H_{10}O_2$	0.68
13.68	Quinuclidine-3-ol	$C_7H_{13}ON$	0.15

Retention time (min)	Designation	Molecular formula	Relative amount (%)
13.83	S-triazolo[4,3-a] pyrazine,3-methyl-	$C_6H_6N_4$	1.31
14.1	2-Hydroxy-3,5-diethyl-5-methylcyclopent-2-en-1-one	$C_{10}H_{16}O_2$	0.55
14.39	1,2-Benzenediol	$C_6H_6O_2$	1.39
14.51	2-Isopropoxyphenol	$C_9H_{12}O_2$	0.68
14.69	Oxazole,4-ethyl-2,5-dimethyl-	$C_7H_{11}ON$	0.95
14.81	2-Methyl-4-(1-methylethyl)-2-cyclohexenone	$C_{10}H_{16}O$	0.57
15.01	Phenol,2,6-dimethoxy-	$C_8H_{10}O_3$	0.22
15.64	1,2-Benzenediol,3-methoxy-	$C_7H_8O_3$	3.53
15.89	1,2-Benzenediol,3-methyl-	$C_7H_8O_2$	1.15
16.27	Pyridine,4-(1-pyrrolidinyl)	$C_9H_{12}N_2$	0.45
16.86	1,2-Benzenediol,4-methyl-	$C_7H_8O_2$	1.73
18.07	Phenol,2,6-dimethoxy-	$C_8H_{10}O_3$	11.67
18.22	Benzenemethanol,3-hydroxy-5-methoxy-	$C_8H_{10}O_3$	1.58
18.65	1,3,5-Triazine-2,4,6(1h,3h,5h)-trione,1,3,5-trimethyl-	$C_6H_9O_3N_3$	2.28
19.07	2,3-Dimethylhydroquinone	$C_8H_{10}O_2$	0.36
19.14	1,3-Benzenediol,4-ethyl-	$C_8H_{10}O_2$	0.27
20.36	1,2,3-Trimethoxybenzene	$C_9H_{10}O_3$	3.11
20.44	Ethanone,1-(2,3,4-trihydroxyphenyl)	$C_8H_8O_4$	1.14
21.41	Ethanone,1-(4-hydroxy-3-methoxyphenyl)-	$C_9H_{10}O_3$	0.18
21.64	8-Quinolinol,5-amino-	C_9H_8ON	0.27
22.21	Benzene,1,2,3-trimethoxy-5-methyl-	$C_{10}H_{14}O_3$	1.52
22.38	2-Propanone,1-(4-hydroxy-3-methoxyphenyl)-	$C_{10}H_{12}O_3$	0.49
22.54	2-(2-Aminoethyl)-4-amino-6-dimethylamino-s-triazine	$C_7H_{14}N_6$	0.14
22.82	4(1H)-Pteridinone,2-amino-	$C_5H_5ON_5$	0.23
23.98	Phenol,2,6-dimethoxy-4-(2-propenyl)-	$C_{11}H_{14}O_3$	0.23

Retention time (min)	Designation	Molecular formula	Relative amount (%)
24.13	Benzene,1,1'-propylidenebis	$C_{15}H_{16}$	0.20
24.53	5-Oxohexanenitrile	C_6H_9ON	0.18
25.1	Butan-2-one,4-(3-hydroxy-2-methoxyphenyl)-	$C_{11}H_{14}O_3$	0.33
25.44	2-Pyrrolidinophenol	$C_{10}H_{13}ON$	0.72
26.26	Phenol,2,6-dimethoxy-4-(2-propenyl)-	$C_{11}H_{14}0_3$	0.40
26.79	Ethanone,1-(4-hydroxy-3,5-dimethoxyphenyl)-	$C_{10}H_{12}O_4$	0.45
27.34	Benzoic acid,4-hydroxy-3,5-dimethoxy-,hydrazide	$C_9H_{12}O_4N_2$	0.14

Table 3. Main components of pyrolysis liquid of particleboard [1].

xylan in hemicelluloses releases acids and aldehydes products, which are not significantly influenced by UF resin. Lignin is the only aromatic constituent in wood with very complicate structures while in UF resin most structures are linear and network linkages. However, many of detected nitrogen compounds show aromatic properties, such as pyridine and pyrimidine. It could be concluded that concerted reactions between UF resin and lignin occur in the pyrolysis process so that lignin is probably the mostly influenced constituent by UF resin in wood.

4. Conclusion

From TG and DTG results, it is found that UF resin in wood-based panels make them to degrade easily at the beginning of thermal treatment but inhibit the degradation of wood-based panels during the second and third stages. Pyrolysis of wood-based panels produces more char and less volatile matters compared to its two components, which indicates that the thermal behavior of poplar particleboard is not just the reflection of the sum of the behavior of poplar and UF resin. The activation energy of both wood and UF resin is higher than the activation energy of wood-based panels during the main pyrolysis stage. The pyrolysis nitrogen gases of wood-based panels were mainly HCN, HCNO, and NH_3.The experimental results showed that the existing form of nitrogen compounds in MDF was a mesh structure of large molecules. The existing forms of nitrogen compounds in MDF char were pyridine and pyrrole. The influence of UF resin on pyrolysis liquid of wood-based panels is mainly on nitrogen compounds. More nitrogenous compounds are detected in the pyrolysis liquid products.

According to the existing research and practice, pyrolysis liquid of wood could improve the physical and chemical properties and fertility of soil and promote plant growth. The pyrolysis liquids of wood-based panels contain a high content of nitric compounds, besides those degraded from the main components of wood. Nitrogen is one of a large number of elements, which plant needs and plays an important role in plant growth. The pyrolysis liquids of wood-based panels with UF resin can also inhibit the growth of mold and bacteria [35]. But the application of pyrolysis liquid of wood-based panels in agriculture is still a new subject, worth exploring.

Acknowledgements

This study was funded by the National Natural Science Foundation of China (31170533).

Author details

Jun Mu* and Zongyuan Lai

*Address all correspondence to: mujun222@sina.com

College of Material and Technology, Beijing Forestry University, Beijing, China

References

[1] Zhang Y, He Z B, Xue L, et al. Influence of a urea–formaldehyde resin adhesive on pyrolysis characteristics and volatiles emission of poplar particleboard. RSC Advances, 2016, 6(16):12850-12861.

[2] Hirata T, Kawamoto S, Okuro A. Pyrolysis of melamine–formaldehyde and urea–formaldehyde resins. Journal of Applied Polymer Science, 1991, 42(12):3147-3163.

[3] Rowell R. The Chemistry of solid wood. Wood Science & Technology, 1985, 19(19):17-18.

[4] Li S J, Chen S H, Mu J. Influence of UF resin on pyrolysis characteristics of biomass components: A thermogravimetric study. International Conference on Biobase Material Science and Engineering (BMSE), 2012, Changsha, China.

[5] Kissinger H E. Variation of peak temperature with heating rate in differential thermal analysis. Journal of Research of the National Bureau of Standards, 1956, 57(4):217-221.

[6] Burnham A K, Braun R L. Global kinetic analysis of complex materials. Energy & Fuels, 1999, 13(1):1-22.

[7] Braun R L, Burnham A K. Analysis of chemical reaction kinetics using a distribution of activation energies and simpler models. Energy & Fuels, 1987, 1(2):153-161.

[8] Anthony D B, Howard J B. Coal devolatilization and hydrogasification. AIChE Journal, 1976, 22(4):625-656.

[9] Lakshmanan C C, White N. A new distributed activation energy model using Weibull distribution for the representation of complex kinetics. Energy & Fuels, 1994, 8(6):1158-1167.

[10] Astarita G. Lumping nonlinear kinetics: Apparent overall order of reaction. AIChE Journal, 1989, 35(4):529-532.

[11] Ferrara F, Orsini A, Plaisant A, et al. Pyrolysis of coal, biomass and their blends: Performance assessment by thermogravimetric analysis. Bioresource Technology, 2014, 171C(2):433-441.

[12] Miura K. A new and simple method to estimate f(E) and ko(E) in the distributed activation energy model from three sets of experimental data. Energy & Fuels, 1995, 9:302-307.

[13] Garcia-perez M, Chaala A, Pakdel H, et al. Characterization of bio-oils in chemical families. Biomass & Bioenergy, 2007, 31(4):222-242.

[14] Evans R J, Milne T A. Molecular characterization of the pyrolysis of biomass. 1. Fundamentals. Energy Fuels, 1987, 1:123-137.

[15] Collard F X, Blin J. A review on pyrolysis of biomass constituents: Mechanisms and composition of the products obtained from the conversion of cellulose, hemicelluloses and lignin. Renewable & Sustainable Energy Reviews, 2014, 38(5):594-608.

[16] Girods P, Dufour A, Rogaume Y, et al. Thermal removal of nitrogen species from wood waste containing urea formaldehyde and melamine formaldehyde resins. Journal of Hazardous Materials, 2008, 159(2-3):210-221.

[17] Feng Y S, Mu J, Chen S H, et al. The influence of urea formaldehyde resins on pyrolysis characteristics and products of wood-based panels. Bioresources, 2012, 7(4):4600-4613.

[18] Li S J, Mu J, Zhang Y. Influence of urea formaldehyde resin on pyrolysis of biomass: A modeling study by TG-FTIR. Spectroscopy & Spectral Analysis, 2014, 34(6):1497-1501(5).

[19] Girods P, Dufour A, Rogaume Y, et al. Comparison of gasification and pyrolysis of thermal pre-treated wood board waste. Journal of Analytical & Applied Pyrolysis, 2009, 85(1):171-183.

[20] Gao N, Li A, Quan C, et al. TG–FTIR and Py–GC/MS analysis on pyrolysis and combustion of pine sawdust. Journal of Analytical & Applied Pyrolysis, 2013, 100(6):26-32.

[21] Li X. Research advance on mechanism of biomass pyrolysis and gasification. Advances in Fine Petrochemicals, 2009, 10(10): 45-50.

[22] Shafizadeh F, Bradbury A G W. Thermal degradation of cellulose in air and nitrogen at low temperatures. Journal of Applied Polymer Science. 1979, 23(5):1431-1442.

[23] Mcgrath T E, Chan W G, Hajaligol M R. Low temperature mechanism for the formation of polycyclic aromatic hydrocarbons from the pyrolysis of cellulose. Journal of Analytical & Applied Pyrolysis, 2003, 66(1):51-70.

[24] Pastorova I, Botto R E, Arisz P W, et al. Cellulose char structure: A combined analytical Py-GC-MS, FTIR, and NMR study. Carbohydrate Research, 1994, 262(1):27-47.

[25] Scheirs J, Camino G, Tumiatti W. Overview of water evolution during the thermal degradation of cellulose. European Polymer Journal, 2001, 37(5):933-942.

[26] Hosoya T, Kawamoto H, Saka S. Pyrolysis behaviors of wood and its constituent polymers at gasification temperature. Journal of Analytical & Applied Pyrolysis, 2006, 78(2):328-336.

[27] Morf P, Hasler P, Nussbaumer T. Mechanisms and kinetics of homogeneous secondary reactions of tar from continuous pyrolysis of wood chips. Fuel, 2002, 81(7):843-853.

[28] Chen S H, Li S J, Mu J. Conversion characteristics of urea formaldehyde resin adhesive in fiberboard through pyrolysis into solid product. Journal of Beijing Forestry University, 2013, 35(4):123-127.

[29] Xu S Y, Chen J J, Cao D R. Analysis of components in wood vinegar. Guangzhou Chemistry, 2006, 31(3):28-31.

[30] Branca C, Paola Giudicianni A, Blasi C D. GC/MS characterization of liquids generated from low-temperature pyrolysis of wood. Industrial & Engineering Chemistry Research, 2003, 42(14):3190-3202.

[31] Yaman S. Pyrolysis of biomass to produce fuels and chemical feedstocks. ChemInform, 2004, 45(31):651-671.

[32] Lappas A A, Samolada M C, Iatridis D K, et al. Biomass pyrolysis in a circulating fluid bed reactor for the production of fuels and chemicals. Fuel, 2002, 81(16):2087-2095.

[33] Mu J, Yu Z M, Zhang D R, et al. Pyrolysis characteristics of disused composite panels and properties of its products. Journal of Beijing Forestry University, 2011, 33(1):125-128.

[34] Mu J, Yu ZM, Liu J J. Carbonized products of disused wood-based panels. Journal of Beijing Forestry University, 2009, 61(3):213-215.

[35] Chen S H, Feng Y S, Mu J, et al. Bacteriostatic characteristics of disused wood-based board pyrolysis condensate liquid. Journal of Beijing Forestry University, 2012, 34(6):131-136.

Chapter 4

Modeling of Pyrolysis in a Stage Scheme of Low-Grade Solid Fuel Gasification

Alexander Kozlov, Anatoly Levin, Denis Svishchev,
Vitaly Shamansky and Alexandre Keiko

Additional information is available at the end of the chapter

http://dx.doi.org/10.5772/67865

Abstract

This paper is concerned with the development of a model of wood pyrolysis in a screw reactor as the first stage of the multistage gasification process. In terms of design, the pilot pyrolyzer represents a recuperative heat exchanger where the heat carrier is represented by a mixture of exhaust and recirculation gases. To prevent clinkering of particles and thermal inhomogeneities, screw-type transportation is used to transport fuel. In order to describe kinetics of pyrolysis and transport of volatiles within the wood particles and their transition to the gas phase, we carried out the studies using a complex of synchronous thermal analysis. The original techniques for the interpretation of measurements were developed for this complex, including the techniques for technical analysis of fuels and identification of detailed kinetics and mechanism of pyrolysis. A detailed numerical modeling of pyrolyzer was performed using the Comsol Multiphysics software, which makes it possible to optimize the design and operating parameters of the pyrolysis process in a screw reactor.

Keywords: pyrolysis of low-grade solid fuel, multistage gasification, screw reactor, CFD modeling, thermal analysis

1. Introduction

One-stage processes of fuel gasification have been developed and thoroughly studied in Russia and other countries throughout the twentieth century. They remain the basis for the creation and design of modern gasification equipment. Numerous studies showed that the one-stage gasification processes had already achieved the limit of perfection beyond which a considerable improvement in their operating characteristics turned out to be impossible or unprofitable.

The gasification technology meeting the contemporary level of technology development should correspond to the following criteria: (1) the gasifier should operate in automatic, mainly unmanned mode,(2) the gasifier operation should be stable and provide insignificant variations in gas composition and flow rate that randomly occur over time,(3) it should produce the minimum amount of liquid and solid waste,(4) the conversion process should be, if possible, little sensitive to changes in the properties of fuel biomass, that is, moisture content and size of particles, and (5) the efficiency of converting fuel chemical energy into gas chemical energy should reach 80–85%.

The staged gasification technology meets the current technical requirements most completely. In the first stage of the process, the allothermal pyrolysis of biomass is organized using producer gas heat and exhaust gases of an engine unit. In this stage, the volatiles (gas and tar) are produced from the fuel. In the second stage, tar is decomposed under the action of heated air and steam outside the fuel bed. Further, the formed steam-gas mixture reacts with the charcoal in the third stage of the process.

2. A review of technologies for multistage solid fuel gasification

One of the first multistage gasifier was developed and patented by Antoine de Lacotte in the 1940s [1, 2]. Oxidative pyrolysis took place in the upper part of fixed-bed reactor. Its gaseous products were taken from the space above the fuel bed and were sent to the external combustion chamber. The oxidation products of the tarred gas returned to the middle part of the fuel bed. One-third of these products were recycled in the pyrolysis stage and two-thirds were involved in the coal-char gasification.

The obtained producer gas was very clean. Diesel internal combustion engine that used this gas could operate without capital maintenance for 30 years [3].

The reactor similar to that designed by Antoine de Lacotte was proposed and patented by the Greek researcher Lamrpos Elefsiniotis [4]. He installed an air supply to the zone of oxidative pyrolysis in the shaft of a gasifier. Such an improvement makes it possible to control the amount of recycled gas coming from the combustion chamber [5]. A downside of the gasifiers invented by de Lacotte and Elefsiniotis is the low efficiency of gasification, which is caused by high heat losses of the external combustion chamber. There is experience of accommodating this chamber inside the fuel bed [6]. In this case, heat coming from the chamber is used in the pyrolysis process and is not dispersed into the environment.

An interesting design of the multistage gasifier is proposed at the Technical University of Denmark [7, 8]. It differs in that the stage of pyrolysis is separated and takes place in a separate screw reactor whose duct is connected to the space above the fuel bed of the downdraft fixed-bed gasifier. In this space, part of pyrolysis gases is oxidized in the air blow. The combustion products come to the coal-char bed in the third stage of the process. A distinctive feature of this multistage process is an active use of the producer gas heat and exhaust gases of the internal combustion engine to heat the air blow and fuel in the screw pyrolyzer. Chemical efficiency of such an allo-autothermal process reaches 93%, whereas in the autothermal conditions it does not exceed 75–80% [9, 10]. Moreover, the tar content in the unrefined gas is at the level of 15 mg/nm^3. Such gas can be used after its conditioning without additional cleaning. The Technical University of Denmark has created a pilot plant Viking with a capacity of 20 kW (*E*). Its scaled-up version with a capacity of 200 kW (*E*) was implemented by the Weiss energy company in Denmark [11]. The Viking gasifier can be used to process fuel with a moisture content of up to 30%. An additional screw reactor of the Weiss cogeneration plant intended for fuel drying makes it possible to convert biomass with a high moisture content (up to 60%). The successful design of the Viking gasifier of the Technical University of Denmark inspired some teams to develop and create similar plants. The Tomas Koch group (company TK Energy) simplified the design of the multistage gasifier by refusing from heating the screw reactor and using a complex system of heat recuperation [12]. The air is blown to the pyrolyzer, thus providing partial fuel combustion and operation of this reactor in the autothermal conditions. The pilot plant with a capacity of 2.3 MW (*t*) is established in the city of Gjol (Denmark).

A pilot plant similar to the Viking gasifier is manufactured by the team from the Institute of Qingdao (China) [13, 14]. They use pelleted biomass as a fuel, and the capacity of the plant makes up 800 kW (*t*). The researchers from Shanghai Jiao Tong University (China) are developing a multistage gasifier with a capacity of 430 kW (*f*) for rice straw processing [15]. In the pyrolyzer of this gasifier, there is a fuel piston instead of screw machine. The researchers from the De Cocody University are studying the work of a screw pyrolyzer embedded in the multistage gasifier [16].

However, the gasifier of the Technical University of Denmark and its analogs have one common drawback. The high degree of biomass combustion is reached only for the sized fuel (chipped wood) prepared from solid species of wood. Such a fuel travels across the screw reactor with minimum destruction of particles, their crushing, and abrasion. Moreover, the char bed in the third stage is effectively and stably gasified. There is no critical packing of the char due to accumulation of coal powder and ash. The processing of the polyfraction or pelleted fuel is characterized by its considerable unburned amount. For example, the degree of fuel carbon conversion in the gasifier of the Jiao Tong University makes up 75–80%. Mass fraction of the lost char residue in the plant of the TK energy reaches 3–10%. Such a considerable amount of unburned fuel leads to a decrease in the gasification efficiency and requires measures for its further processing and utilization.

It is worth noting that to date none of the created plants has been scaled-up and reached the stage of commercial production despite its successful demonstration and a large amount of patents. It is also possible to state that the multistage gasification technology is developed much more slowly than the one-stage gasification technologies in the 1990s, even despite a considerable rise in the fossil fuel prices. There are technical difficulties hindering the widespread adoption of the technology, for example, (1) The number of operating parameters that have an effect on the gasifier operation exceeds 11. This circumstance complicates optimization of its operating conditions. (2) Low temperature of the third stage of the process prevents complete combustion of the residual coal, thus leading to a decline in the gasification efficiency and formation of a great amount of solid waste. (3) Operation of the modern gas-fired equipment (internal combustion engine, fuel cell, gas microturbine, etc.) requires additional removal of tars from gas to the level of 5–100 mg/nm^3. Fine gas cleaning makes the technology much more expensive. Optimization of the second and third stages of the process is required to obtain an almost tar-free gas.

Thus, the development of a promising low-tonnage technology for staged gasification of wood biomass should rest on a detailed research into thermal physics and chemical kinetics of the processes that occur in the equipment components.

In this research, we propose a new concept of staged gasifier, which suggests apart from the use of a screw pyrolyzer the application of a hybrid fixed-bed reactor combining both downdraft and updraft processes. In this case, the tangential air supply is used, which provides good mixing of pyrolysis gases with oxidizer, and, consequently, a high speed and complete decomposition of tar. Such an organization of the process makes it possible to gasify char not only by the entrained flow of incomplete oxidation products of pyrolysis gas obtained in the first stage but also by heated air blow in the process countercurrent with respect to the fuel motion. A block diagram of the developed multistage reactor is presented in **Figure 1**.

This paper presents the results of the research into the first stage of the multistage gasification, that is, allothermal wood pyrolysis in a screw reactor.

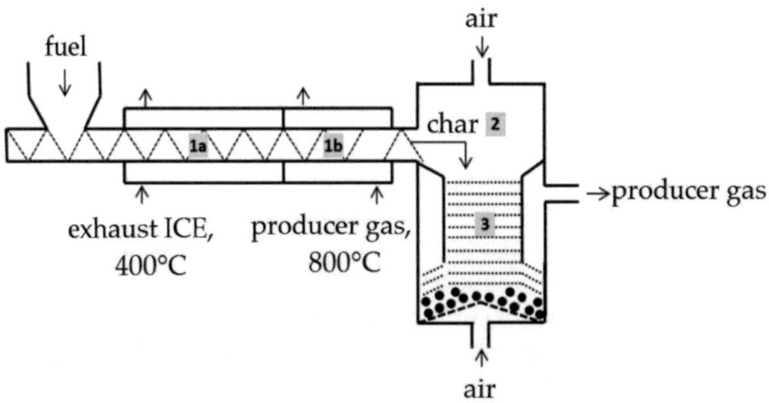

Figure 1. A block diagram of the plant for multistage gasification. Notations: 1—screw pyrolyzer heated by the internal combustion engine (ICE) exhaust gases; 2—space above the fuel bed, producer gas combustion chamber; 3—a two-stage fixed-bed reactor.

3. The methods for CFD modeling of wood biomass pyrolysis process

In the last years, there has been continuous evolution of the calculation models of pyrolysis plants. The research presented in [17] shows a mathematical description of the pyrolysis process. The authors devised a calculation model for the plant designed by them. The model consists of a set of stages of drying, thermal decomposition, interphase transitions, and cooling. Thus, the total duration of the wood feedstock decomposition τ_m process was represented by a set of stages: heating τ_m, drying τ_d, thermal decomposition τ_{td}, and cooling of char coal τ_c. This method allowed the common model to be decomposed into submodels suitable for practical implementation that have different degrees of detail and respective form of equations (ordinary differential or in partial differential). A distinctive feature of this problem statement is the use of a one-dimension representation of distribution of parameters in the space of pyrolyzer chamber. Change in the temperature of gas phase during its passing through the fuel bed was determined from the following equation:

$$\frac{\partial T_1}{\partial \tau} = \frac{\alpha \Delta T F}{(\rho c) V} - w \frac{\partial T}{\partial z}, \quad (1)$$

Solving Eq. (1) and equation of heat conductance for a solid fuel particle made it possible to take into consideration interphase heat exchange:

$$\frac{\partial T_f}{\partial \tau} = \frac{1}{c p x^g} - \frac{\partial}{\partial x_f} \left(x^g \lambda \frac{\partial T_f}{\partial x_f} \right), \quad (2)$$

Moisture content in a fuel particle was determined by the authors [17] from the following expression:

$$\frac{dW}{d\tau} = \frac{\beta(p_p - p)}{\rho V_2}, \quad (3)$$

Solving Eqs. (2) and (3) with the calculation of partial pressure made it possible to determine the duration of a drying stage.

In the same way, the authors of [17] describe all the above-enumerated stages. The only assumption in their model was the absence of nonuniform distribution of parameters in the cross section. To verify their model, the authors of [17] made an experimental bench since in their case, as in many other studies, verification of the constructed mathematical models is based on the unique experimental data. This complicates a comparative analysis of the proposed approaches to the selection of the best description of the processes that occur in similar plants.

A more thorough study on partial gasification, on the basis of a numerical modeling and application of thermogravimetrical analysis, is presented in the recent publications [18, 19]. For example, in the research [20] the authors applied a three-dimensional (3D) representation of the calculated region, which made it possible to more accurately determine variation in temperature profiles and gas motion velocities along the modeled furnace. Thus, a good correspondence was achieved between the calculated data and the experimentally measured process of fuel conversion. The model in [21] included the following equations:

$$r = k^f \prod_{i=1}^{N} c_i^v = A^f T^f \exp(E_a^f / R_g T) \prod_{i=1}^{N} c_i^v, \quad (4)$$

$$\rho C_p \frac{\partial T}{\partial \tau} + \rho C_p u \nabla T = \nabla \cdot (k \nabla T) + Q, \quad (5)$$

$$-n \cdot (-K \nabla T) = \varepsilon (G - \sigma T^4), \quad (6)$$

$$\rho \frac{\partial u}{\partial \tau} + \rho u \cdot \nabla u = \nabla \cdot \left[-pI + \mu \left(\nabla u + (\nabla u)^T \right) - 2/3 \mu (\nabla \cdot u) I \right] + F(a), \quad (7)$$

$$\frac{\partial \rho}{\partial \tau} + \nabla \cdot (\rho u) = 0, \quad (8)$$

The analysis of the numerical modeling results that was carried out by the authors allowed them to make a conclusion on the effect of heat generated in the gasification process on 3d temperature fields and gas velocities inside the furnace. Thus, the detailed three-dimensional representation of an object in the numerical models is a necessary condition for the successful description of complex adjoining processes that occur in the course of gasification.

A thorough analysis of mathematical models was presented in [20]. Large sizes of the plants do not allow us to construct highly detailed aerodynamic models. Therefore, special attention is paid to the development of various closing relationships, in particular the equations of kinetics of chemical reactions. As in the majority of other studies, these authors apply the relationships of the form

$$k_v = A_v \exp\left(-\frac{E_v}{RT_s}\right), \tag{9}$$

$$r_{dry} = k_v \rho Y_{H_2O}, \tag{10}$$

One of the main parts in each similar research is the determination of a required list of conditions that determine the relationship between the reaction rates and parameters of state, for example, temperature. The authors of [21] properly note that such a method can only be applied to certain ranges of temperatures, whereas for high values (above 475K) it gives absolutely unreal estimates of the rate of solid fuel drying. It is possible to conclude that the application of Eqs. (9) and(10) without consideration of conservation laws naturally creates conditions for such problems to arise. Also, the authors of [21] indicate that the CFD models that refer to a continuous porous medium are very attractive in terms of their possible implementation.

Thus, the analysis of the existing studies shows the necessity to consider the detailed modeling of spatially distributed processes of heat-mass exchange, including the equations of kinetics of chemical reactions, considering natural constraints that follow from the laws of conservation. The most promising seems to be the representation of heat-mass exchange equations in a 3D form. In this case, the problem of a correct consideration of closing relationships in such a statement is not an exceptional characteristic of such detailed descriptions due to insufficient information on multi-parameter models that describe pyrolysis process. Thus, each new mathematical model requires verification involving results of mufti-factor experimental studies.

4. Mathematical model

The geometry used for modeling a screw pyrolyzer is presented in **Figure 2**. The interior space is filled with a solid mass by 40%, which should provide easy circulation of the produced gases around the screw. The screw represents a spiral with a wall, 3 mm thick, which is wound on a hollow shaft. The interior space of the pyrolyzer with solid fuel and formed products is separated from heating gases by a wall, 5 mm thick. The screw represents a geometrically complex structure. There are 11 complete spiral turns with a pitch of 139 mm on the hollow

http://dx.doi.org/10.5772/67865

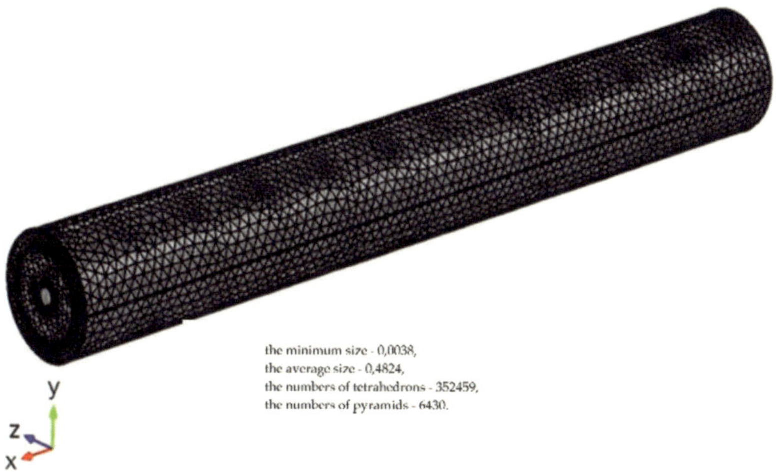

Figure 2. Calculation mesh.

steel shaft. An external diameter of the spiral made up 139 mm, and the shaft diameter was 38.6 mm. In the calculation, we took into account the body of the screw, shaft, and external coating. All the objects are constructed from cylinders. The construction of all these objects caused the emergence of surfaces and regions of small size at the point of their intersection. The total number of components in the calculation mesh equals N. Optimal distribution and sizes of mesh cells were calculated using the Comsol Multiphysics Software on the basis of corresponding algorithms that ensure good convergence for the differential equations applied. The calculation mesh includes different number of cells:

For gas part, it includes hot gas from internal combustion engine and gas inside the screw above the porous medium.

For a surface of solid elements, it includes all solid surfaces.

For the others, it includes solid volumetric parts such as screw, spiral, coating, and so on.

The Comsol Multiphysics Software is used as a simulation environment. Heat exchange in the process of pyrolysis is simulated considering physical properties (porosity, permeability, etc.) of the medium:

$$\frac{\rho}{\varepsilon}\left((u\cdot\nabla)\frac{u}{\varepsilon}\right) = \nabla\cdot\left[-pI + \frac{\mu}{\varepsilon}\left(\nabla u + (\nabla u)^T\right) - \frac{2\mu}{3\varepsilon}(\nabla\cdot u)I\right] - \left(\frac{\mu}{k} + \beta_F|u| + \frac{Q_{\mathrm{br}}}{\varepsilon^2}\right)u + F, \quad (11)$$

Eq. (11) differs from the model in [21] in the consideration of solid fuel porosity ε. In the equation, the changes in masses Q_{br} correspond to the gas masses formed in the course of the solid fuel decomposition. Then, the equation of continuity has the following form:

$$\nabla\cdot(\rho u) = Q_{\mathrm{br}}, \quad (12)$$

Heat exchange in pyrolyzer is calculated using the effective heat transfer coefficient:

$$\rho C_p u\cdot\nabla T + \nabla\cdot(-k_{\text{eff}}\nabla T) = Q, \tag{13}$$

In the area of high-temperature heads, we took into account the radiation heat transfer:

$$q = c_0(T_2^4 - T_1^4), \tag{14}$$

The equation of a diffusion barrier for the plane separating the porous solid body from the gas space in the pyrolyzer is set as follows:

$$\begin{aligned} -nD_{s,i}\nabla c_{i,u} &= \frac{D_{s,i}}{d_s}(c_{i,u} - c_{i,d}) \\ -nD_{s,i}\nabla c_{i,d} &= \frac{D_{s,i}}{d_s}(c_{i,u} - c_{i,d}) \end{aligned}, \tag{15}$$

where d_s is the barrier thickness, mm; D_{si} is the diffusion coefficient, m^2/s. In the calculations, we assumed a bed thickness of 5 mm and a diffusion coefficient of 1e−6 m^2/s.

The boundary conditions chosen to solve the problem are as follows:

- An inlet temperature of fuel is 20°C;
- An inlet temperature of heating gases is 600°C;
- A velocity of fuel flow in pyrolyzer is 1 mm/s;
- An inlet velocity of heating gases is 0.1–1 m/s.

5. Modeling results

5.1. Calculation of the fields of temperatures and gas flow velocities

The calculated fields of temperature, velocities, densities of media, streamlines, and other derivative data were obtained by using the mathematical model of heat and mass transfer processes in the screw pyrolyzers by using the Comsol Multiphysics software. **Figure 3** presents the calculated temperature values in the vertical sections along the pyrolyzer and also the values of formed gas flow velocity in the orthogonally directed planes. The calculation example demonstrates a sufficiently high degree of fuel heating and an adequate pattern of gas flow over the porous medium around the screw.

Figure 4 demonstrates successful heating of solid fuel providing the condition for its presence at the temperature of above 500°C for 20 min at the fuel motion velocity of 0.001 m/s.

5.2. Parametric calculations

The parametric calculations are an essential part of the optimization study which is intended for determination of the most promising stepwise gasification technology of low-grade solid fuel. The parametric calculations can be made on the stable calculation model allowing the variation of parameters in the solution, that is, boundary conditions, over a sufficiently wide range. The Comsol Multiphysics software makes it possible to calculate steady-state conditions specifying different laws of change in the boundary conditions.

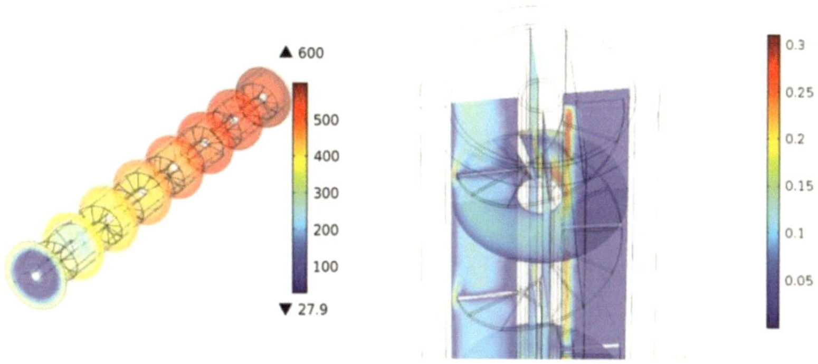

Figure 3. The example of calculations performed to determine the fields of temperatures and the motion velocities of pyrolysis products. The heating gas velocity at the inlet section is w=0.1 m/s.

The developed calculation model of the screw pyrolyzer was tested on the basis of the following parametric studies:

A. the heating gas temperature at the pyrolyzer inlet was set from 770 to 910 K with a step of 20 K;

B. the heating gas velocity at the pyrolyzer inlet was set from 0.1 to 0.9 m/s with an interval of 0.2 m/s.

The results of testing (see **Figures 5** and **6**) confirmed the solution reliability and stability with practically full convergence of numerical solutions and as a result the equal time spent on each calculation variant. The other variants of parametric studies are quite possible and seem to be the least problematic from the viewpoint of numerical solution stability, since the current tasks in solving the numerical problems are aimed first of all at the achievement of stability to the initial and boundary conditions.

The parametric calculations with the changing heating gas flow rate reveal an asymptote in the solutions of temperature distribution; as far as starting with 0.3 m/s, the calculated values do

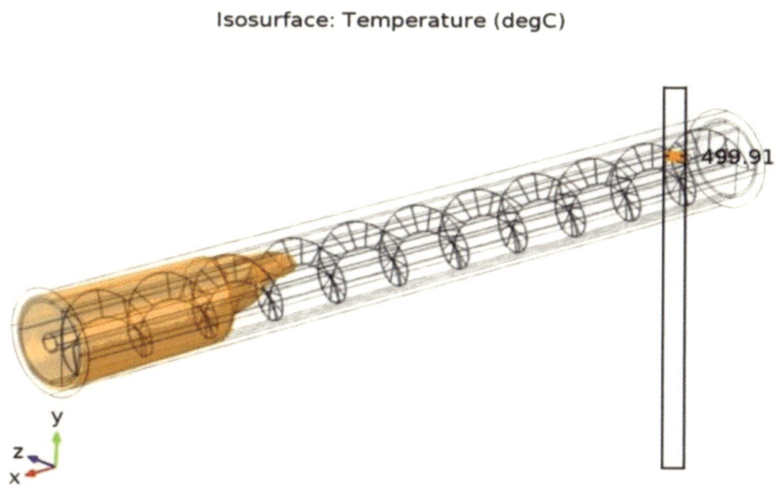

Figure 4. The surface formed by the isotherm at 500°C.

not change considerably at further increase of gas velocity values (**Figure 5**). This result enables the supposition about the existing optimal ratio between the working parameters of the developed stepwise gasification scheme in the forthcoming optimization study.

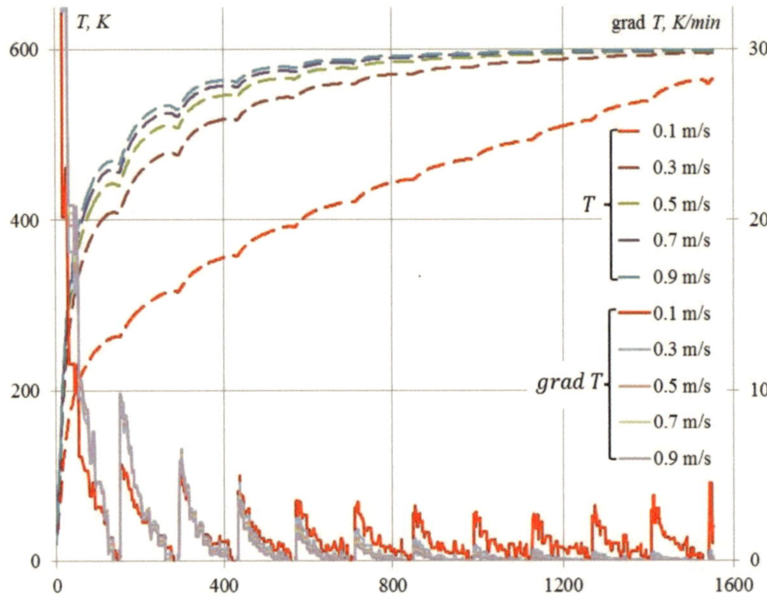

Figure 5. Distribution of the temperature and its gradient in the pyrolyzer space for different heating gas flow rates.

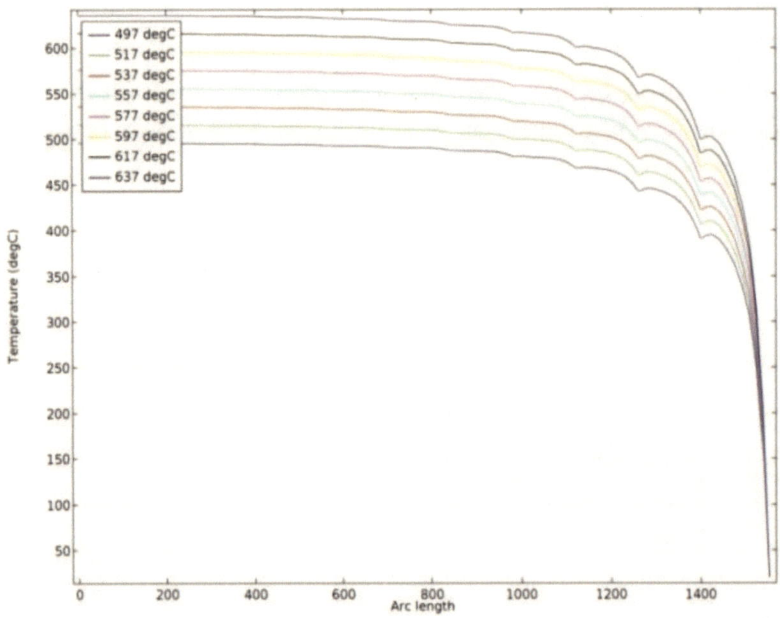

Figure 6. Temperature distribution in the pyrolyzer space for different assigned values of heating gas temperature.

5.3. Analysis of the calculated values of gas flow velocity

Figure 7 illustrates a screwed character of the gas streamlines along the pyrolyzer axis which arranges the mixing of formed fractions (steam, etc.) in the gas space. As a result, the efficiency of the heating process and the uniformity of temperature fields are secured, avoiding the zones with irregular reaction rates.

The calculated fields of the gas phase velocity u in the longitudinal cross section of the screw reactor are presented in **Figure 8**. The highest values of u are observed over the surface of filled low-grade solid fuel.

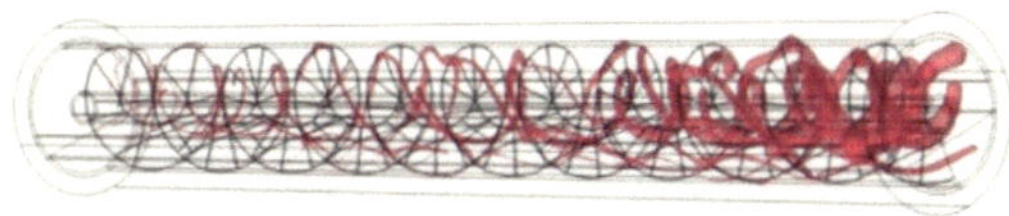

Figure 7. The calculated streamlines of the formed gases in the process of low-grade solid fuel heating in the pyrolyzer. The color indicates the calculated temperature values.

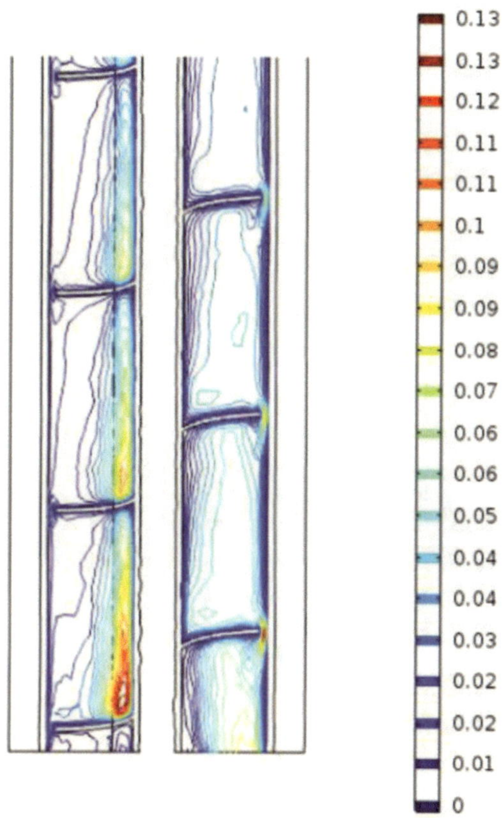

Figure 8. The calculated isolines of velocity.

Markedly distinct velocity fields and as a consequence the diffusion coefficients characterizing transfer of the gas phase from the porous mass to the free space can be formed for different operating conditions.

Figure 9 presents instantaneous values of gas flow velocities through the surface of the filled solid fuel mass. The effect of heating gas temperature on the diffusion processes of formed gas yield was analyzed. The calculations show a negligible effect of the boundary conditions on motion of the media in the region modeled by the diffusion thin bed. This fact suggests feasibility of using the constant coefficients to describe gas transition from the porous medium to the space free for gas flow in the screw reactor. The maximum change in the velocity was equal to ~1% at the change of the heating gas temperature by 140 K at the pyrolyzer inlet. Thereby, the general distribution type of values in the velocity field did not change.

5.4. Application of the model for distribution of the formed substance concentrations

The validity of applying the stationary thermophysical dependences to the determination of tree properties in the pyrolysis process was analyzed. For this purpose, the temperature

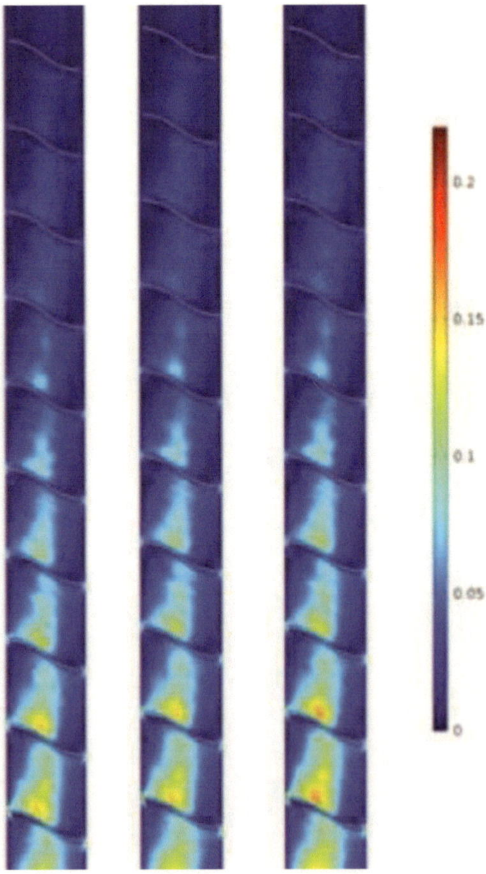

Figure 9. The fields of gas flow velocity on the surface.

gradient values in the fuel medium along the motion of solid particles were determined. Taking into account the velocity of fuel mass motion with respect to the pyrolyzer, it is possible to compare the temperature gradient with the heating rates of experimental samples. The conclusion confirmed feasibility of using the equations of chemical kinetics which were determined at $dT/d\tau = 30$ K/min.

The applied approach to determination of the fields of concentrations of substances formed in the course of substance reaction demonstrated satisfactory results of the calculation example (**Figure 10**). The highest concentration of steams formed during drying is observed in the zone of intensive heating—in the initial part of the screw reactor, and as they flow along the reactor the concentrations level off in the sections.

Because of inhomogeneity of heating, the steams formed in the process of solid fuel mass heating substantially differ in concentrations on the initial part of the screw reactor (**Figure 9**). **Figure 9** shows that the share of formed steams essentially depends on the operating factors. For example, more steam is formed inside the porous bed at the lower values of heating gas temperature at the pyrolyzer inlet. Because of the obvious nonphysical character of these results, the following explanation is advisable: in the available model of fraction formation, there are no constraints on the quantity of substances formed. As was mentioned in the work [21], the direct application of the Arrhenius-modified models leads to the similar effects. The correct analysis of physical constraints requires more strict descriptions of the laws of concentration distribution which are based not only on the values of gas temperatures and velocities. One of the methods to solve this problem is to specify the initial concentration of associated water. Then, the equations describing change in concentrations will have the following form:

Figure 10. The calculated fields of steam concentrations.

$$\begin{cases} \nabla\cdot(-D_1\nabla c_1) + u\cdot\nabla c_1 = R_1 \\ \nabla\cdot(-D_2\nabla c_2) + u\cdot\nabla c_2 = R_1{}' \end{cases} \tag{16}$$

where indices 1 and 2 correspond to concentrations of the steams and associated water, respectively.

5.5. Analysis of contribution of radiation heat transfer

Optimization calculations even for stationary mathematical models with concentrated presentation of the description of processes need huge computational resources. It is not surprising that the numerical calculation of the three-dimensional fields of velocity, temperature, and others by using the server Xeon CPU E5606 (8 cores, 256 gigabyte of the random access memory) of one condition required 24.5 working hours. Therefore, it is topical to reduce the mathematical model by the addition of assumptions allowing the number of equations to be decreased. From this point of view, the statement that the contribution of radiation heat transfer is unessential seems to be the most attractive. The identical operation conditions of the screw pyrolyzer were calculated considering the radiation heat transfer and without it. The calculation results (**Figure 11**) demonstrate invariability of the key isotherm and the temperature field. The contribution of radiation heat transfer, which amounted to less than 5% of the total energy exchange balance, was also estimated. If the radiation heat transfer is excluded from the general model, the computation time can substantially be decreased from 24.5 working hours of the server with the indicated configuration up to 30–40 min.

At the same time, it should be noted that the conclusion about inessential contribution of radiation heat transfer to the whole heat exchange contradicts the results of the work [21], which poses the question on deeper experimental study of this problem.

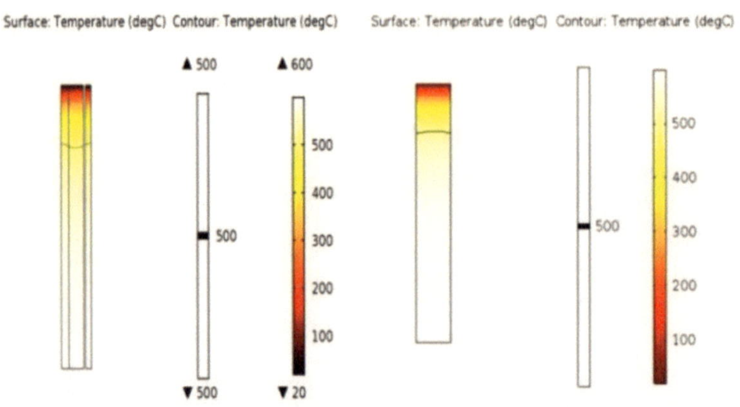

Figure 11. The calculation results of temperature field and isotherm position at T=500°C by the models, which account for the radiation heat transfer (a) and without it (b).

http://dx.doi.org/10.5772/67865

6. Conclusion

The model of heat transfer and aerodynamics of the pyrolyzer was developed for express optimization calculations. The temperature distribution in the pyrolyzer reactor was obtained in a wide range of operating conditions. Moreover, it is confirmed that in the considered thermal conditions the process of fuel conversion proceeds to completion. Inclusion of the kinetic block in the model of heat transfer made it possible to determine the concentration fields of the formed substances (steam, etc.). Numerical experiments confirmed the possibility for the application of this model to optimization calculations.

Possible use of stationary experimental dependences for the determination of the rate of change in solid fuel density, yield of volatiles, and so on is shown for the range of considered working conditions. The results of parametric study demonstrated the existence of the asymptote for the optimal amount of heating gases which ensures the first stage of pyrolysis.

Acknowledgements

Results of the study were obtained using the Unique Scientific Plant "High-temperature circuit." This work was carried out at the Melentiev Energy Systems Institute and financially supported by the Russian Science Foundation (project number 16-19-10227).

Nomenclature

A	pre-exponent factor (1/s)
c_p	specific heat capacity of the gas mixture (KJ/kgK)
D_{si}	diffusion coefficient (m^2/s)
E	activation energy (J/mol)
F	surface area (m^2)
k	kinetic rate constant (1/s)
P	gas pressure (Pa)
p	partial pressure (Pa)
Q	heat transfer to wood chips by convection and radiation (W/m^3)
R	universal gas constant ~8.314 (J/molK)
r	reaction or process rate (kg/m^3s)
T	temperature (K)
τ	time (s)
u	velocity vector (m/s)
V	volume (m^3)
W	moisture (%)
x	coordinate (m)

Greek letters

α	heat transfer coefficient (W/m^2K)
β	mass transfer coefficient (kg/m^2sPa)
ε	porosity (–)
λ	thermal conductivity (W/m K)
μ	air viscosity (kg/ms)
p	density (kg/m^3)
σ	normal stress tensor (Pa)
ω	flow rate (m/s)

Subscripts

f	fuel
g	gas
s	solid phase

Author details

Alexander Kozlov*, Anatoly Levin, Denis Svishchev, Vitaly Shamansky and Alexandre Keiko

*Address all correspondence to: kozlov@isem.irk.ru

Melentiev Energy Systems Institute, Siberian Branch of the Russian Academy of Sciences, Irkutsk, Russian Federation

References

[1] Lacotte A. D. (1942). Procédé de gazéification totale des combustibles à haute teneur en matières volatiles. Pat. FRT51403 19401120.

[2] Societe Distibois (1951). Method for the gasification of fuel having high contents of volatile matter. Pat. FRX648691 19391121.

[3] Kaupp A. (1984). Small Scale Gas Producer-Engine Systems. Vieweg Teubner Verlag.

[4] Elefsiniotis L. (2008). Three-Stage Gasifier, Fixed Bed, Which Has Buffer Zone of Gaseous Flow Between Pyrolysis Zone and Combustion Zone. Pat. WO2007GR000172 0070306.

[5] Antonopoulos I.S., Karagiannidis A., Elefsiniotis L., Perkoulidis G., Gkouletsos A. (2011). Development of an innovative 3-stage steady-bed gasifier for municipal solid waste and biomass. Fuel Processing Technology, 92(12), 2389–2396. DOI 10.1016/j.fuproc.2011.08.016.

[6] Susanto H., Beenackers A. A. (1996). A moving-bed gasifier with internal recycle of pyrolysis gas. Fuel, 75(11), 1339–1347. DOI 10.1016/0016-2361(96)00083-X.

[7] Henriksen U., Ahrenfeldt J., Jensen T. K., Gøbel B., Bentzen J. D., Hindsgaul C., Sørensen L. H. (2006). The design, construction and operation of a 75kW two-stage gasifier. Energy, 31(10), 1542–1553. DOI 10.1016/j.energy.2005.05.031

[8] Gøbel B., Henriksen U. B., Ahrenfeldt J., Jensen T. K., Hindsgaul C., Bentzen J. D., Sørensen L. H. (2004). Status-2000 Hours of operation with the Viking gasifier. Grafica Lito.

[9] Hofmann P., Schweiger A., Fryda L., Panopoulos K. D., Hohenwarter U., Bentzen, J. D., Kakaras E. (2007). High temperature electrolyte supported Ni-GDC/YSZ/LSM SOFC operation on two-stage Viking gasifier product gas. Journal of Power Sources, 173(1), 357–366. DOI 10.1016/j.jpowsour.2007.04.073.

[10] Svishchev D. A., Kozlov A. N., Donskoy I. G., Ryzhkov A. F. (2016). A semi-empirical approach to the thermodynamic analysis of downdraft gasification. Fuel, 168, 91–106. DOI 10.1016/j.fuel.2015.11.066.

[11] Bentzen J. D., Hummelshøj R., Henriksen U. B., Gøbel B., Ahrenfeldt J., Elmegaard B. (2004, January). Upscale of the two-stage gasification process. In World Conference and Technology Exhibition on Biomass for Energy and Industry: Grafica Lito-Florence.

[12] Babu S. P. (2006). IEA Bioenergy Agreement Task 33: Thermal Gasification of Biomass Work Shop No. 1: Perspectives on Biomass Gasification.

[13] He T., Han D., Wu J., Li J., Wang Z., Wu J. (2015). Simulation of biomass gasification and application in pilot plant. Energy Technology, 3(2), 162–167.

[14] Wang Z., He T., Qin J., Wu J., Li J., Zi Z., Sun L. (2015). Gasification of biomass with oxygen-enriched air in a pilot scale two-stage gasifier. Fuel, 150, 386–393. DOI 10.1016/j.fuel.2015.02.056.

[15] Su Y., Luo, Y. (2009, March). Experiment on rice straw gasification in a two-stage gasifier. In Power and Energy Engineering Conference, 2009. APPEEC 2009. Asia-Pacific (pp. 1–4). IEEE.

[16] Fassinou W. F., Van de Steene L., Toure S., Volle G., Girard P. (2009). Pyrolysis of *Pinus pinaster* in a two-stage gasifier: Influence of processing parameters and thermal cracking of tar. Fuel Processing Technology, 90(1), 75–90. DOI 10.1016/j.fuproc.2008.07.016.

[17] Grachev A.N., Safin R.G. Valeev I.A. (2006), Pyrolysis of waste wood processing enterprises. Chemistry and Chemical Engineering, 49, 10, 104–108.

[18] Wang G., Li, W., Li, B., Chen, H. (2008). TG study on pyrolysis of biomass and its three components under syngas. Fuel, 87(4), 552–558. DOI 10.1016/j.fuel.2007.02.032.

[19] Han F., Meng A., Li Q., Zhang Y. (2016). Thermal decomposition and evolved gas analysis (TG-MS) of lignite coals from Southwest China. Journal of the Energy Institute, 89(1), 94–100. DOI 10.1016/j.joei.2015.01.007.

[20] Meng X., de Jong W., Badri F. S., Benito P., Basile F., Verkooijen A. H. (2012). Combustion study of partially gasified willow and DDGS chars using TG analysis and COMSOL modeling. Biomass and Bioenergy, 39, 356–369. DOI 10.1016/j.biombioe.2012.01.032.

[21] Khodaei H., Al-Abdeli Y. M., Guzzomi F., Yeoh G. H. (2015). An overview of processes and considerations in the modelling of fixed-bed biomass combustion. Energy, 88, 946–972. DOI 10.1016/j.energy.2015.05.099.

Chapter 5

Comprehensive Two-Dimensional Gas Chromatography and Its Application to the Investigation of Pyrolytic Liquids

Gabriela Pereira da Silva Maciel,
Juliana Macedo da Silva, Mozart Daltro Bispo,
Laiza Canielas Krause, Rosângela Assis Jacques,
Claudia Alcaraz Zini and Elina Bastos Caramão

Additional information is available at the end of the chapter

http://dx.doi.org/10.5772/68077

Abstract

The chapter presents basic principles of one-dimensional gas chromatography (1D-GC) and comprehensive two-dimensional gas chromatography (GC × GC) related to the main advantages of the two-dimensional technique, as well as its application to the study of organic compounds in liquids derived from coal, mainly through pyrolysis and extraction. It also shows the investigation of compounds contained in bio-oils obtained from biomass through pyrolysis, using GC × GC. Advances in scientific knowledge related to the composition of these complex matrices are shown through different examples of GC × GC analyses, such as the identification of trace compounds that would not be perceived by 1D-GC, organized patterns of elution of structurally related compounds that help their identification, etc. Examples shown make it clear that GC × GC is the technique of choice to elucidate composition of these complex matrices.

Keywords: comprehensive two-dimensional gas chromatography, GC × GC, pyrolysis, bio-oil, coal tar, coal bitumen, mass spectrometry

1. Introduction

Gas chromatography (GC) or one-dimensional gas chromatography (1D-GC) is a mature technique that offers high peak capacity and complementary data provided by different detectors. It has been successfully applied to several areas of knowledge, which confirms its maturity as an analytical technique. However, several matrices are so complex that 1D-GC may not be capable of separating their components, such as petroleum and derivatives, natural aromas, perfumes, foods and beverages, environmental samples, etc. One aspect that makes it even more difficult is the presence of compounds in trace concentrations that co-elute with other components of interest or with matrix interferents that are present in larger amounts in a sample. Such a situation is commonly seen in complex matrices as the ones formerly mentioned. It is especially cumbersome when the trace compound plays an important role in the sample under investigation, and some examples may be mentioned on this regard: nitrogen and sulfur compounds in petroleum, coal tar and bitumen, aroma active compounds in wine aroma, allergens in perfumes, etc. Misidentification, wrong quantification of important compounds may arise from such a situation and consequently false results that might lead to wrong decisions. In addition, high-quality mass spectra with no interferences are a difficult goal to achieve in 1D-GC, although specific detectors may be of help when some chemical groups are being target.

Several trials have been made to improve chromatographic resolution (R_s) in 1D-GC, according to the following equation:

$$R_s = \frac{\sqrt{N}}{4}\left(\frac{\alpha - 1}{\alpha}\right)\left(\frac{k}{1+k}\right) \tag{1}$$

where N is the number of plates, α is the separation factor of a pair of solutes, and k is the retention factor of the most retained solute [1].

One alternative is to extend the length of the chromatographic column, however, as R_s is a function of the quadratic root of N; this strategy ends up in time-consuming analyses, larger peak base width and consequently higher detection limits. A classical illustration is the analysis of gasoline made by Berger in a 550 m long chromatographic column (1.3 million effective plates) where 970 compounds were separated, although not all compounds of this sample were resolved. Analysis time was eleven hours, and this example makes it clear why such a time-consuming approach is not welcome [2].

Alternatively, it is possible to employ smaller column diameters, which are associated with high carrier gas flow and higher oven heating rates in fast GC. However, resolution observed in fast GC chromatograms is usually similar to conventional ones, except that analysis time is shorter [3, 4]. In summary, with the before mentioned equation of chromatographic resolution in mind, it seems that the most effective way of improving resolution is to modify the separation factor (α), and this can be accomplished using multidimensional gas chromatography, with different stationary phases. Two conditions define multidimensional separation techniques: (1) the components of the sample subjected to two different mechanisms, and (2) components should remain separated until the end of the separation process [5–7]. Two-dimensional gas chromatography (2D-GC) may be classified as heart cutting (GC-GC) and as comprehensive (GC × GC). In GC-GC, only some parts of the effluent from the first chromatographic dimension (^{1}D) are introduced in the second dimension (^{2}D). It is of special interest when only target components are the focus of the study, but not when the whole sample is under investigation. The length of the second column is conventional, and runs in this dimension are independent of the analysis in ^{1}D. The longer the number of chromatographic band cuts, the longer the analysis time, as every cut has to undergo a further separation in the second dimension, adding extra time to the complete analytical process. In particular, care

http://dx.doi.org/10.5772/68077

should be taken, as cuts should not be too long, otherwise compounds that have already been separated in ^{1}D may be mixed again before reaching the ^{2}D column [8]. In GC × GC, all the effluent from ^{1}D or a representative part of it is lead to ^{2}D, while the separation achieved in the first dimension should be maintained. The analysis time is the same as the one of ^{1}D run, and second dimension runs occur at same time in which ^{1}D analysis is running [9]. The goal of this chapter is to present some fundamental aspects of GC × GC and the advantages of its application to the investigation of pyrolytic liquids from coal and biomass, as well as of bitumen.

2. Comprehensive two-dimensional gas chromatography (GC × GC)

Comprehensive two-dimensional gas chromatography (GC × GC) started in 1991, due to the brilliant contribution of Professor John Philips and his research group [10]. Even though it is a relatively young technique, it has already experienced several stages of development and is maturing in a fast pace. The first applications of GC × GC were related to petrochemical samples [11, 12]; however, other applications have followed these preliminary ones, such as forensic samples [13], environmental samples [14] food samples [15, 16], petroleum and derivatives [17] etc. The fractions coming from ^{1}D [typically 15–60 m of length, 0.25–0.32 mm of internal diameter (^1dc) and 0.1–1 μm of film thickness (^1df)] are sampled and focused in the modulator and further transferred to ^{2}D. The second column is shorter [commonly 0.5–2 m of length, 0.1–0.18 mm of internal diameter (^2dc) and 0.1 μm of film thickness (^2df)] to allow fast separation of the sequential chromatographic bands.

Figure 1(A) schematically illustrates a general GC × GC equipment. The stationary phases of both columns should be of different natures and should be chosen according to the dimensionality of the sample [18]. If the column in the ^{1}D is of nonpolar nature, the stationary phase of the second column should take profit of another mechanism of separation, such as, polar interactions (e.g., π-π interactions, among others). This type of column set is designated "conventional" and was the most common in the early applications of GC × GC. When the ^{1}D is polar and the ^{2}D contains a nonpolar stationary phase, the column set is called "inverse" set. Modulation is the key process of GC × GC, and the modulator is located in between the two columns, at the end of the ^{1}D or in the beginning of the ^{2}D. **Figure 1(B)** illustrates the modulation process with a cryogenic liquid. The modulator provides sampling and concentration of the chromatographic bands eluting from ^{1}D and leads these bands to a fast separation in ^{2}D [18].

The modulator collects the chromatographic band eluting from the ^{1}D during a preset period of time (**Figure 1(B)** B1). The modulator chosen for illustrative purposes in the example presented in **Figure 1(B)** is a double jet cryogenic modulator. When the modulator is actuated (**Figure 1(B)** B1), a cooled nitrogen gas jet is directed to a small area of the modulator. During this period of time, the chromatographic band undergoes compression and is narrowed due to the low temperature of the refrigerating fluid, and the narrower band is introduced in the ^{2}D as a short pulse (**Figure 1(B)** B2). Therefore, the analytes are concentrated enhancing the chromatographic analysis sensitivity and resolution. The chromatographic band release is provided by a hot jet in the same region of the capillary or only due to the heating provided by the chromatographic oven. At the end of a modulation process, a new fraction of ^{1}D effluent

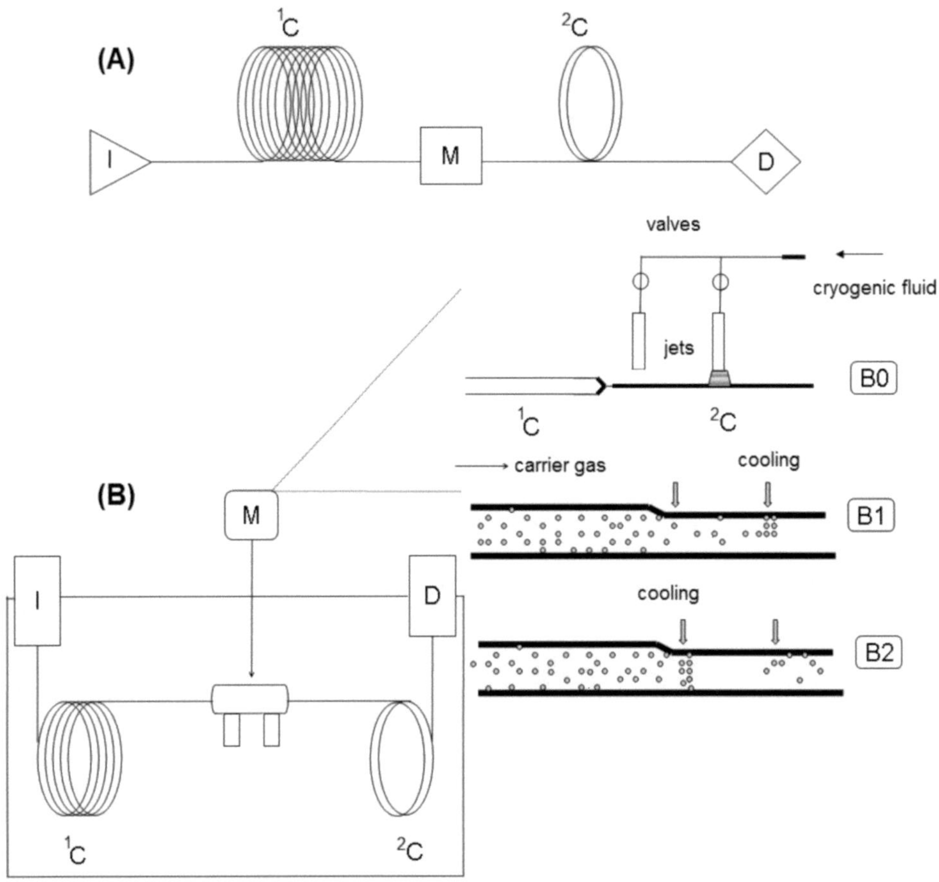

Figure 1. Representative schemes of a GC × GC system. (A) Main parts of a GC × GC equipment. (B) Illustration of a double jet cryogenic modulation system (adapted from Dalluge et al. [19]).

enters the modulator (the flow from ^{1}D is continuous), while the former faction undergoes a fast separation in the ^{2}D column. This process of sampling, focusing, and release of the continuous chromatographic bands eluting from ^{1}D occurs during the whole GC × GC analysis. A complete cycle of modulation is called period of modulation (P_M), and this time period is equal to the ^{2}D run time. At the end, the compounds that elute from ^{2}D pass through the detector and a series of short ^{2}D chromatograms are obtained. The technique is called "comprehensive" because all the effluent coming from ^{1}D is transferred to the second dimension.

Modulators are classified according to its operating principle in thermal and flow modulators. Thermal modulators are the ones where a positive and/or negative difference in temperature is employed to achieve modulation [20–22]. The example given refers to a thermal modulator that employs a cryogenic system. It provides high efficiency of band focusing and further fast release of chromatographic bands [23]. There are several different designs of cryogenic modulators that employ refrigerating liquids (nitrogen or carbon dioxide), such as the one with a cold double jet (example given above) [8], the one with four jets (two cold and two hot jets) [24] the loop modulator [25] etc. Carbon dioxide is an efficient cooling liquid; however, if

http://dx.doi.org/10.5772/68077

compounds under study are too volatile (less than six carbons), liquid nitrogen is considered to be more appropriate [23]. Even though cryogenic modulators are well known for its high efficiency, flow modulators are simple and low cost devices, as they do not require the use of a cryogenic fluid. Flow modulators have been the focus of intense research in the last few years and are becoming an alternative, especially for routine analyses. A more complete discussion on this subject is available in review and research articles elsewhere [21, 22, 24].

In order to preserve the separation obtained in the first dimension, sampling fractions should not be longer than ¼ of the widths of the peak in ^{1}D [26, 27] in ^{2}D are in the range of 2–8 s and so, runs are essentially isothermal. In contrast, the ^{1}D run time commonly stays between 30 and 120 min. The two columns may be located in one oven or the second column may be in a second one. A second oven provides higher flexibility regarding temperature control. Heating rate in the first column is usually low (1–5°C min^{-1}) in order to provide at a reasonable amount of modulations per ^{1}D peak aiming to maintain the separations achieved in ^{1}D. For example, if peak width is 18 s (^1w_b), an appropriate P_M should not be higher than 6 s [23, 26]. Peak widths in the second dimension (^2w_b) range from 50 to 600 ms, and detection requires a frequency of acquisition from a minimum of 50–100 Hz [19, 28, 29]. Several detectors meet these requirements, and mass spectrometric detectors are especially appropriate to be used with GC × GC, as they provide information on the identity of separated components. However, several selective detectors have also been employed with GC × GC, and the interested reader can find review papers that deal with this subject [30].

Figure 2 illustrates the whole GC × GC process steps. The result of the analysis is a great number of short chromatograms obtained from the ^{2}D continuous analyses (step 1. Modulation). A large peak representing a co-elution of three components is presented to exemplify the

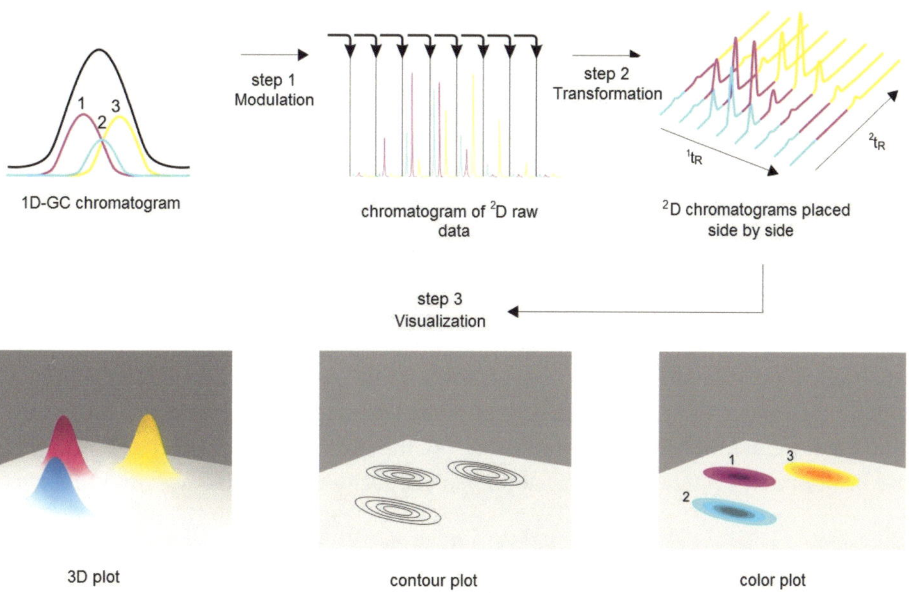

Figure 2. Representative scheme of the process of data generation and visualization of GC × GC chromatograms (basis for this figure is Dallüge et al. [31]).

modulation process. A raw chromatogram is obtained as the sum of all ²D chromatograms. Each color line represents one chromatographic peak (pink, blue, and yellow, respectively). The transformation of the raw data in a two-dimensional (2D) chromatogram (^1t_R vs. ^2t_R) is performed via software (step 2. Transformation). In addition, three-dimensional (3D) chromatograms are also obtained, associating raw data, the start of modulation process, P_M, injection time, and the signals given through the detector. In the third step (Visualization) shown in **Figure 2**, a 3D-plot is presented, and the intensity of the colors represents the intensity of peak signals. A projection of the top view of the 3D-plot gives rise to 2D-plots on the plane, such as color plots. In contour plots, each line represents a different and specific intensity of the chromatographic signal.

Some advantages of the use of GC × GC in relation to 1D-GC and GC-GC should be mentioned [32]:

(a) Higher peak capacity, which provides better chromatographic separation among analytes and among analytes and matrix interferents.

(b) Enhancement in detectability because of narrower chromatographic bands coming from the modulation process.

(c) All sample are submitted to both separation processes.

(d) Analysis time to complete both separations is the same as the one used for ¹D separation.

(e) Ordered elution of structurally related compounds may help increasing reliability of the identification process, and fingerprint provided by chromatograms may be used for pattern recognition.

(f) Two sets of retention data also increase reliability in compound identification process.

The concept of peak capacity was introduced by Giddings in 1967 and is defined as the maximum number of compounds that can be placed side by side in the separation space (chromatogram), having a predetermined resolution in a certain period of retention time, according to Eq. (2) [33].

$$n_c = \frac{L}{w_b R_s} \tag{2}$$

where L is the column length, w_b is the base peak width, and R_s is resolution.

The theoretical peak capacity of a system is the sum of the individual peak capacities of each column. In the case of GC-GC, where several "cuts" (heartcuts) are made and these chromatographic fractions are lead to other (s) columns (s), the contribution of every secondary column should be summed up. Therefore, the total peak capacity of a GC-GC system will be n_{ctot}.

$$n_{ctot} \cong \sum n_{ci} = m \times \bar{n}_c \tag{3}$$

where n_{ci} represents peak capacity of each individual column, m is the number of columns employed, and $\bar{n}$ is the mean peak capacity of the columns.

In a GC × GC system, theoretical peak capacity is approximately the product of the peak capacities of the individual columns:

$$n_{ctot} \cong n_{cD} \times n_{cD} \quad (4)$$

Figure 3 represents the peak capacities of one chromatographic column (**Figure 3(A)**), of a heart cut two-dimensional system (**Figure 3(B)** GC-GC) and of a comprehensive two-dimensional gas chromatography system (GC × GC) [6, 34]. This representative drawing makes it clear that the theoretical peak capacity of GC × GC is higher than the one of 1D-GC and then

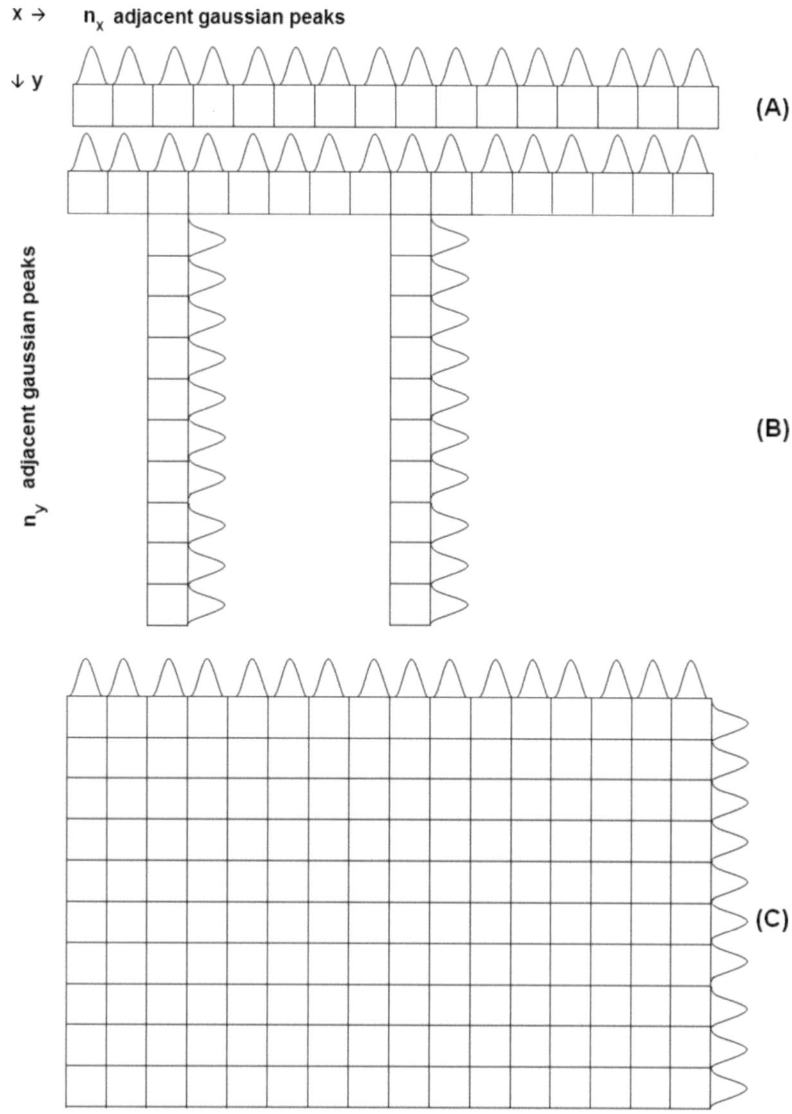

Figure 3. Representative illustration of theoretical peak capacities: (A) 1D-GC; (B) GC-GC; (C) GC × GC (based on Giddings [6]).

any other two-dimensional instrumental arrangement. However, this great peak capacity can only be achieved if there is no correlation between the separation mechanisms of the two columns, and they are considered independent of each other. In this case, the system may be called orthogonal [31]. In practice, according to statistical considerations, n_{ctot} will always be lower than the maximum possible, but an appropriate choice of the stationary phases will help to take the best of GC × GC peak capacity [35]. However, most importantly is that the dimensionalities of the sample and of the system should be combined in order to get the best, not only of the occupation of the 2D space, but also of the possible organized pattern that structurally related compounds may present. The number of independent variables that should be specified in relation to the identification of sample components refers to its dimensionality (double bonds, aromaticity, polar groups, etc.). Orderly elution in the 2D space can be obtained out of orthogonality conditions, as far as the dimensions of the sample and of the system are appropriately combined [36, 37].

3. Application of GC × GC to the investigation of coal derived liquids

3.1. The importance of coal

Mineral coal represents an important source of energy worldwide and several aspects, such as its abundance, widespread geographical availability, storage near major centers, and ease of transportation favor its use [38]. It consists of hydrocarbons and other components that contain heteroatoms, such as nitrogen and sulfur, originated from decomposition of organic materials (trunks, roots, branches, and leaves of trees). Coal may be classified according to its maturity (carbon content and physicochemical properties) which depends on temperature and pressure conditions during its formation. The higher the carbon content the more mature the coal is considered: peat (~45%), lignite (~46–60%), sub-bituminous coal (~42–52%), bituminous coal (~46–86%), and anthracite (>86%). Among them, sub-bituminous coal is the most employed as fuel [39]. Several applications of coal may be listed, such as its use in power plants for energy generation (main use), for steel production, and cement manufacturing [40]. In addition, it is used for residential energy needs, alumina refineries, paper manufacturers, chemical and pharmaceutical industries, and for specific products, such as activated carbon, carbon fiber, and silicon metal [41].

According to the 2015 annual report of the International Energy Agency (IEA), 10 countries produce more than 100 million tons/year (Mt/y), and China leads the ranking (3527 Mt), followed by United States (813 Mt), Indonesia (691 Mt) and Australia (509 Mt). Nowadays, India has become the largest importer, while Australia is the major exporter of coal in the world [42]. However, production of coal has been decreasing in the last years, due to several reasons, such as the economic changes in China and new environmental policies all over the world. China has adopted new hydraulic, nuclear, wind and solar energy instead of coal, in many cases due to concerns about environmental effects involved in all the processes included in the coal production, its products and derivatives [43]. More than 70 pollutant compounds are generated in the combustion of coal. In addition, the polluting potential can be seen in

http://dx.doi.org/10.5772/68077

different stages of the process of production (mining, production itself, use, and disposal of residues) and requires investments in green technologies to minimize risks of contamination and to avoid environmental accidents [38]. The main environmental effects of coal utilization are closely linked to the type of coal and its chemical composition, such as carbon, sulfur, and nitrogen content. During combustion, carbon monoxide (CO), sulfur dioxide (SO_2), and nitrogen oxides (NOx) are emitted, among other pollutants [44]. Due to these environmental problems, coal upgrading has been prioritized. Several different processes of coal upgrading can be performed, such as moisture removal, de-mineralization, and removal of harmful constituents (e.g., sulfur, nitrogen, and heavy metals) [45–51]. Consequently, it is important to know the composition of coal and of its derivatives in order to develop environmentally friendly processes and uses of these materials.

3.2. Coal derivatives

Liquids derived from coal are obtained mainly through pyrolysis, liquefaction, and solvent extraction. When coal is heated to high temperatures (500–1000°C) in an inert environment, 20–40% of the original solid is released into gas or volatiles. This process is called pyrolysis or devolatilization of coal. Coal is also converted into liquid fuel either directly or indirectly, via liquefaction [52]. Direct coal liquefaction, industrialized in the western countries of the past century, is considered as an effective way to produce transportation fuels and chemicals [53]. Liquefaction employs severe conditions of temperature and pressure to achieve a significant productivity. However, coal extracts may also be obtained with milder conditions, such as extraction with solvents, where compounds that are less strongly linked to the carbonaceous matrix are obtained. Solvent extraction may be performed in a relatively high temperature environment (370–480°C), where the matrix can swell with the solvent and overcome the van der Waals forces allowing the soluble material to be extracted. A nonhydrogen-donor solvent can be used, such as anthracene oil, in the absence of hydrogen [52]. Other diverse techniques may be employed for coal extraction in analytical scale, and some of them may be cited: Soxhlet extraction [54, 55], ultrasound extraction (UE) [56–58], and supercritical fluid extraction (SFE) [55, 59–61].

Coal tar is the liquid produced in pyrolysis, and it is also a byproduct of coal carbonization/gasification. It contains complex chemicals that are otherwise difficult to acquire from other fossil fuels such as gas and oil [62]. During thermal processes, several chemical reactions occur: cracking, dehydration, isomerization, rearrangements, aromatization, condensation, etc., and the amount of coal tar produced with pyrolysis is around 2.5–4% upon the weight of coal. Heavier molecular weight compounds are broken in smaller molecules under a partially or totally oxygen free atmosphere [63]. The pyrolytic oil may be employed as an alternative fuel and in fuel mixtures, and in this case, it should present properties that are similar to fossil fuels [64]. For that purpose, it is important to remove oxygen and other heteroatoms during pyrolysis [63]. In general, a wide variety of aromatic chemicals including polycyclic aromatic hydrocarbons (PAH), phenols, and heterocyclic compounds (containing sulfur and nitrogen atoms) exist in coal tar, depending on the reaction temperature and the nature of original coals [62]. Its complex compositions make coal tars an important feedstock for value-added substances and materials [65], although its content in carcinogenic compounds is a concern [66].

Quinolines and derivatives are important constituents of pharmacologically active synthetic compounds and have been applied for the treatment of malaria [67], HIV [68] and as corrosion inhibitors for steel [68]. Antitumor, antibacterial, anti-inflammatory, psychotropic and anti-histaminic properties have been attributed to many alkaloids, whose structures are based on carbazoles [69].

Pyridine is widely employed as starting material to produce agrochemicals, as, for example, paraquat and diquat, chlorpyrifos, and antifungal agents such as the zinc salt of pyrithione and pharmaceuticals. It is also used as solvent in organic chemistry along with methyl-pyridines [70]. Carbazoles have also been used as photoconductors, semiconductors, and, due to their light-emitting properties, they are interesting organic tools for physics experiments [69, 71]. Approximately 100% of the carbazole, quinoline and thiophene compounds are derived from coal tar [64].

Therefore, an environmentally friendly use of coal tar demands evaluation of its molecular features. A deep and wide characterization of physicochemical and molecular characteristics of coal and coal tars is fundamental to clarify their complex composition. An appropriate analytical approach helps to improve coal processing in order to obtain higher added value products, as well as enlighten the possible environmental impacts of coal (or coal tar) during combustion, pyrolysis and other industrial processes [64, 65].

3.3. Characterization of coal tar

Characterization of coal is usually made through the analyses of the liquids that are derived from coal using extraction, pyrolysis, and liquefactions. Several research works have been published on this matter, dealing with molecular mass distribution and characterization of molecular structure of coal liquid derivatives. Molecular weight and functional groups distributions are carried out by size exclusion chromatography [72] and Fourier transform infrared spectroscopy (FTIR) [73], respectively. The ratio of aromatic/saturated compounds and the distribution of heteroatoms has been determined by X-ray photoelectron spectroscopy (XPS) [74], magnetic resonance spectroscopy (NMR) [75], and X-ray absorption spectroscopy (XAS) [76]. Although these analytical techniques provide satisfactory information about coal tar characterization, analyses in molecular level are still a challenge.

Extracts of liquefaction and coal tar were analyzed by laser desorption (LD)-MS and matrix-assisted laser desorption/ionization (MALDI)-MS for the determination of the molecular mass (MM) distribution, and results were compared to size exclusion chromatography (SEC) [77]. In the SEC, the MM distribution of the coal tars with a carbon content of up to 89% was almost identical and tending to decrease for higher rank coals. For coal-derived liquefaction extracts up to 87% carbon, there is a constant increase in MM with increasing rank and a steady decline for coals with more than 87% of carbon. Analyzes carried out with MALDI showed that liquefaction extracts were similar among them, with a little increase in MM with the elevation of the carbon content; however, coal tar contained a higher MM distribution. SEC has been shown to underestimate the MM range when compared to (MALDI)-MS results.

Zhu et al. developed MS analytical methods using gas chromatography/mass spectrometry (GC/MS), time-of-flight MS coupled with an atmospheric solids analysis probe (ASAP-TOFMS) and electrospray ionization (ESI)-Orbitrap MS for the characterization of a petroleum ether (designated as PE-HTCT) of a high-temperature coal tar (HTCT) [66]. GC/MS detected hydrocarbons and molecules with low molecular weight and polarity, while mid-polarity compounds were investigated by PE-HTCT, without any sample pretreatment, through a rapid *in-situ* ASAP-TOFMS analysis. (ESI) Orbitrap MS was employed to characterize heteroatom-containing compounds of PE-HTCT. The combination of the three MS-based analytical methods provided an overview of the components present in the PE-HTCT. Sixty-seven percent of the species identified on PE-HTCT using GC/MS were hydrocarbons, but ASAP-TOFMS and ESI-Orbitrap MS detected only a small portion of them. These two analytical techniques were better choices for identification of polar compounds and heteroatom-containing compounds.

Gas chromatography with mass spectrometric detection (GC/MS) is commonly employed for the analysis of volatile fractions of liquids derived from coal [65, 66, 78–82] including also the use of an analytical pyrolyzer as injector [83]. However, other detectors are also employed, such as flame ionization detector (FID) [84–86] and atomic emission detector (AED) [87].

There are a considerable number of published research papers that report the occurrence of complex mixtures of organic compounds in coal tar. Major compounds are polycyclic aromatic hydrocarbons (PAH), and heterocyclic compounds containing oxygen, nitrogen, and sulfur are present as minor constituents [64]. Even though one-dimensional gas chromatography (1D-GC) is a well-established analytical technique for the analysis of coal tar, it presents drawbacks because the matrix is extremely complex, containing a high number of compounds, which result in an unresolved complex mixtures in many cases. Compounds present in higher amounts are together with trace level compounds (such as sulfur compounds) that are important in terms of health, environment, and process, which makes the analysis of such target trace compounds an extremely difficult task [63, 88]. Due to the already reported advantages of GC × GC, it is the technique of choice for the analysis of coal tar. However, in comparison with reports on bio-oil from biomass, the ones published on this matter are scarce [63, 89–92].

3.4. Application of GC × GC to liquids derived from coal

Researchers have highlighted the advantages of GC × GC for the analysis of coal-derived liquids, especially the possibility of resolving hundreds of compounds, in comparison with 1D-GC. The advantage of a mass spectrometric detector coupled to GC × GC, the structured elution order of structurally related compounds and the information about two sets of retention data also help identifying unknown compounds.

The parallel between pyrolytic oils from biomass (bio-oil) and coal (coal tar) can be seen in their complex composition as coal tar contains linear and branched saturated hydrocarbons, aromatic hydrocarbons and several compounds containing heteroatoms as well as bio-oils. Characterization of liquids from coal liquefaction has been mainly performed by GC × GC/TOFMS [92–94]. Applications of GC × GC/TOFMS to coal tar are also reported, such as, the use of chemical signatures of coal tars of different manufacture gas plants (MGP), where coal tar is a byproduct [90]. Coal tar is a common occurrence at MGP sites, some of which are severely affected and environmental remediation of polluted sites is a major challenge

[95–98]. A GC × GC method provided minimal sample pretreatment and a highly accurate and detailed chemical signature [89, 93]. The analytical method developed specifically for coal tar was different from the methods already reported for conventional forensic analysis of oil-spill in scientific literature [89, 99] and nitrogen containing compounds in coal tar, which are minor components of this matrix [91]. Some examples are given in the following texts.

3.4.1. Sulfur compounds in bitumen and tar

The removal of organic sulfur compounds (OSC) from coal is essential to provide an environmentally acceptable use for coal. Combustion of sulfur compounds leads to the emission of SOx gases, which are major contributors to acid rain [100]. In addition, OSC promote equipment corrosion [101] and can be deleterious to human health, contributing to respiratory diseases, such as emphysema, pneumoconiosis, asthma, and bronchitis [102]. It is also well known that organically bonded sulfur is difficult to remove from coal with the available technology. Development of new technological approaches for coal desulfurization depends on knowledge about the different organic sulfur compounds present in coals. Extraction of organic compounds from coal may be performed through a variety of processes, and among them, pyrolysis is considered as one of the most efficient thermal treatments in terms of yield of extracted material [99]. Pyrolysis causes breakage and rearrangement of several chemical bonds, generating artifacts. Conversely, process such as extraction is softer, and in this case, only compounds weakly bonded to the carbonaceous matrix are extracted. Bitumen is the name given to a coal extract. If the objective is to identify the compounds that are originally present in coal matrix, extraction processes are an appropriate choice. Several techniques of extraction may be employed to obtain bitumen, such as Soxhlet extraction, ultrasonic extraction (UE), and supercritical-fluid extraction (SFE) [100, 102].

Selective sulfur detectors have been used for the analyses of OSC in several matrices (petroleum and derivatives, coal tar and pitch, garlic, wine and beer volatiles, etc), as they reduce analysis time by eliminating cumbersome and time-consuming sample preparation procedures that can cause contamination or loss of analytes. A review about this subject has been reported by Wardencki and Zygmunt [103]. Among them, some may be listed: flame-selective flame detector and (FPD), double flame FPD [103], photometric detector of (PFPD) [104], sulfur chemiluminescence detector (SCD) [103, 105], and the atomic emission detector (AED) [106]. On the other hand, these detectors also present some drawbacks, as the lack of structural information, the quenching effect due to co-elution with hydrocarbons, water or CO_2 for FPD, among others. As far as the authors are aware, there is no paper reported for the analysis of OSC via GC × GC with a sulfur selective detector, although an analytical method has been developed for diesel samples and used to quantify sulfur impurities in petroleum fractions, which are also complex samples.

Machado et al. [99] employed three different techniques of extraction (ASE, UE, and SFE) to obtain bitumen from coal. The largest number of compounds (3347) was obtained through ASE, while extracts were analyzed by GC × GC/TOFMS. The analyses made by 1D-GC/qMS in scan mode resulted in 93 tentatively identified compounds, although none of the target OSC could be identified (**Figure 4A**). The following classes of OSC were found: thiophenes, benzothiophenes, naphthothiophenes, dibenzothiophenes, benzonaphthothiophenes, and

http://dx.doi.org/10.5772/68077

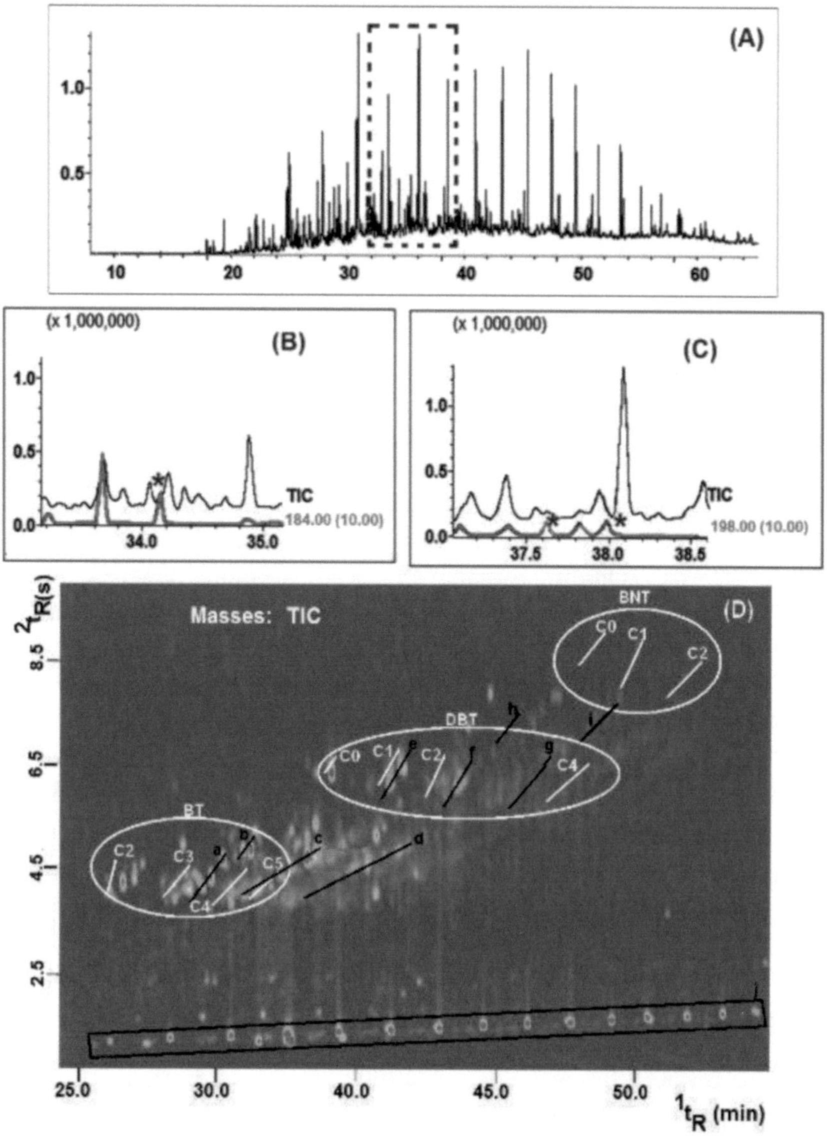

Figure 4. 1D-GC/qMS of the ASE bitumen: (A) scan mode; (B) EIM mode, with *m/z* 184 as the chosen ion (bellow line); (C) EIM mode, with *m/z* 198 as the chosen ion (bellow line). (D) GC × GC/TOFMS color plot of ASE bitumen. White lines - sulfur compounds: benzothiophenes (BT), dibenzothiophenes (DBT), benzonaphthothiophenes (BNT); number of methyl groups attached to compounds (CX). Black lines – hydrocarbons: a, naphthalenes with three methyl groups; b, *H*-xanthenes; c, naphthalenes with four methyl groups; d, naphthalenes with five methyl groups; e, phenanthrenes with one methyl group; f, phenanthrenes with two methyl groups; g, phenanthrenes with three methyl groups; h, pyrenes with one methyl group; i, pyrenes with two methyl groups; j, aliphatic hydrocarbons;. (adapted from Machado et al. [99]).

thioxanthenes, reaching 150 OSC. **Figure 4**A illustrates the image pattern formed by the peaks in the two-dimensional separation space (2D), which is an important characteristic of 2D-chromatograms that enhances the reliability of the compound identification process, as their distribution in the chromatogram is due to their physicochemical characteristics. Molecules are separated according to their number of carbon atoms, and this phenomenon is known as roof tile effect. Examples of roof tile effect are common in 2D-chromatograms of petroleum and petroleum-derived samples [17] and were also observed in coal bitumen.

The use of extracted ion-monitoring tool (EIM) made it possible to identify six alkylated and nonalkylated dibenzothiophenes (DBT). Co-elutions of polycyclic aromatic sulfur heterocycles (PASH) with polycyclic aromatic hydrocarbons (PAH) are well reported in scientific literature due to the occurrence of the same ions in their mass spectra and are clearly shown in the following examples. A dashed red line shows one region of the chromatogram where some compounds elute (**Figure 4A**) and the asterisks mark the chromatographic peaks under discussion in **Figure 4B** and **C**: peaks corresponding to *m/z* 184 (DBT or naphtothiophenes, NT) and *m/z* 198 (C1-DBT), respectively, where C1 designates one alkyl group with one carbon atom. The peak corresponding to ion 184 (**Figure 4B**, pink line) was observed in scan mode (**Figure 4B**, black line), as a frontal shoulder of major intensity and was identified as phenyl ethyl ether (retention time 34.2 min). The earlier eluting peak that contains the same characteristic ion does not correspond to an OSC, but to methylnaphthalene. However, a careful look in the GC × GC/TOFMS shows that an OSC elutes in this same region and that the sole use of EIM of 1D-GC/qMS may lead to erroneous results, such as misidentification, where OSC remains unnoticed. A further consequence of co-elutions is the overestimation of a PAH (or a matrix interferent peak) that co-elutes with an OSC leading to error in quantitative analysis. It is important to highlight that both chemical classes (PAH and PASH) are toxic. In some cases, PASH may even show higher mutagenicity than PAH. This is the case of the mutagenicity of 2- and 3-aminodibenzothiophene is higher than the one of benzo*[a]*pirene [107]

An alternative to get around this analytical problem is to submit the sample to preseparation steps, in order to minimize the presence of interferents. However, every time new stages are added to an analytical procedure, higher probability of contamination and loss of analytes are also added. Consequently, the analytical method becomes longer. The adoption of direct injection of the sample in a GC × GC/TOFMS or the use of a simple pretreatment step before its chromatographic analysis seem to be more appropriate, as it eliminates the already mentioned analytical drawbacks. In a similar way, **Figure 1**C presents two out of the four possible C1-DBT isomers that were identified (two asterisks indicate the t_R of each one of them) and the co-elution of the major peak of methyl anthracene (C1-anthracene, t_R 38.0 min, asterisk more to the right) with a methyl-dibenzothiophene (C1-DBT). The blue line shows the *m/z* 198 (EIM), and the black line designates the current derived from all detected ions (scan mode).

OSC in coal tar were also investigated, and 1D-GC co-elutions of PASH with PAH and phenols were also elucidated in GC × GC/TOFMS analyses [89]. Some of them were reported for the first time, such as the co-elution of trimethyl thiophene and phenol. The co-elution of C4-BT and trimethyl-naphthalene in coal tar and the separation of these compounds by GC × GC/TOFMS served as an illustration of many other co-elutions. In addition, thioxanthene was tentatively identified for the first time in coal tar, and the quality of mass spectra of compounds was noticeably enhanced with GC × GC/TOFMS in comparison with 1D-GC/qMS.

http://dx.doi.org/10.5772/68077

3.4.2. Nitrogen containing compounds in coal tar

The determination of the forms of nitrogen in coal is attracting increasing attention because of attempts to identify relationships between fuel nitrogen functionality and NOx production in coal combustion. Previous interest in nitrogen functionality in coals had arisen from efforts to understand the chemical transformations occurring during coal coking [108]. Nitrogen polyaromatic compounds (NPAC) are present in fossil fuels as a complex mixture. Percentages of NPAC are usually less than 0.5% in crude oil and around 1–2% in shale oil and coal tar [49]. While conventional petroleum-derived fuels are low in nitrogen, typical liquid fuels derived from coal have much higher nitrogen contents and give rise to significant amounts of NO, during combustion. The conversion of nitrogen (1% or more) from coal to NO, even at low combustion temperatures, is a serious environmental problem [109]. Over the past few years, it has been recognized that many of the nitrogen-containing compounds found in coal-derived liquids are both carcinogenic and mutagenic, particularly in the presence of other polycyclic aromatic compounds (PAC). Several basic nitrogen compounds are toxic and the aza-heterocycles and primary aromatic amines are suspected of carcinogenic activity. In general, neutral nitrogen compounds are supposed to be less toxic; however, several some dibenzocarbazoles have been already reported as carcinogens [49]. In addition, the combustion of nitrogen rich fuels is known to release many NPAC and NO, into the atmosphere [110]. Nitrogen-containing compounds play an important role in linking sub-units of the macromolecular coal structure through hydrogen bonding and N-linkages. Consequently, analysis of these compounds in liquids derived from coal is important as nitrogen-containing compounds are known to deactivate catalysts usually employed in the upgrading of primary products of coal liquefaction [74].

The N-containing compounds can be classified in neutral pyrrolic and basic pyridinic types, according to their chemical characteristics. The neutral nitrogen species comprise carbazole, pyrrole, indole, and their alkylated and benzo analogues; while basic nitrogen species include pyridines, quinolines, benzoquinolines and their alkylated analogues [30, 110]. N-compounds occur as minor components in coal tar, mainly as pyridinic, pyrrolic, and some aromatic primary amines [111].

Gas chromatography coupled to different selective detectors, such as the chemiluminescence detector (NCD), has been used for the analysis of NPAC in bituminous coal [112]. Burchill et al. used nitrogen and phosphorus detector (NPD) for the analysis of NPAC in coal tar products [49, 113]. The hyphenation of GC × GC with nitrogen chemiluminescence detection (GC × GC-NCD) has been reported to achieve the identification of various N-compounds in diesel fuel [114–116], while NPD was employed for the analysis of N-containing compounds in heavy gas oil [30]. To date, the use of GC × GC with selective detector for the analysis of nitrogenous compounds in coal tar has not been reported.

Nitrogen detection is subject to interference with the hydrocarbons of the matrix [91]. Therefore, different techniques of isolation of N-compounds from coal tar are traditionally carried out by aqueous acid extraction [49, 50] precipitation with gaseous hydrogen chloride [117], solvent extraction with methanol [118], and cation-exchange chromatography [119, 120].

Informative detectors, such as the most commonly used mass spectrometers (quadrupole, ion trap), are not sensitive enough to achieve detection of N-compounds at low concentrations. Moreover, they do not provide selective unambiguous extraction, identification and quantification of N-compounds in complex matrices such as coal tar. Speciation of NPAC in coal tar by GC × GC has already been reported in the literature using time-of-flight mass spectrometry detector (TOFMS). Silva et al. [91] applied GC × GC/TOFMS to determine NPAC in a Brazilian sub-bituminous coal tar without any previous prefractionation. One hundred and twelve NPAC were tentatively identified, and twelve NPAC were identified through injection with analytical standards. Pyridines and quinolines were the major basic N-compounds, whereas carbazoles were the major neutral ones. An important contribution of the GC × GC was the identification of all possible isomers of methyl quinolines, methyl pyridines and methyl carbazoles, being seven, three and five compounds, respectively. In addition, it was possible to identify nitriles, which is a very difficult task to perform using direct injection in GC/MS, as they are present in trace levels [121] 1D-GC offers poor separation between neutral and basic NPAC, mainly for the pairs acridine/carbazole and indole/quinoline, which cannot be resolved by mass detectors such as quadrupole, and ion trap [114]. However, the identification of these compounds was achieved with GC × GC/TOFMS due to differences in the second dimension retention time and with the help of spectral deconvolution.

Chromatographic separation of basic N-compounds is still also a challenge because pyridines, anilines, quinolines and acridines overlap in the first dimension [116]. Therefore, a preliminary fractionation of hydrocarbons, oxygen, nitrogen, and sulfur compounds must be performed in order to provide greater sensitivity and resolution, whenever one-dimensional techniques are employed. However, the separation of these classes has been achieved with no pretreatment, but with direct analysis with GC × GC/TOFMS, as presented in **Figure 5** [91]. The separation of quinolines from polyaromatic hydrocarbons and anilines from phenols is another challenge, which is exemplified in **Figure 5**.

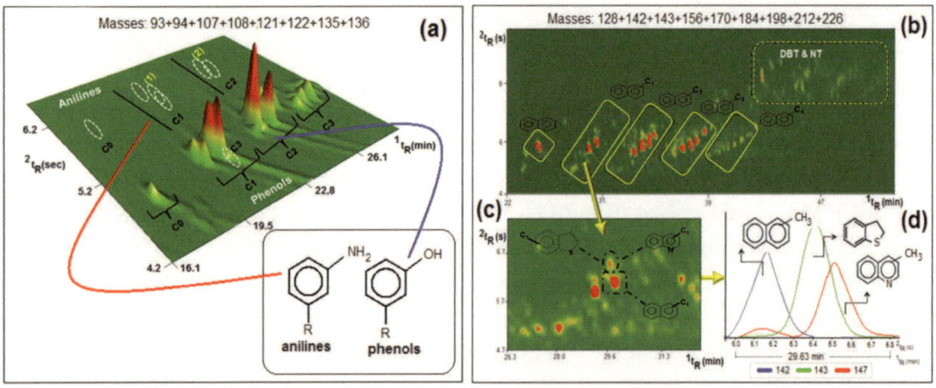

Figure 5. Demonstration of the peak capacity and resolution of the GC × GC/TOFMS: (a) 3D-Diagram (EIM mode) from selected region of anilines (C0–C3) and phenols (C0–C3); (b) 2D-Diagram (EIM mode) highlighting the region of elutions of alkyl naphthalenes (N), dibenzothiophenes (DBT) and naphthothiophenes (NT); (c) detail of the diagram showing the separation of three compounds and (d) reconstructed chromatogram with the specific ions of each of the three compounds shown in (b), including their structures (adapted from Silva et al. [91]).

Extrography is one of the methods of sample preparation used in literature [121] that allows separation among aliphatics, aromatics and neutral NPAC, whereas all phenols still elute in the fraction containing the basic NPAC. However, **Figure 5(a)** shows that it was possible to adequately separate phenols from anilines (basic compounds) in the second dimension of GC × GC. **Figure 5(b)** also clearly presents the roof tile effect related to alkyl naphthalenes (N), dibenzothiophenes (DBT) and naphthothiophenes (NT), as well as the separation between an aromatic hydrocarbon and two other compounds in the second dimension (^{2}D) (one PAH and one PASH), although co-elution of PAH, PASH and NPAC compounds is evident (**Figure 5(c) and (d)**). In this case, detection and tentative identification of the three co-eluting compounds was possible due to spectral deconvolution (**Figure 5(d)**).

4. Application of GC × GC to the investigation of pyrolytic liquids from biomass

4.1. The importance of biomass and bio-oil

Environmental problems fueled by the combustion of fossil fuels, the future decline in oil production due to the depletion of reserves, as a precursor of fossil fuels biomass is composed of carbon and can be an alternative to the substitution or complementation of fossil fuels with many applications [122].

Biomass can be used in energy applications for the production of heat and transport fuels. The conversion of biomass into fuels and chemicals, such as bio-plastics, biofertilizers and bio-polyesters, can share the demand with those of petrochemical origin, avoiding the depletion of fossil reserves in several regions of the world. A valorization of the entire content of biomass in bio refineries inspired by the model of those exploited for petroleum products constitutes one of the key issues [122–124].

Biomass is a renewable source of energy that could mitigate pollution of fossil fuels. It is abundant as agricultural residues and forest waste, whose thermochemical decomposition, under optimal conditions produces valuable gases, condensates and char. Char is used in fertilizers, while the condensed liquid, pyrolysis oil or bio-oil, serves as the mother liquor for numerous value-added chemicals and prospective engine-fuel applications [122]. Bio-oils are complex mixtures of water and hundreds of organic compounds that belong to different chemical classes, such as acids, aldehydes, ketones, alcohols, esters, anhydro-sugars (such as levoglucosan), furans, phenols, guaiacols, syringols, nitrogen-containing compounds, as well as large molecular oligomers (holocellulose-derived anhydro-oligosaccharides and lignin-derived oligomers [125–128]. Bio-oil compounds may be used for several different applications, such as phenolic compounds as raw material in chemical industries for the production of phenolic resins, among other uses [122]. Benzenediols such as catechol and hydroquinone have many applications as antivirus and as antioxidant for the production of rubbers, inhibitor of vinyl polymerization reactions, and also in topical application in skin whitening [123, 124, 129]. Guaiacol can be used as an indicator in several experiments involving enzymes, due to its characteristic color change according to reaction medium. In addition, it has been also applied as antiseptic, expectorant, and anesthetic [130]. Furthermore, ketones are applied in chemical syntheses, as, for example, furanones are precursors in the synthesis of brominated derivative fungicidal coatings and are currently obtained only by extraction of marine algae [131, 132].

4.2. Application of GC × GC to pyrolytic liquids derived from biomass

Chemical analysis and complete characterization are absolutely essential to gather the necessary data that provide information for research into bio-oil. The chemicals encountered in bio-oil depend on the composition of biomass, from which it is produced, as well as on the process parameters used during pyrolysis. Investigation of the molecular composition of bio-oil is mandatory for ensuring its proper characterization and for determining applications for which it is appropriate [125–127]. Considering the great variety of biomasses that can be used in the pyrolysis process, as well as the high complexity of bio-oils, due to the large number of components and chemical classes present, the characterization of different bio-oils, as well as their differentiation, presents itself as an analytical challenge [125, 126]. Kanaujia et al. [127] reviewed the most used analytical techniques to characterize the bio-oils of which we can mention liquid chromatography (LC), gel permeation chromatography (GPC), infrared spectroscopy (IR), GC, among others that were addressed according to their corresponding technical characteristics. GC/MS is an important tool to evaluate bio-oil profiles, helping to determine its subsequent final use, as well as for monitoring upgrade processes.

Hyphenated techniques such as heart-cutting multidimensional GC (MDGC) [133] and comprehensive two-dimensional gas chromatography (GC × GC) have gained much attention for the analysis of bio-oils in the last years [134–136]. GC × GC coupled to mass spectrometry (MS) proved to be a powerful technique for the separation of complex matrices and is ideal for characterizing compounds with similar chemical structures [137–139]. Similarly to what has been seen for the complex liquids derived from coal, one of the main advantages of GC × GC is an enormous increase in the resolving power when compared to 1D-GC [140]. Additionally, different well-ordered groups that are characterized by distinct patterns can be distinguished in the two-dimensional plane, providing important information about the chemical structures of the molecules of interest [134, 135].

The first studies that employed GC × GC for the analysis of bio-oils were conducted by Marsman and Sfetsas. Marsman et al. [141–143] used GC × GC-FID and GC × GC/TOFMS systems to tentatively identify roughly 248 and 368 compounds in bio-oil from beech (*Fagussylvatica*) and hydrodeoxygenated beech bio-oil (using ruthenium as catalyst), respectively. The authors also classified the compounds according to their chemical function into nine groups (acids, aldehydes, ketones, furans, guaiacols, syringols, sugars, alkyl phenols, alkyl-benzenes). The major compounds found in beech bio-oil were: hydroxymethyl furfural, furanone, furfural, mequinol and butanediol. Sfetsas et al. [144] have also analyzed the constituents of three bio-oils, where GC × GC was employed with FID and TOFMS detection systems. In this study, approximately 96 compounds (area >0.3%) were tentatively identified, and they were classified as acids, esters, aldehydes, ketones, hydrocarbons, aromatic hydrocarbons, phenols, sugars, and other nonclassified compounds. The major components were acetic acid, levoglucosane, and hydroxypropanone. Several other researchers have also reported the composition of different bio-oils.

http://dx.doi.org/10.5772/68077

Moraes et al. have used GC × GC for the qualitative characterization of bio-oils derived from the intermediate pyrolysis of various agro industrial residues such as orange bagasse [145], peach pit [146, 147], rice husk [147], and sugar cane straw [135]. Analysis of bio-oil from orange bagasse [145] was performed, without any pretreatment, by GC × GC-FID and GC × GC/TOFMS. A hundred sixty-seven compounds from different chemical classes (acids, aldehydes, ketones, phenols, esters, ethers and some nitrogen-compounds) were tentatively identified, and only twenty-six among them presented a chromatographic area percentage above 1%. Another research work of the same research group has shown the presence of approximately 500 compounds detected in each one of the bio-oils from peach core and rice husk. A conventional set of columns was employed in GC × GC/TOFMS, and the major part of compounds were ketones, phenols, alcohols, ethers, acids, aldehydes sugar derivatives and hydrocarbons [147].

Migliorini et al. [146] performed a comparative study between GC/qMS and GC × GC/TOFMS for peach pit bio-oil. The results showed 51 and 220 tentatively identified components, respectively, confirming the superiority of the multidimensional technique. The ordered distribution of the compounds in the 2D space allowed the identification of all the isomers of C1 to C4 alkyl phenols. Compounds such as alcohols, aldehydes, anhydrides, ketones, esters, ethers and phenols were present in the chromatograms of the two analytical systems employed, whereas phenols shared the major chromatographic area percent. However, GC × GC/TOFMS provided tentative identification of carboxylic acids, hydrocarbons and sugar derivatives such as levoglucosan that were not found in 1D-GC/MS analyses.

Moraes et al. [147] studied the similarities and differences between peach pit and rice husk bio-oils, using GC × GC/TOFMS with a conventional set of columns. The results showed similar qualitative and semi-quantitative composition for both bio-oil samples, with the presence of ketones, phenols, alcohols, ethers, acids, sugar derivatives, aldehydes and hydrocarbons. A hundred and six and 223 compounds were tentatively identified in rice husk and peach pit bio-oils, respectively. The bio-oil of sugar cane straw was also investigated using a similar analytical approach. Furfural and hexenoic acid were major compounds in this bio-oil, and a total of 123 compounds were tentatively identified.

Araújo et al. [148] evaluated volatile organic compounds that were released from biopitch (obtained via the distillation of *Eucalyptus* sp. bio-oil) and its polyurethane films. Biopitch was used in place of polyols in the polyurethane synthesis. The chemical characterization was performed in a system using solid phase microextraction (SPME) and GC × GC/TOFMS. Although these researchers had the expectation of finding aromatic compounds in the heavier fraction of *Eucalyptus* biopitch, they verified the presence of phenols with an aromatic ring, which differentiates this biopitch from those obtained from fossil materials (coal and oil). These results have shown that the Eucalyptus biopitch has potential for the development of "green" materials.

Mango seed waste represents an important environmental problem in Brazil due to the large amounts produced in the industrial processing of the corresponding fruit. Pyrolysis is an attractive alternative for the use of the residual seeds. Lazzari et al. [149] have investigated the pyrolytic liquid of mango seed waste (tegument and almond) using GC × GC/TOFMS. More than 100 compounds were tentatively identified in each bio-oil, through comparison of experimental and reported linear temperature programmed retention indices (LTPRI) and mass spectra. Major chemical classes in terms of chromatographic are percentage were ketones (20.6%), acids (16.8%), and hydrocarbons (7.2%) in the almond bio-oil, while the bio-oil of the tegument was rich in phenols (32.6%) and ketones (22.9%).

Almeida et al. [150] reported qualitative and semi-quantitative characterization of bio-oil generated from the residual fiber obtained from the shells of coconut (*Cocos nucifera* L. var. Dwarf). Chemical characterization of the bio-oil and the aqueous phase was performed by GC/qMS and GC × GC/TOFMS. Forty-two compounds were identified in the organic extract of the aqueous phase and 81 in the bio-oil, using GC/qMS, while GC × GC/TOFMS provided identification of 68 and 95 for the same samples, respectively. Alkyl phenols were predominant in both phases, and aldehydes, ketones, and phenols were also present in both pyrolytic products. The chromatographic area percent of phenols in bio-oil was 59 and 12% for aldehydes (mainly furfural), using GC/qMS, while aqueous phase extract rendered 77% for phenols. The composition of pyrolytic bio-oil and aqueous phase from shells of coconut indicates that these products may find potential use in industry, which also means that environmental impact due to wrong disposal of coconut shells may also be minimized.

The vegetable oil (VO) from the seeds and the bio-oil (BO) from the pyrolysis of the residual cakes from *Crambe abyssinica* were characterized with GC/qMS and GC × GC/TOFMS by Onorevoli et al. [151]. Three residual cakes were pyrolyzed: one coming from mechanical pressing extraction, another from Soxhlet and a third from the compressed propane extraction (CPE, subcritical state). Procedures resulted in the identification of 195, 307, and 361 compounds, respectively, in the bio-oils of residual cakes. While the BO presented complex composition, with esters, ketones, ethers, phenols, alcohols, acids, hydrocarbons, and nitrogen compounds, VO were basically composed of fatty acids, mainly erucic and oleic acid.

The amount of nitrogen compounds was higher in GC × GC/MS analysis than in GC/MS, and several co-elutions were found in 1D-GC that may explain the incorrect identification of the nitrogen containing analytes, which explains this difference. **Figure 6** shows one illustrative example through the separation of heptenoic acid, *p*-cresol and C_2-aniline, which were identified in the CPE bio-oil sample by GC × GC/TOFMS (^1t_R = 15.30 min and ^2t_R = 3.74 s, 4.76 and 4.80 s, respectively). *P*-cresol was the only one compound identified by GC/MS in this corresponding retention time, with a chromatographic area percent of 12.0%. This example shows that a direct analysis of this bio-oil by GC/MS, without previous fractionation steps can lead to erroneous results due to co-elutions.

The castor seed cake is a solid residue generated from castor oil extraction during the pressing of castor seeds. In 2011, world production of castor seed was 2,767,548 tons [3], generating approximately 1,383,774 ton of waste. Based on this fact, Silva et al. [152] used slow pyrolysis from castor seed cakes to obtain bio-oil and determine its chemical composition by GC × GC/TOFMS. The calorific value of the castor bio-oil was found to be 37.5 MJ kg^{-1}, similar to that of petroleum-derived fuels (43–46 MJ kg^{-1}). In addition, a large variety of compounds have been tentatively identified in this bio-oil, such as hydrocarbons (paraffins, olefins and aromatics), nitrogen compounds (nitriles, anilines, quinolines, pyridine, indoles, pyrazines, pyrroles, carbazols and acridines), oxygen compounds (phenols, acids, ketones, esters, and furans) and sulfur compounds (disulfides and thiophenes), in a total of 408 tentatively identified compounds.

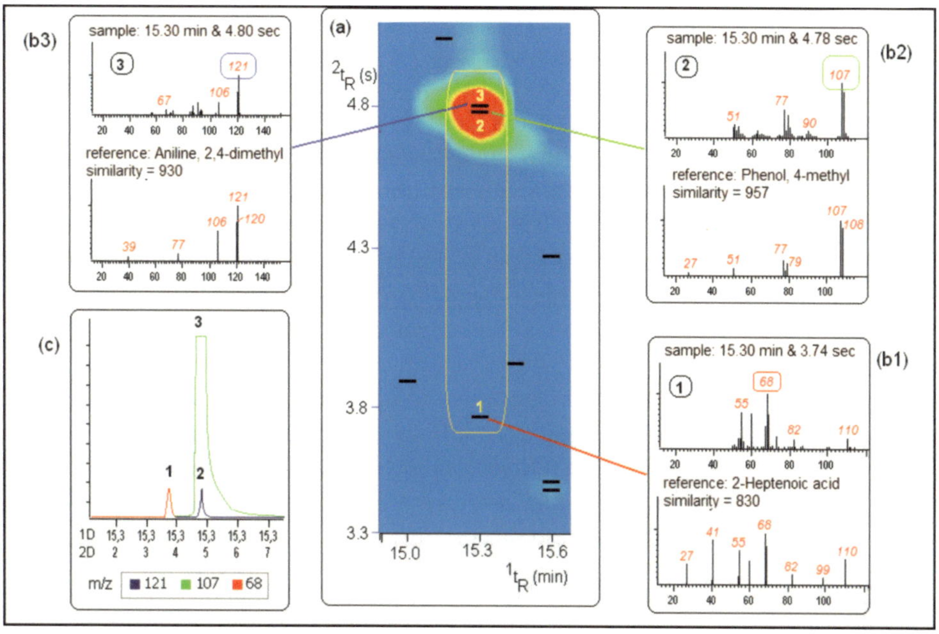

Figure 6. Peaks that co-eluted in GC/MS but were separate in GC × GC/TOFMS. (a) Part of the 2D Diagram of the CPE bio-oil sample; (b) mass spectra of peaks 1–3 and their respective comparisons with library mass spectra; (c) reconstructed chromatogram of the peaks: peak 1 = heptenoic acid; peak 2 = C_1 phenol; and peak 3 = C_2 aniline (adapted from Onorevoli et al. [151]).

Torri et al. [136] compared the composition of bio-oils from hardwood (*Eucalyptus sp.*) and softwoods (*Picea abies*) wood residues produced in large scale in pulp and paper industries, using fast and intermediate pyrolysis. The bio-oil characterization was performed by GC/qMS and GC × GC/TOFMS. The GC × GC analysis allowed the detection of dimethoxy-phenols, tracing potential markers of hardwood bio-oil, which might co-elute in 1D-GC with some methoxy-phenols and benzenediols. The separation and identification of toxic aromatic polycyclic hydrocarbons by GC × GC/TOFMS, which may also co-elute in 1D-GC with other bio-oil components, have raised the possibility of erroneous identification/quantification of these toxic components by 1D-GC due to the environmental and health risks they pose.

Tessarolo et al. [153] used GC/MS and GC × GC/TOFMS for the chemical characterization of bio-oils from palm fruit bunch and pine wood chips prepared via flash pyrolysis. Compounds such as furanones, benzenediols, phenols, indanones, cyclopentenones, and alkylpyridines were identified. Analyses with GC/MS resulted in the detection of 166 and 129 compounds in bio-oils of empty palm fruit bunch and pine wood chips, respectively. However, GC × GC/TOFMS allowed the detection of 631 and 857 analytes, respectively. Here, also, the superior performance of GC × GC/TOFMS was clearly demonstrated with the use of a medium polarity stationary phase in the second dimension (BPX-50) and a nonpolar column in the first dimension (DB-5). The major classes of compounds in the two bio-oil samples were furanones, ketones, cyclopentenones, benzenediols, furans, phenols, methoxy- and dimethoxy-phenols and anhydrous sugars. Different composition profiles due to the biomass sources were found;

aldehydes, esters, and pyridines were obtained from empty palm fruit bunch bio-oil, while alcohols and cyclopentanediones were found in pine wood chips bio-oil. Negative-ion electrospray ionization Fourier transform ion cyclotron resonance mass spectrometry (ESI(−)-FT-ICR MS) was also used to analyze the same bio-oils [154]. Some low-oxygen content classes not detected by ESI-FT-ICR MS were identified by GC × GC/TOFMS (classes from O0 to O8 with a C3 to C14). The high chromatographic resolution power of GC × GC/TOFMS with the high dynamic *m/z* range of ESI-FT-ICRMS allowed the identification of volatile, nonvolatile and polar compounds of bio-oils. The same research group [155] has employed GC × GC/TOFMS, FT-ICR-MS and hydrogen nuclear magnetic resonance (^{1}H NMR) to characterize bio-oils from catalytic pyrolysis of pine wood and sugar cane bagasse. The bio-oils of both biomasses presented similar composition, differing only in the proportions found in the classes of predominant compounds. The main groups of compounds identified were acids, ketones, phenols, oxygenated heterocyclics and aromatic hydrocarbons. An increase in hydrocarbons was obtained, whenever ZSM-5 (Zeolite Socony Mobil-5) was employed, as would be expected. The three analytical techniques used were of fundamental importance for an efficient chemical characterization of the bio-oils produced.

GC × GC with a quadrupole mass spectrometric detector (qMS) was used in the research works of da Cunha et al. [156] and Schneider et al. [157]. Da Cunha et al. have performed the fractionation of sugar cane straw bio-oil, using a silica column with pressurized liquids. A hundred sixty-six compounds were tentatively identified in the polar fraction resulting from the fractionation process (acids, aldehydes, ketones, esters, phenols, ethers, alcohols, and sugar derivatives), while the nonpolar fraction provided mainly aromatic hydrocarbons, aliphatic, cyclic and olefinic compounds [156]. Schneider et al. [157] analyzed polar compounds extracted with an alkaline solution from the *lignocell* (sawdust from forest timber) bio-oil. The results obtained in the GC × GC/qMS system allowed the characterization of 130 compounds from different chemical classes (phenols, ethers, ketones, aldehydes, acids, alcohols, and aromatic hydrocarbons). Only thirty-five compounds were tentatively identified by GC/MS, and they belonged to the following classes: phenols, aldehydes, ketones, ethers, and acids [157]. Maciel et al. [158] also studied the pyrolysis of sugarcane straw by GC × GC/TOFMS, with a focus on the composition of the aqueous phase. Organic solvents were employed to extract the components of the aqueous phase, and the composition of the organic extracts was similar to the one found in bio-oil, except that the aqueous phase extracts were richer in phenols, such as *ortho, meta* and *para* cresols. The use of GC × GC/qMS has shown high efficiency for the analyses of bio-oils, and it is an interesting choice for this type of analyses, as its cost is lower than the one of GC × GC/TOFMS.

5. Final comments

The basic principles of 1D-GC and 2D-GC (GC-GC and GC × GC) have been presented, and the increased resolution, sensitivity, peak capacity, mass spectral quality, and the organized distribution of compounds that are chemically similar, in the GC × GC chromatograms, have also been highlighted. A comparison of the application of 1D-GC and GC × GC to complex samples such as coal tar, coal bitumen, and pyrolytic bio-oil has shown that the special characteristics of GC × GC provide benefits, such as speciation of difficult to separate trace compounds, such as sulfur compounds in coal tar and bitumen and nitrogen-containing compounds in coal tar. The number of compounds tentatively identified in coal-derived liquids and bio-oils is higher with GC × GC than the ones found with 1D-GC, due to higher GC × GC peak capacity and resolution. In addition, organized distribution of structurally related compounds in the 2D-plots is of help in the process of identification of compounds. Even though 1D-GC may be an important tool for process monitoring (pyrolysis, upgrading, etc.) as it serves to evaluate chromatographic profiles and help to determine the subsequent use of produced compounds, it can also lead to misleading results due to its insufficient performance when dealing with complex samples. Sample pretreatment before GC/MS analysis is a strategy to improve analytical performance; however, it adds one-step to the analytical process, which becomes more time-consuming, laborious, and prone to contamination. Selective detectors are also alternatives to improve 1D-GC performance whenever specific compounds are target, although they do not provide a whole view of the sample composition. The before mentioned GC × GC analytical benefits can be of especial importance, whenever trace compound analysis is performed because of environmental and/or health issues, which is the case of the presence of sulfur and nitrogen compounds in pyrolytic or extraction liquids from coal or biomass residues. Mass spectrometric detector adds information to GC × GC analysis, especially when spectral deconvolution (TOFMS) is employed for compounds that partially co-elute and GC × GC/qMS is a good choice for a less costly GC × GC/MS analysis. A more comprehensive and detailed analysis of coal derived liquids and also of bio-oils is of importance, as far as these liquids are potential alternative sources of high added-value products and also as alternative fuels. The same knowledge is important for the development of environmentally safe technologies for the production of coal derived liquids and pyrolytic liquids of biomass residues, which are of interest to the chemical industry. These liquid samples are extremely complex, contain a great number of different compounds and chemical classes and are alto very diverse due to the original variety of coals and residual biomasses, as well as due to the different parameters that may be employed during production and upgrading processes. This variability in terms of raw material input and process poses a great analytical challenge, which is addressed to a great extent by GC × GC/TOFMS, without the need of previous laborious sample pretreatment, or multiple chromatographic steps, while keeping analysis time in the range of a 1D-GC analysis. However, as bio-oils and coal derived liquids are extremely complex samples, other instrumental techniques should also be employed along with GC × GC/TOFMS to achieve a total characterization of these samples, such as liquid chromatography with mass spectrometric detector, infrared spectroscopy, among others.

Acknowledgements

Authors would like to thank CNPq, CAPES, FINEP and PETROBRAS for the financial support to this work through scholarships, grants, and direct funding.

Author details

Gabriela Pereira da Silva Maciel[1], Juliana Macedo da Silva[2], Mozart Daltro Bispo[3], Laiza Canielas Krause[3,4], Rosângela Assis Jacques[1,4], Claudia Alcaraz Zini[1,4] and Elina Bastos Caramão[1,3,4*]

*Address all correspondence to: elina@ufrgs.br

1 Institute of Chemistry, Federal University of Rio Grande do Sul (UFRGS), Porto Alegre, Rio Grande do Sul, Brazil

2 Liberato Salzano Vieira da Cunha, Federal Technical School, Novo Hamburgo, Rio Grande do Sul, Brazil

3 Tiradentes University, Aracaju, Sergipe, Brazil

4 INCT-E&A, Brazil

References

[1] Skoog DA, Holler FJ, Crouch SR. Principles of instrumental analysis. Belmont, CA: Thomson Brooks/Cole, 2007. p. 1039. ISBN-10: 0495012017

[2] Berger TA. Separation of a gasoline on an open tubular column with 1.3 million effective plates. Chromatographia. 1996;**42**:63-71. DOI: 10.1007/BF02271057

[3] Cramers CA, Janssen HG, Van Deursen MM, Leclercq PA. High-speed gas chromatography: An overview of various concepts. Journal of Chromatography A. 1999;**856**:315-29. DOI: 10.1016/S0021-9673(99)00227-7

[4] Shellie R, Marriott P. Opportunities for ultra-high resolution analysis of essential oils using comprehensive two-dimensional gas chromatography: A review. Flavour and Fragrance Journal. 2003;**18**:179-91. DOI: 10.1002/ffj.1225

[5] Giddings JC. Two-dimensional separations: concept and promise. Analytical Chemistry. 1984;**56**:1258A. DOI: 10.1021/ac60252a025

[6] Giddings JC. Concept and comparison in multidimensional separation. Journal of High Resolution Chromatography. 1987;**10**:319-23.

[7] Bertsch W. Two-dimensional gas chromatography: concepts, instrumentation, and applications—part 1: Comprehensive two-dimensional gas chromatography. Journal of High Resolution Chromatography. [Internet]. 1999;**22**:647-65.

[8] Gordon BM, Uhrig MS, Borderging MF, Chung HL, Coleman WM, Elder JF. Analysis of flue-cured tobacco essential oil by hyphenated analytical techniques. Journal of Chromatographic Science. 1988;**26**:174-80. DOI: 10.1093/chromsci/26.4.174

[9] Liu Z, Sirimanne SR, Patterson DG, Needham LL, Phillips JB. Comprehensive two-dimensional gas chromatography for the fast separation and determination of pesticides extracted from human serum. Analytical Chemistry. 1994;**66**:3086-92. DOI: 10.1021/ac00091a016

[10] Liu Z, Phillips JB. Comprehensive two-dimensional gas chromatography using an on-column thermal modulator interface. Journal of Chromatographic Science. 1991;**29**:227-31. DOI: 10.1093/chromsci/29.6.227

[11] Beens J, Blomberg J, Schoenmakers PJ. Proper tuning of comprehensive two-dimensional gas chromatography (GC × GC) to optimize the separation of complex oil fractions. Journal of High Resolution Chromatography. 2000;**23**:182-8. DOI: 10.1002/(SICI)1521-4168(20000301)23:3<182::AID-JHRC182>3.0.CO;2-E

[12] Blomberg J, Schoenmakers P, Brinkman UAT. Gas chromatographic methods for oil analysis. Journal of Chromatography A. 2002;**972**:137-73. DOI: 10.1016/S0021-9673(02)00995-0

[13] Sampat A, Lopatka M, Sjerps M, Vivo-Truyols G, Schoenmakers P, van Asten A. Forensic potential of comprehensive two-dimensional gas chromatography. Trends in Analytical Chemistry. 2016;**80**:345-63. DOI: 10.1016/j.trac.2015.10.011

[14] Nolvachai Y, Kulsing C, Marriott PJ. Pesticides analysis: Advantages of increased dimensionality in gas chromatography and mass spectrometry. Critical Reviews in Environmental Science and Technology. 2015;**45**:2135-73. DOI: 10.1080/10643389.2015.1010431

[15] Cordero C, Kiefl J, Schieberle P, Reichenbach SE, Bicchi C. Comprehensive two-dimensional gas chromatography and food sensory properties: Potential and challenges. Analytical and Bioanalytical Chemistry. 2015;**407**:169-91. DOI: 10.1007/s00216-014-8248-z

[16] Almstetter MF, Oefner PJ, Dettmer K. Comprehensive two-dimensional gas chromatography in metabolomics. Analytical and Bioanalytical Chemistry. 2012;**402**:1993-2013. DOI: 10.1007/s00216-011-5630-y

[17] Von Mühlen C, Zini CA, Caramão EB, Marriott PJ. Applications of comprehensive two-dimensional gas chromatography to the characterization of petrochemical and related samples. Journal of Chromatography A. 2006;**1105**:39-50. DOI: 10.1016/j.chroma.2005.09.036

[18] Marriott P, Shellie R. Principles and applications of comprehensive two-dimensional gas chromatography. TrAC Trends in Analytical Chemistry. 2002;**21**:573-83. DOI: S0165-9936(02)00814-2

[19] Dallüge J, Beens J, Brinkman UAT. Optimization and characterization of comprehensive two-dimensional gas chromatography with time-of-flight mass spectrometric detection (GC × GC – TOF MS). Journal of Separation Science2002;**25**:201-14. DOI: 1615-9306/2002/0403-0201

[20] Mostafa A, Edwards M, Górecki T. Optimization aspects of comprehensive two-dimensional gas chromatography. Journal of Chromatography A. 2012;**1255**:38-55. DOI: 10.1016/j.chroma.2012.02.064

[21] Krupcik J, Gorovenko R, Spánik I., Sandra P, Armstrong DW. Flow-modulated comprehensive two-dimensional gas chromatography with simultaneous flame ionization and quadrupole mass spectrometric detection. Journal of Chromatography A. 2013;**1280**:104-11. DOI: 10.1016/j.chroma.2013.01.015

[22] Tranchida PQ, Purcaro G, Dugo P, Mondello L, Purcaro G. Modulators for comprehensive two-dimensional gas chromatography. TrAC Trends in Analytical Chemistry. 2011;**30**:1437-61. DOI: 10.1016/j.trac.2011.06.010

[23] Adahchour M, Beens J, Vreuls RJJ, Brinkman UAT. Recent developments in comprehensive two-dimensional gas chromatography (GC × GC). I. Introduction and instrumental set-up. TrAC Trends in Analytical Chemistry. 2006;**25**:438-54. DOI: 10.1016/j.trac.2006.03.002

[24] Machado ME, Bregles LP, de Menezes EW, Caramão EB, Benvenutti E V, Zini CA. Comparison between pre-fractionation and fractionation process of heavy gas oil for determination of sulfur compounds using comprehensive two-dimensional gas chromatography. Journal of Chromatography A. 2013;**1274**:165-72. DOI: 10.1016/j.chroma.2012.12.002

[25] Blomberg J, Riemersma T, Zuijlen M Van, Chaabani H. Comprehensive two-dimensional gas chromatography coupled with fast sulphur-chemiluminescence detection: Implications of detector electronics. Journal of Chromatography A. 2004;**1050**:77-84. DOI: 10.1016/j.chroma.2004.07.105

[26] Seeley J V. Theoretical study of incomplete sampling of the first dimension in comprehensive two-dimensional chromatography. Journal of Chromatography A. 2002;**962**:21-7. DOI: 10.1016/S0021-9673(02)00461-2

[27] Murphy RE, Schure MR, Foley JP. Effect of Sampling Rate on Resolution in Comprehensive Two-Dimensional Liquid Chromatography. Analytical Chemistry [Internet]. 1998;**70**:1585-94. DOI:10.1021/ac971184b

[28] Lee AL, Lewis AC, Bartle KD, McQuaid JB, Marriott PJ. A comparison of modulating interface technologies in comprehensive two- dimensional gas chromatography (GC × GC). Journal of Microcolumn Separations. 2000;**12**:187-93.

[29] Górecki T, Harynuk J, Panić O. The evolution of comprehensive two-dimensional gas chromatography (GC x GC). Journal of Separation Science. 2004;**27**:359-79. DOI: 10.1002/jssc.200301650

[30] von Mühlen C, Khummueng W, Zini CA, Caramão EB, Marriott PJ. Detector technologies for comprehensive two-dimensional gas chromatography. Journal of Separation Science. 2006;**29**:1909-21. DOI: 10.1002/jssc.200500443

[31] Dallüge J, Beens J, Brinkman U a T, Dalluge J, Beens J, Brinkman U a T. Comprehensive two-dimensional gas chromatography: A powerful and versatile analytical tool. Journal of Chromatography A. 2003;**1000**:69-108. DOI: 10.1016/S0021-9673(03)00242-5

[32] Phillips JB, Beens J. Comprehensive two-dimensional gas chromatography: A hyphenated method with strong coupling between the two dimensions. Journal of Chromatography A. 1999;**856**:331-47. DOI: 10.1016/S0021-9673(99)00815-8

[33] Giddings JC. Maximum number of components resolvable by gel filtration and other elution chromatographic methods. Analytical Chemistry. 1967;**39**:1027-8. DOI: 10.1021/ac60252a025

[34] Giddings JC. Unified Separation Science. New York: John Wiley & Sons; 1991. p. 352. ISBN: 978-0-471-52089-4

[35] Davis JM. Statistical theory of spot overlap in two-dimensional separations. Analytical Chemistry. 1991;**63**:2141-52. DOI: 10.1021/ac00019a014

[36] Giddings JC. Sample dimensionality : A predictor of order-disorder in component peak distribution in multidimensional separation. Journal of Chromatography A. 1995;**703**:3-15. SSDI: 0021-9673(95)00249-9

[37] Ryan D, Morrison P, Marriott P. Orthogonality considerations in comprehensive two-dimensional gas chromatography. Journal of Chromatography A. 2005;**1071**:47-53. DOI: 10.1016/j.chroma.2004.09.020

[38] Xia XH, Chen B, Wu XD, Hu Y, Liu DH, Hu CY. Coal use for world economy: Provision and transfer network by multi-region input-output analysis. Journal of Cleaner Production. 2017;**143**:125-44. DOI: 10.1016/j.jclepro.2016.12.142

[39] Speight J. The Chemistry and Technology of Coal; 3th ed. Florida: CRC Press; 2012. p. 845. ISBN: 9781439836460

[40] Osborne D, editor. The Coal Handbook: Towards Cleaner Production; 1st ed. Australia: Woodhead Publishing; 2013. p. 776. ISBN: 9780857094223

[41] Schernikau L. Economics of the International Coal Trade: The Renaissance of Steam Coal; Netherlands: Springer; 2010. p. 250. DOI: 10.1007/978-90-481-9240-3

[42] International Energy Agency. Excerpt from world energy balances coal information (2016 edition). [Internet]. 2015. Available from: https://www.iea.org/publications/freepublications/publication/KeyWorldEnergyTrends.pdf [Accessed: 2016-01-03]

[43] International Energy Agency. Excerpt from coal information (2016 edition). [Internet]. 2015. Available from: http://www.iea.org/publications/freepublications/publication/KeyWorldEnergyTrends.pdf. [Accessed: 2016-01-05]

[44] Markandya A, Wilkinson P. Electricity generation and health. Lancet. 2007;**370**:979-90. DOI: 10.1016/S0140-6736(07)61253-7.

[45] Charutawai K, Ngamprasertsith S, Prasassarakich P. Supercritical desulfurization of low rank coal with ethanol/KOH. Fuel Processing Technology. 2003;**84**:207-16. DOI: 10.1016/S0378-3820(03)00056-0

[46] Karaca H, Ceylan K, Olcay A. Catalytic dissolution of two Turkish lignites in tetralin under nitrogen atmosphere: Effects of the extraction parameters on the conversion. Fuel. 2001;**80**:559-64. DOI: 10.1016/S0016-2361(00)00119-8

[47] Karaca S, Akyürek M, Bayrakçeken S. The removal of pyritic sulfur from Aşkale lignite in aqueous suspension by nitric acid. Fuel Processing Technology. 2003;**80**:1-8. DOI: 10.1016/S0378-3820(02)00026-7

[48] Ratanakandilok S, Ngamprasertsith S, Prasassarakich P. Coal desulfurization with methanol/water and methanol/KOH. Fuel. 2001;**80**:1937-1942. PII: S0016-2361(01)00047-3

[49] Burchill P, Herod AA, Pritchard E. Investigation of nitrogen compounds in coal tar products. 1. Unfractionated materials. Fuel. 1983;**62**:11-9. DOI: 10.1016/0016-2361(83)90245-4

[50] Burchill P, Herod AA, Pritchard E. Estimation of Basic Nitrogen Compounds in Some Coal Liquefaction Products. Journal of Chromatography A. 1982;**246**:271-95. DOI: 0021-9673/82/0000-0000/S0275

[51] Gonsalvesh L, Marinov SP, Stefanova M, Carleer R, Yperman J. Organic sulphur alterations in biodesulphurized low rank coals. Fuel. 2012;**97**:489-503. DOI: 10.1016/j.fuel.2012.02.015

[52] Akash BA. Thermochemical Liquefaction of Coal. International Journal of Thermal and Environmental Engineering. 2013;**5**:51-60. DOI: 10.5383/ijtee.05.01.006

[53] You Q, Wu SY, Wu YQ, Huang S, Gao JS, Shang JX, et al. Product distributions and characterizations for integrated mild-liquefaction and carbonization of low rank coals. Fuel Processing Technology. 2017;**156**:54-61. DOI: 10.1016/j.fuproc.2016.09.022

[54] Li W, Ye C, Feng J, Xie K. Influence of Column Chromatography and Soxhlet Extraction on the Composition of Coal Pyridine-Soluble. Chinese Journal of Analytical Chemistry. 2006;**34**:905-10. ISBN: 3516018453

[55] Kolak JJ, Burruss RC. The use of solvent extractions and solubility theory to discern hydrocarbon associations in coal, with application to the coal-supercritical CO2 system. Organic Geochemistry. 2014;**73**:56-69. DOI: 10.1016/j.orggeochem.2014.05.002

[56] He W, Liu Z, Liu Q, Shi L, Shi X, Wu J, et al. Behavior of radicals during solvent extraction of three low rank bituminous coals. Fuel Processing Technology. 2017;**156**:221-7. DOI: 10.1016/j.fuproc.2016.10.029

[57] Tang Z, Zhai C, Zou Q, Qin L. Changes to coal pores and fracture development by ultrasonic wave excitation using nuclear magnetic resonance. Fuel. 2016;**186**:571-8. DOI: 10.1016/j.fuel.2016.08.103

[58] Zhong X, Wang M, Dou G, Wang D, Chen Y. Structural characterization and oxidation study of a Chinese lignite with the aid of ultrasonic extraction. Journal of the Energy Institute. 2015;**88**:398-405. DOI: 10.1016/j.joei.2014.11.004

[59] Vishal V. In-situ disposal of CO_2 : Liquid and supercritical CO_2 permeability in coal at multiple down-hole stress conditions. Journal of CO_2 Utilization. 2017;**17**:235-42. DOI: 10.1016/j.jcou.2016.12.011

[60] Yi-hui D, Hang C, Dong-fei W, Wei-guang MA, Jin-feng W, De-ping XU, et al. Supercritical fluid extraction and fractionation of high-temperature coal tar. Journal of Fuel Chemistry and Technology. 2010;**38**:140-43.

[61] Dariva C, de Oliveira J, Vale MGR, Caramão EB. Supercritical fluid extraction of a high-ash Brazilian coal : Extraction with pure ethanol and isopropanol and their aqueous solutions. Fuel. 1997;**76**:585-91. PII: SOO16-2361(97)00060-4

[62] Benhabib K, Faure P, Sardin M, Simonnot MO. Characteristics of a solid coal tar sampled from a contaminated soil and of the organics transferred into water. Fuel. 2010;**89**:352-9 DOI:10.1016/j.fuel.2009.06.009

[63] Rathsack P, Otto M. Classification of chemical compound classes in slow pyrolysis liquids from brown coal using comprehensive gas-chromatography. Fuel. 2014;**116**:841-9. Available from: DOI: 10.1016/j.fuel.2013.05.100

[64] Li C, Suzuki K. Resources, properties and utilization of tar. Resources, Conservation and Recycling. 2010;**54**:905-15. DOI: 10.1016/j.resconrec.2010.01.009

[65] Borwitzky H, Schomburg G. Separation and identification of polynuclear rromatic compounds in coal tar by using glass capillary chromatography including combined gas vhromatography-mass spectrometry. Journal of Chromatography. 1979;**170**:99-124.

[66] Zhu JL, Fan X, Wei XY, Wang SZ, Zhu TG, Zhou CC, et al. Molecular characterization of heteroatomic compounds in a high-temperature coal tar using three mass spectrometers. Fuel Processing Technology. 2015;**138**:65-73. DOI: 10.1016/j.fuproc.2015.04.020

[67] Kaur K, Jain M, Reddy RP, Jain R. Quinolines and structurally related heterocycles as antimalarials. European Journal of Medicinal Chemistry. 2010;**45**:3245-64. DOI: 10.1016/j.ejmech.2010.04.011

[68] Ebenso EE, Obot IB, Murulana LC. Quinoline and its derivatives as effective corrosion inhibitors for mild steel in acidic medium. International Journal of Electrochemical Science. 2010;**5**:1574-86. ISSN: 14523981 14523981

[69] Caruso A, Voisin-Chiret AS, Lancelot JC, Sinicropi MS, Garofalo A, Rault S. Efficient and simple synthesis of 6-aryl-1,4-dimethyl-9H-carbazoles. Molecules. 2008;**13**:1312-20. DOI: 10.3390/molecules13061312

[70] Kroschwitz JI. Pyridine and pyridine derivatives. In: Kirk-Othmer Encyclopedia of Chemical Technology EDN. 4th ed. New Jersey: John Wiley & Sons Inc; 2006. p. 1084. ISBN: 978-0-471-48496-7

[71] Samuel P, Maity S, Khan S, Roy SC. New opportunities for research in coal derived chemicals. Journal of Scientific & Industrial Research. 2008;**67**:1051-8. ISSN: 00224456

[72] Hausler DW, Taylor LT. Size exclusion chromatography of organically bound metals and coal-derived materials with inductively coupled plasma atomic emission spectrometric detection. Analytical Chemistry. 1981;**53**:1227-31. DOI: 10.1021/ac00231a022

[73] Seshadri S, Young C, Cronauer C. Characterization of coal liquids by ^{13}C n.m.r.and FT-i.r.spectroscopy - fractions of oils of SRC-I and asphaltenes and preasphaltenes of. Fuel. 1985;**64**:22-8. 0016-2361/85/01002247$3.00

[74] Bartle KD, Perry DL, Wallace S. The functionality of nitrogen in coal and derived liquids: An XPS study. Fuel Processing Technology. 1987;**15**:351-61. DOI: 10.1016/0378-3820(87)90057-9

[75] Alien DT, Petrakis L, Grandy DW, gavalas GR, Gates BC. Determination of functional groups of coal-derived liquids by n.m.r. and elemental analysis. Fuel. 1984;**63**:803-9. DOI: 10.1016/0016-2361(84)90071-1

[76] Kasrai M, Brown JR, Bancroft GM, Tan KH, Chen JM. Characterization of sulphur in coal from sulphur L-edge XANES spectra. Fuel. 1990;**69**:411-4. DOI: 10.1016/0016-2361(90)90306-B

[77] Herod AA, Li CZ, Parker JE, John P, Johnson CAF, Smith GP. Characterization of coal by matrix-assisted laser desorption ionization mass spectrometry.I. the argonne coal samples. Rapid Communications in Mass Spectrometry. 1994;**8**:808-14. DOI: 10.1002/rcm.1290081004

[78] Zhang D, Wang S, Ma X, Tian Y. Interaction between coal and distillation residues of coal tar during co-pyrolysis. Fuel Processing Technology. 2015;**138**:221-7. DOI:10.1016/j.fuproc.2015.06.002

[79] Li XH, Ma JS, Li LL, Li BF, Feng J, Turmel W. Semi-coke as solid heat carrier for low-temperature coal tar upgrading. Fuel Processing Technology. 2016;**143**:79-85. DOI: 10.1016/j.fuproc.2015.11.013

[80] Luo K, Zhang C, Zhu S, Bai Y, Li F. Tar formation during coal pyrolysis under N2 and CO2 atmospheres at elevated pressures. Journal of Analytical and Applied Pyrolysis. 2016;**118**:130-5. DOI: 10.1016/j.jaap.2016.01.009

[81] Du ZY, Wang X, Zhang ZH, Feng J, Qin YH, Zhang B. Evolution properties of cellulose- and lignin-derived pyrolysis tars after interacting with coal chars. Journal of Analytical and Applied Pyrolysis. 2016;**122**:332-41. DOI: 10.1016/j.jaap.2016.09.008

[82] Oña-Ruales JO, Sharma AK, Wise SA. Identification and quantification of seven fused aromatic rings $C_{26}H_{14}$ peri-condensed benzenoid polycyclic aromatic hydrocarbons in a complex mixture of polycyclic aromatic hydrocarbons from coal tar. Journal of Chromatography A. 2016;**1442**:83-93. DOI: 10.1016/j.chroma.2016.02.082

[83] Kong X, Bai Y, Yan L, Li F. Catalytic upgrading of coal gaseous tar over Y-type zeolites. Fuel. 2016;**180**:205-10. DOI:10.1016/j.fuel.2016.03.101

[84] Tursun Y, Xu S, Wang G, Wang C, Xiao Y. Tar formation during co-gasification of biomass and coal under different gasification condition. Journal of Analytical and Applied Pyrolysis. 2015;**111**:191-9. DOI: 10.1016/j.jaap.2014.11.012

[85] Montiano MG, Fernández AM, Díaz-Faes E, Barriocanal C. Tar from biomass/coal-containing briquettes. Evaluation of PAHs. Fuel. 2015;**154**:261-7. DOI: 10.1016/j.fuel.2015.03.067

[86] Nguyen M, Berndt C, Reichel D, Krzack S, Meyer B. Pyrolysis behaviour study of a tar- and sulphur-rich brown coal and GC-FID/MS analysis of its tar. Journal of Analytical and Applied Pyrolysis . 2015;**115**:194-202. DOI: 10.1016/j.jaap.2015.07.014

[87] Bermejo J, Fernández AL, Prada V, Granda M, Menéndez R. Monitoring the synthesis of new pitches from coal tar and its fractions by chromatography and related techniques. Journal of Chromatography A. 1999;**849**:507-19. DOI: 10.1016/S0021-9673(99)00608-1

[88] Marinov SP, Stefanova M, Stamenova V, Carleer R, Yperman J. Sulphur functionality study of steam pyrolyzed "mequinenza" lignite using reductive pyrolysis technique coupled with MS and GC/MS detection systems. Fuel Processing Technology. 2005;**86**:523-34. DOI: 10.1016/j.fuproc.2004.04.003

[89] Machado ME, Caramão EB, Zini CA. Investigation of sulphur compounds in coal tar using monodimensional and comprehensive two-dimensional gas chromatography. Journal of Chromatography A. 2011;**1218**:3200. DOI: 10.1016/j.chroma.2010.11.077

[90] Gauchotte-Lindsay C, Richards P, McGregor LA, Thomas R, Kalin RM. A one-step method for priority compounds of concern in tar from former industrial sites: Trimethylsilyl derivatisation with comprehensive two-dimensional gas chromatography. Journal of Chromatography A. 2012;**1253**:154-63. DOI: 10.1016/j.chroma.2012.06.093

[91] Silva JM da, Machado ME, Maciel GPS, Dal Molin D, Caramão EB. Speciation of nitrogen-containing compounds in an unfractionated coal tar sample by comprehensive two-dimensional gas chromatography coupled to time-of-flight mass spectrometry. Journal of Chromatography A. 2014;**1373**:159-68. DOI: 10.1016/j.chroma.2014.11.004

[92] Hamilton JF, Lewis AC, Millan M, Bartle keith D, Herod AA, Kandiyoti R. Comprehensive two-dimensional gas chromatography coupled to time-of-flight mass spectrometry of coal liquids produced during a coal liquefaction process. Energy Fuels. 2007;**21**:286-94. DOI: 10.1021/ef060366i

[93] Mcgregor LA, Gauchotte-lindsay C, Nic N, Thomas R, Daly P, Kalin RM. Ultra resolution chemical fingerprinting of dense non-aqueous phase liquids from manufactured gas plants by reversed phase comprehensive two-dimensional gas chromatography. Journal of Chromatography A. 2011;**1218**:4755-63. DOI: 10.1016/j.chroma.2011.05.045

[94] Herod AA, Zhuo Y, Kandiyoti R. Size-exclusion chromatography of large molecules from coal liquids, petroleum residues, soots, biomass tars and humic substances. Journal of Biochemical and Biophysical Methods. 2003;**56**:335-61.

[95] Birak PS, Miller CT. Dense non-aqueous phase liquids at former manufactured gas plants: Challenges to modeling and remediation. Journal of Contaminant Hydrology. 2009;**105**:81-98. DOI: 10.1016/S0165-022X(03)00070-8

[96] Costa HJ, White keith A, Ruspantini JJ. Distinguishing PAH Background and MGP Residues in Sediments of a Freshwater Creek. Environmental Forensics. 2004;**5**:171-82. DOI: 10.1080/15275920490495909

[97] Gaines RB, Frysinger GS, Reddy CM, Nelson RK. Wang Z, Stout S, editors. Oil Spill Environmental Forensics: Fingerprinting and Source Identification; 1st ed. London: Academic Press; 2006. p. 620. ISBN:9780123695239

[98] Mcgregor LA, Gauchotte-Lindsay C, Daéid NN, Thomas R, Kalin RM. Multivariate statistical methods for the environmental forensic classification of coal tars from former manufactured gas plants. Environmental Science & Technology. 2012;**46**:3744-52. DOI: 10.1021/es203708w

[99] Machado ME, Fontanive FC, Oliveira JV De, Caramão EB, Zini CA. Identification of organic sulfur compounds in coal bitumen obtained by different extraction techniques using comprehensive two-dimensional gas chromatography coupled to time-of-flight mass spectrometric detection. Analytical and Bioanalytical Chemistry. 2011;**401**:2433-44. DOI: 10.1007/s00216-011-5171-4

[100] Schlosberg RH, editor. Chemistry of Coal Conversion; 1st ed. New York: Plenum Press; 1985. p. 336. ISBN 0306419742

[101] Barooah PK, Baruah MK. Sulphur in Assam coal. Fuel Processing Technology. 1996;46:83-97. DOI: 10.1016/0378-3820(95)00058-5

[102] Borah D. Oxidation of high sulphur coal. Part 2. Desulphurisation of organic sulphur by hydrogen peroxide in presence of metal ions. Fuel. 2001;80:1475-88. DOI: 10.1016/S0016-2361(01)00002-3

[103] Wardencki W, Zygmunt B. Gas chromatographic sulphur-sensitive detectors in environmental analysis. Analytica Chimica Acta. 1991;255:1-13. DOI: 10.1016/0003-2670(91)85080-C

[104] Stee LLP Van, Beens J, Vreuls RJJ, Brinkman UAT. Comprehensive two-dimensional gas chromatography with atomic emission detection and correlation with mass spectrometric detection : principles and application in petrochemical analysis. Biomolecular Analysis and Spectroscopy, 2003;**1019**:89-99. DOI: 10.1016/S0021-9673(03)01301-3

[105] Eckerttilotta SE, Hawthorne SB, Miller DJ. Comparison of commercially available atomic emission and chemiluminescence detectors for sulfur-selective gas-chromatographic detection. Journal of Chromatography. 1992;591:313-23. DOI: 10.1016/0021-9673(92)80249-t

[106] GerbersmannC,LobinskiR,AdamsFC.Determinationofvolatilesulfurcompoundsinwater samples, beer and coffee with purge and trap gas chromatography-microwave-induced plasma atomic emission spectrometry. Analytica Chimica Acta. 1995;316:93-104. SSDI: 0003-2670(95)00344-4

[107] Nishioka M, Campbell RM, West WR, Smith PA, Booth GM, Lee ML. Determination of aminodibenzothiophenes in a coal liquid. Analytical Chemistry. 1985;57:1868-71. DOI: 10.1021/ac00286a017

[108] Buckley AN. Nitrogen functionality in coals and coal-tar pitch determined by X-ray photoelectron spectroscopy. Fuel Processing Technology. 1994;**38**:165-179. SSDI 0378-3820(93)E0106-0

[109] Solomon PR, Colket B. Evolution of fuel nitrogen in coal devolatilization. Fuel. 1978;**57**:749-755. doi.org/10.1016/0016-2361(78)90133-3

[110] Mitera J. Separation compounds and identification in coal-tar pitch. Fuel. 1989;68: 596-600.

[111] Burchill P, Herod AA, Mahon JP, Pritchard E. The class separation of nitrogen compounds in coal tars. Journal of Chromatography. 1983;281:109-24. doi.org/10.1016/S0021-9673(01)87871-7

[112] Wornat MJ, Sarofim AF, Longwell JP, Lafleur AL. Effect of pyrolysis conditions on the composition of nitrogen-containing polycyclic aromatic compounds from a bituminous coal. Energy & Fuels.1988;2:775-82. DOI: 10.1021/ef00012a009

[113] Burchill P, Herod AA. Investigation of nitrogen compounds tar products.2. Basic fractions in coal. Fuel. 1983;62:20-9. doi.org/10.1016/0016-2361(83)90246-6

[114] Adam F, Bertoncini F, Brodusch N, Durand E, Thiébaut D, Espinat D. New benchmark for basic and neutral nitrogen compounds speciation in middle distillates using comprehensive two-dimensional gas chromatography. Journal of Chromatography A. 2007;1148:55-64. DOI: 10.1016/j.chroma.2007.01.142

[115] Dutriez T, Borras J, Courtiade M, Thiébaut D, Dulot H, Bertoncini F. Challenge in the speciation of nitrogen-containing compounds in heavy petroleum fractions by high temperature comprehensive two-dimensional gas chromatography. Journal of Chromatography A. 2011;1218:3190-9. DOI: 10.1016/j.chroma.2010.10.056

[116] Lissitsyna K, Huertas S, Quintero LC, Polo LM. Novel simple method for quantitation of nitrogen compounds in middle distillates using solid phase extraction and comprehensive two-dimensional gas chromatography. Fuel. 2013;104:752-7. DOI: 10.1016/j.fuel.2012.08.054

[117] Bodzek D, Krzyianowska T, Marzec A. Heterocompounds present in asphaltenes from various products of coal hydrogenation. Fuel. 1979;58:196-202. doi.org/10.1016/0016-2361(79)90117-0

[118] Kodera Y, Mite Y, Nakayama T. Solvent extraction from coal liquids of nitrogen compounds. Fuel. 1991;70:765-9. doi.org/10.1016/0016-2361(91)90076-M

[119] Burchill P, Herod AA, Mahon JP, Pritchard E. Comparison of methods for the Isolation of basic nitrogen compounds from coal tars. Journal of Chromatography. 1983;265:223-38. doi.org/10.1016/S0021-9673(01)96719-6

[120] Wu JC, Sung H, Lin Y, Lin S. Removal of tar base from coal tar aromatics employing solid acid adsorbents. Separation and Purification Technology. 2000;21:145-53. PII:SI383-5866(00)00198-2

[121] Granda M, Merhdez R, Bernad P, Bermejo J. Efficiency of extrography in the fractionation of coal-derived oils. Fuel. 1993;72:397-403. DOI: 10.1016/0016-2361(93)90061-6

[122] Bu Q, Lei H, Zacher AH, Wang L, Ren S, Liang J. A review of catalytic hydrodeoxygenation of lignin-derived phenols from biomass pyrolysis. Bioresource Technology. 2012;124:470-7. DOI: 10.1016/j.biortech.2012.08.089

[123] Kong Y, Chen X, Wang W, Chen Z. A novel palygorskite-modified carbon paste amperometric sensor for catechol determination. Analytica Chimica Acta. 2011;688:203-7. DOI: 10.1016/j.aca.2011.01.007

[124] Barbosa LCA, Maltha CRA, Demuner AJ, Pinheiro PF, Varejão JOS, Montanari RM. Synthesis and evaluation of antimicrobial activity ofhalogenated furanones and compounds analogues to nostoclides. Quim Nova. 2010;33:2020-6.

[125] Isahak WNRW, Hisham MWM, Yarmo MA, Hin TY. A review on bio-oil production from biomass by using pyrolysis method. Renewable & Sustainable Energy Reviews. 2012;16:5910-23. DOI: 10.1016/j.rser.2012.05.039

[126] Collard FX, Blin J. A review on pyrolysis of biomass constituents: Mechanisms and composition of the products obtained from the conversion of cellulose, hemicelluloses and lignin. Renewable & Sustainable Energy Reviews. 2014;38:594-608. DOI: 10.1016/j.rser.2014.06.013

[127] Kanaujia PK, Sharma YK, Garg MO, Tripathi D, Singh R. Review of analytical strategies in the production and upgrading of bio-oils derived from lignocellulosic biomass. Journal of Analytical and Applied Pyrolysis. 2014;105:55-74. DOI: 10.1016/j.jaap.2013.10.004

[128] Kanaujia PK, Sharma YK, Agrawal UC, Garg MO. Analytical approaches to characterizing pyrolysis oil from biomass. TrAC Trends in Analytical Chemistry. 2013;42:125-36. DOI: 10.1016/j.trac.2012.09.009

[129] Ziyatdinova GK, Gainetdinova AA, Budnikov GK. Reactions of Synthetic Phenolic Antioxidants with Electrogenerated Titrants and Their Analytical Applications. Journal of Analytical Chemistry. 2010;**65**: 929-934. DOI: 10.1134/S1061934810090078

[130] Surmont R, Verniest G, Kimpe N De. Short Synthesis of the Seed Germination Inhibitor 3,4,5-Trimethyl-2(5H)-furanone. The Journal of Organic Chemistry. 2010;75:5750-3. DOI: 10.1021/jo1010476

[131] Romero D, Traxler MF, López D, Kolter R. Antibiotics as signal molecules. Chemical Reviews. 2011;111:5492-505. DOI: 10.1021/cr2000509

[132] Al-bataineh SA, Britcher LG, Griesser HJ. XPS characterization of the surface immobilization of antibacterial furanones. Surface Science. 2006;600:952-62. DOI: 10.1016/j.susc.2005.12.028

[133] Fullana A, Contreras JA, Striebich RC, Sidhu SS. Multidimensional GC/MS analysis of pyrolytic oils. Journal of Analytical and Applied Pyrolysis. 2005;**74**(1-2):315-26. DOI: 10.1016/j.jaap.2004.11.036

[134] Djokic MR, Dijkmans T, Yildiz G, Prins W, Van Geem KM. Quantitative analysis of crude and stabilized bio-oils by comprehensive two-dimensional gas-chromatography. Journal of Chromatography A. 2012;1257:131-40. DOI: 10.1016/j.chroma.2012.07.035

[135] Moraes MSA, Georges F, Almeida SR, Damasceno FC, Maciel GPS, Zini CA, Caramão EB. Analysis of products from pyrolysis of Brazilian sugar cane straw. Fuel Processing Technology. 2012;101:35-43. DOI: 10.1016/j.fuproc.2012.03.004

[136] Torri IDV, Paasikallio V, Faccini CS, Huff R, Caramão EB, Sacon V, Zini CA. Bio-oil production of softwood and hardwood forest industry residues through fast and intermediate pyrolysis and its chromatographic characterization. Bioresource Technology. 2016;200:680-90. DOI: 10.1016/j.biortech.2015.10.086

[137] Mondello L, Tranchida PQ, Dugo P, Dugo G. Comprehensive two-dimensional gas chromatography-mass spectrometry: A review. Mass Spectrometry Reviews. 2008;27:101-24. DOI: 10.1002/mas.20158

[138] Marriott PJ, Chin ST, Maikhunthod B, Schmarr HG, Bieri S. Multidimensional gas chromatography. TrAC Trends in Analytical Chemistry. 2012;34:1-20. DOI: 10.1016/j.trac.2011.10.013

[139] Tranchida PQ, Franchina FA, Dugo P, Mondello L, Introduction I. Comprehensive two-dimensional gas chromatography-mass spectrometry : recent evolution and current trends. Mass Spectrometry Reviews. 2016;35:524-34. DOI: 10.1002/mas.21443

[140] Shen DK, Gu S. The mechanism for thermal decomposition of cellulose and its main products. Bioresource Technology. 2009;100:6496-504. DOI: 10.1016/j.biortech.2009.06.095

[141] Marsman JH, Wildschut J, Evers P, de Koning S, Heeres HJ. Identification and classification of components in flash pyrolysis oil and hydrodeoxygenated oils by two-dimensional gas chromatography and time-of-flight mass spectrometry. Journal of Chromatography A. 2008;1188:17-25. DOI: 10.1016/j.chroma.2008.02.034

[142] Marsman JH, Wildschut J, Mahfud F, Heeres HJ. Identification of components in fast pyrolysis oil and upgraded products by comprehensive two-dimensional gas chromatography and flame ionisation detection. Journal of Chromatography A. 2007;1150:21-7. DOI: 10.1016/j.chroma.2006.11.047

[143] Windt M, Meier D, Marsman JH, Heeres HJ, de Koning S. Micro-pyrolysis of technical lignins in a new modular rig and product analysis by GC-MS/FID and GC × GC-TOFMS/FID. Journal of Analytical and Applied Pyrolysis. 2009;85:38-46. DOI: 10.1016/j.jaap.2008.11.011

[144] Sfetsas T, Michailof C, Lappas A, Li Q, Kneale B. Qualitative and quantitative analysis of pyrolysis oil by gas chromatography with flame ionization detection and comprehensive two-dimensional gas chromatography with time-of-flight mass spectrometry. Journal of Chromatography A. 2011;**1218**:3317-25. DOI: 10.1016/j.chroma.2010.10.034

[145] Moraes MSA, Bortoluzzi JH, Migliorini MV, Zini CA, Caramão EB. Comprehensive two-dimensional gas chromatography applied to the qualitative analysis of major compounds of bio-oil from pyrolysis of orange pulp. Scientia Chromatographica. 2011;3:301-14. DOI: 10.4322/sc.2011.018

[146] Migliorini MV, Silvana M, Moraes A, Machado ME, Caramão EB. Caracterization of phenols in bio-oil from the pyrolysis of peach core by GC/MS e GC × GC/TOFMS. Scientia Chromatographica. 2013;5:47-65. DOI: 10.4322/sc.2013.006

[147] Moraes MSA, Migliorini MV, Damasceno FC, Georges F, Almeida S, Zini CA, Caramão, EB. Qualitative analysis of bio oils of agricultural residues obtained through pyrolysis using comprehensive two dimensional gas chromatography with time-of-flight mass spectrometric detector. Journal of Analytical and Applied Pyrolysis. 2012;98:51-64. DOI: 10.1016/j.jaap.2012.05.007

[148] Araújo RCS, Pasa VMD, Marriott PJ, Cardeal ZL. Analysis of volatile organic compounds in polyurethane coatings based on Eucalyptus sp.bio-oil pitch using comprehensive two-dimensional gas chromatography (GC × GC). Journal of Analytical and Applied Pyrolysis. 2010;88:91-7. DOI: 10.1016/j.jaap.2010.02.012

[149] Lazzari E, Schena T, Primaz CT, Maciel GPS, Machado ME, Cardoso CAL, Jacques RA, Caramão EB. Production and chromatographic characterization of bio-oil from the pyrolysis of mango seed waste. Industrial Crops and Products. 2016;83:529-36. DOI: 10.1016/j.indcrop.2015.12.073

[150] Almeida TM, Bispo MD, Cardoso ART, Migliorini M V., Schena T, Campos MCV, Machado ME, López JÁ, Krause LC, Caramão EB. Preliminary studies of bio-oil from fast pyrolysis of coconut fibers. Journal of Agricultural and Food Chemistry. 2013;61:6812-21. DOI: 10.1021/jf401379s

[151] Onorevoli B, Machado ME, Dariva C, Franceschi E, Krause LC, Jacques RA, Caramão EB. A one-dimensional and comprehensive two-dimensional gas chromatography study of the oil and the bio-oil of the residual cakes from the seeds of Crambe abyssinica. Industrial Crops and Products. 2014;52:8-18. DOI: 10.1016/j.indcrop.2013.09.034

[152] Silva RVS, Casilli A, Sampaio AL, Ávila BMF, Veloso MCC, Azevedo DA, Romeiro GA. The analytical characterization of castor seed cake pyrolysis bio-oils by using comprehensive GC coupled to time of flight mass spectrometry. Journal of Analytical and Applied Pyrolysis. 2014;106:152-9. DOI: 10.1016/j.jaap.2014.01.013

[153] Tessarolo NS, Luciana RM, Silva RSF, Azevedo DA. Chemical characterization of bio-oils using comprehensive two-dimensional gas chromatography with time-of-flight mass spectrometry. Journal of Chromatography A. 2013;1279:68-75. DOI: 10.1016/j.chroma.2012.12.052

[154] Tessarolo NS, Silva RC, Vanini G, Pinho A, Romão W, de Castro EVR, Azevedo DA. Assessing the chemical composition of bio-oils using FT-ICR mass spectrometry and comprehensive two-dimensional gas chromatography with time-of-flight mass spectrometry. Microchemical Journal. 2014;117:68-76. DOI: 10.1016/j.microc.2014.06.006

[155] Tessarolo NS, Silva RVS, Vanini G, Casilli A, Ximenes VL, Mendes FL, Pinho AR, Romão W, Castro EVR, Kaiser CR, Azevedo DA. Characterization of thermal and catalytic pyrolysis bio-oils by high-resolution techniques : 1H NMR, GC × GC-TOFMS and FT-ICR MS. Journal of Analytical and Applied Pyrolysis. 2016;117:257-67. DOI:10.1016/j.jaap.2015.11.007

[156] da Cunha ME, Schneider JK, Brasil MC, Cardoso CA, Monteiro LR, Mendes FL, Pinho A, Jacques RA, Machado ME, Freitas LS, Caramão EB. Analysis of fractions and bio-oil of sugar cane straw by one-dimensional and two-dimensional gas chromatography with quadrupole mass spectrometry (GC × GC/qMS). Microchemical Journal . 2013;110:113-9. DOI: 10.1016/j.microc.2013.03.004

[157] Schneider JK, da Cunha ME, dos Santos AL, Maciel GPS, Brasil MC, Pinho AR, Mendes FL, Jacques RA, Caramão EB. Comprehensive two dimensional gas chromatography with fast-quadrupole mass spectrometry detector analysis of polar compounds extracted from the bio-oil from the pyrolysis of sawdust. Journal of Chromatography A. 2014;1356:236-40. DOI: 10.1016/j.chroma.2014.06.053

[158] Maciel GPS, Machado ME, Barbará JA, Molin DD, Caramão EB, Jacques RA. Biomass and Bioenergy GC x GC/TOFMS analysis concerning the identification of organic compounds extracted from the aqueous phase of sugarcane straw fast pyrolysis oil. Biomass and Bioenergy. 2016;85:198-206. DOI: 10.1016/j.biombioe.2015.11.009

Section 2

Innovative Pyrolysis Processes

Chapter 6

Microwave-Assisted Pyrolysis of Biomass for Bio-Oil Production

Yaning Zhang, Paul Chen, Shiyu Liu, Liangliang Fan,
Nan Zhou, Min Min, Yanling Cheng, Peng Peng,
Erik Anderson, Yunpu Wang, Yiqin Wan,
Yuhuan Liu, Bingxi Li and Roger Ruan

Additional information is available at the end of the chapter

http://dx.doi.org/10.5772/67442

Abstract

Microwave-assisted pyrolysis (MAP) is a new thermochemical process that converts biomass to bio-oil. Compared with the conventional electrical heating pyrolysis, MAP is more rapid, efficient, selective, controllable, and flexible. This chapter provides an up-to-date knowledge of bio-oil production from microwave-assisted pyrolysis of biomass. The chemical, physical, and energy properties of bio-oils obtained from microwave-assisted pyrolysis of biomass are described in comparison with those from conventional pyrolysis, the characteristics of microwave-assisted pyrolysis as affected by biomass feedstock properties, microwave heating operations, use of exogenous microwave absorbents, and catalysts are discussed. With the advantages it offers and the further research and development recommended, microwave-assisted pyrolysis has a bright future in production of bio-oils that can effectively narrow the energy gap and reduce negative environmental impacts of our energy production and application practice.

Keywords: microwave-assisted pyrolysis, bio-oil, biomass, pyrolysis temperature, microwave power, pyrolysis time, feedstock characteristics, microwave absorbent, reaction catalyst

1. Introduction

The oil crisis in the mid-1970s drove the price hike of crude oil and hence the hot pursue of alternative energy resources. This led to the development of pyrolysis technologies, which have the potential to achieve bio-oil yields of 70–80 wt.% of biomass [1, 2] or as high as 80–95 wt.% [3]. If the 1.08 × 10^5 Mtoe (Million Tonnes of Oil Equivalent) of annually available biomass [4] is converted to bio-oils through pyrolysis technologies with an average bio-oil conversion ratio of 75 wt.% [5], this would result in an annual bio-oil production of 8.1 × 10^4 Mtoe, effectively narrowing the large energy gap. On the other hand, the value-added bio-oil produced from biomass offers the following advantages: (a) CO_2/GHG (greenhouse gas) neutral, (b) negligible or zero SO_x emissions, (c) lower NO_x emissions, (d) biodegradable, (e) locally produceable, (f) renewable, and (g) sustainable [3, 6, 7]. Bio-oil obtained from biomass can therefore also significantly reduce the negative environmental impacts.

Pyrolysis is typically defined as a thermochemical decomposition of biomass at medium to high temperatures in the absence of oxygen. The main operating parameters in pyrolysis are heating rate, pyrolysis temperature, and residence time [7]. Depending on these main operating parameters, the pyrolysis process can be generally classified into three subclasses as slow pyrolysis, fast pyrolysis, and flash pyrolysis. Conventional slow pyrolysis is characterized by heating rate of <1°C/s, temperature range of 300–700°C, and residence time of >450 s. Fast pyrolysis is characterized by heating rate of 10–300°C/s, temperature range of 550–1250°C, and residence time of 0.5–20 s. Flash pyrolysis is characterized by heating rate of >1000°C/s, temperature range of 800–1300°C, and residence time of <0.5 s [4, 7–9]. Generally, fast pyrolysis and flash pyrolysis are more promising alternative approaches than slow pyrolysis because of the fact that they can convert a wide range of biomass feedstock for higher bio-oil production [7, 9].

Recently, a new pyrolysis technique, microwave-assisted pyrolysis (MAP), was developed [9, 10] and it has drawn serious attention because of its advantages over the traditional electrical heating methods. For conventional electrical heating methods, heat is transferred from high-temperature gas to the fuel particle surface through convection mechanism and it is then further transferred from the outside surface to the inside core through conduction mechanism. A temperature gradient from outside to inside of the feedstock particle is formed because of the poor thermal conductivity of the feedstock material, and the released volatile diffuses from the inside core to the outside surface through a higher temperature region (**Figure 1**). For the microwave heating method, microwave penetrates the feedstock particle and microwave energy is transformed into heat inside the particle. Because of the heat loss effect of particle surface, heat constantly accumulates inside the feedstock and is transferred outwards. A temperature gradient from inside to outside of feedstock particle is formed also because of the poor thermal conductivity of the feedstock material, and the released volatile diffuses from the inside core to the outside surface through a lower temperature region (**Figure 1**).

The unique heating mechanisms of microwave heating make microwave-assisted pyrolysis (MAP) have many advantages over the traditional electrical heating pyrolysis. **Table 1** shows the characteristics of microwave-assisted heating and conventional electrical heating. Because of these advantages, microwave-assisted pyrolysis of biomass was widely studied for bio-oil production.

http://dx.doi.org/10.5772/67442

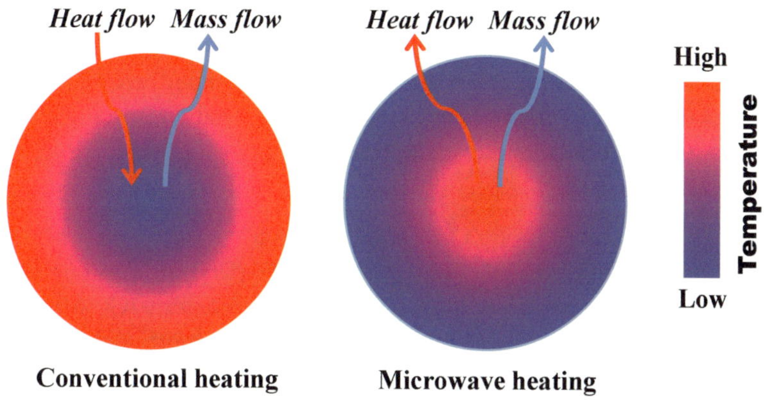

Figure 1. Schematic diagram of temperature distribution, heat transfer, and mass transfer in the conventional heating and microwave heating [9].

Microwave-assisted heating	Conventional electrical heating
Conversion of energy	Transfer of energy
In-core volumetric and uniform heating at molecular level	Superficial heating through conduction, convection, and radiation
Hot spot	No hot spot
Rapid and efficient	Slow, inefficient, limited
Higher electricity conversion efficiency	Lower electricity conversion efficiency
Selective	Nonselective
Dependent on material's properties	Less dependent
Precise and controlled heating	Less controllable
Process flexible	Less flexible
Equipment portable	Less portable
Lower contaminants	Higher contaminants
Lower thermal inertia and faster response	Higher thermal inertia and slower response

Table 1. Comparison between microwave-assisted heating and conventional electrical heating [8, 9].

2. Bio-oil from conventional and microwave-assisted pyrolysis of biomass

Bio-oil is dark brown and free-flowing organic liquid. The synonyms for bio-oil include pyrolysis oil, pyrolysis liquid, bio-crude oil, wood oil, wood distillate, wood liquid, liquid smoke, pyroligneous acid, pyrolytic tar, and so on [3]. Bio-oil obtained from microwave-assisted pyrolysis of a biomass is a mixture of different molecular size components derived from depolymerization and fragmentation of the biomass, it therefore has different composition, properties, and characteristics as compared with diesel oils and petroleum oils [2].

2.1. Chemical composition

Pyrolysis bio-oil is a complex mixture of more than 300 compounds. In addition to water and solids, many chemical compounds with distinct functional groups are found in bio-oils, including hydrocarbons, ketones, esters, aldehydes, phenols, alcohols, furans, acids, sugars, and the others.

Functional groups, e.g., aromatics by O-H stretching, esters by C=O stretching, etc., can be detected using Fourier transform infrared spectroscopy (FT-IR) [11] while the compositions are usually determined by using gas chromatography-mass spectrometry (GC-MS). **Table 2** shows the frequency ranges (relating to wavelength ranges) and FTIR functional groups of pyrolysis bio-oil compounds. As a given functional group (e.g., O-H stretching and C=O stretching) may

Frequency range (cm^{-1})	Functional groups
3475	Alcohol O-H stretching
3400	H-bonded OH stretching
3078, 3020	Terminal vinyl C-H stretching
2964, 2865	Methyl C-H asymmetric and symmetric stretching
2957, 2872	Methyl C-H stretching
2931, 2853	Methylene C-H asymmetric and symmetric stretching
1850–1750	Aromatic combination bands
1710	Ketone C=O
1640	Alkenyl C=C stretching
1496, 1457	C=C-C aryl ring stretching
1407, 1366	Phenol or tertiary alcohol O-H bending
1279	Primary or secondary O-H in-plane bending
1240	Aromatic ethers, aryl-O stretching
1238, 1193	Phenol C-O stretching
1202, 1083, 1028, 1020	Aromatic C-H in-plane bending
1115	Tertiary alcohol C-O stretching
1080	Secondary alcohol C-O stretching
1050, 1039	Primary alcohol C-O stretching
1020, 965–880	Aromatic bending
840, 774, 755, 748, 731	Aromatic C-H out-of-plane bending
694	C-C ring bending
615	Alcohol O-H out-of-plane bending

Table 2. FT-IR functional groups of pyrolysis bio-oil compounds [12, 13].

indicate more than one chemical compound, the FT-IR analysis therefore may have difficulties in determining different compounds in the bio-oils, limiting the applications of FT-IR.

Compared with FT-IR, GC-MS is more widely used in determining specific chemical compounds in pyrolysis bio-oils. **Table 3** shows the numbers, molecular weights, and area percentages of the bio-oil compounds detected by GC-MS for microwave-assisted pyrolysis of biomass. **Table 4** shows some of the detailed bio-oil compounds obtained from microwave-assisted pyrolysis of various biomass feedstocks. Among the various bio-oil compounds, hydrocarbons, phenols, and furans are generally regarded as high value chemicals. However, the others are regarded as undesirable products because they contain higher oxygen contents, which would increase the instability of bio-oil, decrease the heating value of bio-oil, and cause the other drawbacks [18].

2.2. Physical properties

The physical properties of bio-oil mainly include moisture content, solid content, viscosity, density, pH, and so on.

The moisture content of pyrolysis bio-oil mainly comes from two sources: (a) the water in the raw feedstock and (b) the water produced from the dehydration reactions occurred during pyrolysis process. The moisture content of bio-oil can vary in a wide range (7.86–69.19 wt.%) depending on the feedstock characteristics and process conditions [23, 28, 29]. As water is miscible with the oligocellulosic-derived compounds because of the solubilizing effect of polar hydrophilic compounds (acids, alcohols, hydroxyaldehydes, and ketones), the presence of water in bio-oil would reduce the oil viscosity and improve the flow characteristics, which are beneficial to the combustion process. However, it may also lower the heating values of bio-oils, thereby increasing the ignition delay and decreasing the combustion rate [28].

Class	Compound	Molecular weight	Percentage (% area)	Source
Hydrocarbons	>100	78–368	93.2	[14]
Ketones	~70	84–384	17.9	[15]
Esters	>50	86–508	59.4	[15]
Aldehydes	>20	44–182	83.0	[16]
Phenols	~40	94–194	74.8	[17]
Alcohols	~30	46–386	12.9	[18]
Furans	~20	72–172	26.1	[19]
Acids	>30	60–284	53.8	[20]
Sugars	>2	162–180	47.7	[21]
Others[a]			90.0	[13]

Note: [a] The others mainly include acetamide, propanamide, pyridine, etc. [22, 23].

Table 3. Bio-oil compounds detected by GC-MS for microwave-assisted pyrolysis of biomass.

Compound	Formula	Corn stover[a]	Corn stover[b]	Corn stover[c]	Macro algae[d]	Micro algae[e]	Organic waste[f]	Bio-waste[g]
Hydrocarbons								
Toluene	C_7H_8				1.2		6.9	
Styrene	C_8H_8	2.0–2.7						
Naphthalene	$C_{10}H_8$	11.0–21.4	0.8–4.3		3.4			
Ketones								
Cyclopentanone	C_5H_8O				0.3			
1,3-Cyclopentanedione, 2-methyl-	$C_6H_8O_2$		2.2–4.0					
Benzaldehyde, 4-methyl-	C_8H_8O		0–3.6					
Esters								
Gamma-butyrolactone	$C_4H_6O_2$			2.2–2.7		2.0		
Oxacyclopentadec-6-en-2-one	$C_{14}H_{24}O_2$		8.7–16.0					
Hexadecanoic acid, methyl ester	$C_{17}H_{34}O_2$						17.9	
Aldehydes								
2-Furaldehyde (furfural)	$C_5H_4O_2$			4.3–79.6				
5-Methyl-2-furancarboxaldehyde	$C_6H_6O_2$		0.7–1.9					
Phenols								
Phenol	C_6H_6O	4.3–17.8	6.4–9.2	3.6–6.7	6.0	9.5		2.0
Phenol, 4-methyl-	C_7H_8O		3.5–6.1	1.4–3.5	9.0			
Phenol, 4-ethyl-	$C_8H_{10}O$	17.0–20.5		2.1–3.5				
Phenol, 4-ethyl-2-methoxy-	$C_9H_{12}O_2$	4.3–7.0	2.2–2.9					
Alcohols								
2-Furanmethanol	$C_5H_6O_2$			1.8–4.2		1.7		
Lyratol	$C_{10}H_{16}O$		0.7–1.3					

http://dx.doi.org/10.5772/67442

Compound	Formula	Corn stover[a]	Corn stover[b]	Corn stover[c]	Macro algae[d]	Micro algae[e]	Organic waste[f]	Bio-waste[g]
Furans								
Tert-butyl methyl carbonate	$C_6H_{12}O_3$		0–1.4					
Benzofuran	C_8H_6O	0.2–2. 7						
Benzofuran, 2,3-dihydro-	C_8H_8O		0–1.2	3.1–9.6				
Benzofuran, 4,7-dimethyl-	$C_{10}H_{10}O$	0–3.8			0.5			
Acids								
Acetic acid	$C_2H_4O_2$		0.5–1.1					0.8
Hexadecanoic acid	$C_{16}H_{32}O_2$			0–2.3				0.8
Sugars								
Levoglucosan	$C_6H_{10}O_5$			4.2–7.7				27.9
Glucopyranose	$C_6H_{12}O_6$							1.3
Others								
Acetamide	C_2H_5NO					8.2		
Pyridine	C_5H_5N				2.4			
Pyrazine, methyl-	$C_5H_6N_2$				0.8	2.9		
Benzyl nitrile	C_8H_7N					3.0		
Hexadecanamide	$C_{16}H_{33}NO$				1.0			

[a] Ref. [24].
[b] Ref. [25].
[c] Ref. [16].
[d] Ref. [22].
[e] Ref. [23].
[f] Ref. [26].
[g] Ref. [27].

Table 4. Compounds of bio-oil obtained from microwave-assisted pyrolysis of biomass.

The solid content of pyrolysis bio-oil mainly comes from the fly chars and ashes in the condensable gases during pyrolysis process, and it is of significant importance with respect to the particulate emissions during combustion process because it can wear the fuel system, block the filter, and clog the fuel nozzle [30]. Generally, the larger the particle size and the higher the particle amount, the more serious the solid content problem is [28]. For engine and boiler applications, the solid content of bio-oil should be controlled within 1% [31].

The viscosity of pyrolysis bio-oil can vary over a wide range from as low as 10 cp to as high as 10,000 cp depending on its temperature and can be even higher when the bio-oil is stored in poor conditions for longer periods. The viscosity of pyrolysis bio-oil is also affected by water content, process configuration, process parameters, feedstock characteristics, storage conditions, and storage periods [1]. Engine companies are concerned about the viscosity of pyrolysis bio-oil, because high viscosity of bio-oil could cause excessive fuel injection pressure during the engine warm-up stage. On the other hand, engines would be starved for fuel at low temperatures because the fuel would move slowly through the filters or lines due to high viscosities [32].

The low pH values of pyrolysis bio-oils are mainly attributed to the substantial amounts of organic acids, e.g., acetic acid and formic acid, which make the bio-oil be corrosive to common construction materials such as carbon steel and aluminum [2, 28]. This corrosiveness is especially severe when the pyrolysis bio-oil is at elevated temperatures and increased water contents. In this case, polyolefins are usually used as alternative construction materials due to their resistance to the corrosiveness of bio-oils [2].

Table 5 shows the main physical properties of the microwave-assisted pyrolysis bio-oil, conventional electrical heating pyrolysis bio-oil, diesel oil, and petroleum oil. Both microwave-assisted pyrolysis bio-oil and conventional electrical heating pyrolysis bio-oil have similar physical properties, whereas they are rather different from those of the diesel oil and petroleum oil. This is due to the fact that the pyrolysis bio-oils are produced from mainly the depolymerization and fragmentation reactions of the three key construction blocks (cellulose, hemicellulose, and lignin) of biomass [2]. Although microwave-assisted pyrolysis bio-oil and conventional

Property	Unit	Pyrolysis bio-oil		Diesel oil[c]	Petroleum oil[d]
		Microwave[a]	Conventional[b]		
Moisture	wt.%	15.2	4.5–43.0	–	0.1
Solid content	wt.%	0.22	0.1–3.0	–	0.1
Dynamic viscosity	mPa·s	60 at 50°C	40–100 at 50°C	1.6–2.3 at 50°C	180 at 40°C
Density	g/mL	1.25	0.91–1.29	0.83–0.84	0.94
pH	–	2.87	2.3–5.5	–	–

[a] Ref. [28].
[b] Ref. [2].
[c] Ref. [32].
[d] Ref. [31].

Table 5. Main physical properties of different oils.

electrical heating pyrolysis bio-oil may have similar physical properties, the physical properties may be significantly different due to the different pyrolysis mechanisms [19].

2.3. Energy properties

The distinctive characteristics of the elemental compositions for both microwave-assisted pyrolysis bio-oil and conventional electrical heating pyrolysis bio-oil are that they have higher oxygen content (29–40%) whereas lower carbon content (54–60%) as compared with diesel oil and petroleum oil (**Table 6**). Oxygen is present in most of the oil compounds that have been identified in the bio-oils (more than 300), and the distributions of these compounds are strongly dependent on biomass characteristics (approximate analysis and ultimate analysis) and operation conditions (pyrolysis temperature, heating rate, and residence time) [2].

As oxygen itself is not a combustible element, the high oxygen content of a bio-oil would result in a lower HHV and also a lower LHV of the bio-oil (**Table 6**). The HHV of bio-oil can be measured directly by a bomb calorimeter, and it can also be estimated through the following empirical correlations [33, 34]:

$$\mathrm{HHV} = 0.3491\,\eta_C + 1.1783\,\eta_H + 0.1005\,\eta_S - 0.1034\,\eta_O - 0.0151\,\eta_N - 0.0211\,\eta_{ash} \quad (1)$$

$$\mathrm{HHV} = 0.3382\,\eta_C + 1.4428(\eta_H - 0.125\,\eta_O) \quad (2)$$

where HHV is the higher heating value of bio-oil (MJ/kg); η_C is the weight percentage of carbon bio-oil (%); η_H is the weight percentage of hydrogen in bio-oil (%); η_S is the weight percentage of sulfur in bio-oil (%); η_O is the weight percentage of oxygen in bio-oil (%); η_N is the weight percentage of nitrogen in bio-oil (%); η_{ash} is the weight percentage of ash in bio-oil (%). The LHV of bio-oil can then be estimated through the following relationship [35]:

$$\mathrm{HHV} = \mathrm{LHV} + 21.978\,\eta_H \quad (3)$$

Property	Unit	Pyrolysis bio-oil		Diesel oil[c]	Petroleum oil[d]
		Microwave[a]	Conventional[b]		
Elemental analysis					
C	wt.%	60.66	54–58	86.23–86.31	85
H	wt.%	7.70	5.5–7.0	13.14–13.27	11
N	wt.%	2.02	0–0.2	–	0.3
S	wt.%	0.15	–	0.034–0.039	–
O	wt.%	29.4	35–40	–	1.0
HHV	MJ/kg	17.51	14–19	42.7–43.0	40

[a] Ref. [28].
[b] Ref. [2].
[c] Ref. [32].
[d] Ref. [31].

Table 6. Main chemical properties of different oils.

where LHV is the lower heating value of bio-oil (MJ/kg).

It is worth noting that microwave-assisted pyrolysis bio-oil may have significantly higher HHV (e.g., 31.13 MJ/kg) than electrical heating pyrolysis bio-oil (e.g., 12.31 MJ/kg) because the microwave-assisted pyrolysis bio-oil usually has much lower oxygen content (e.g., 25.78 wt.% vs. 56.10 wt.%) [36].

3. Moderate microwave-assisted pyrolysis

3.1. The process

Figure 2 shows the schematic diagram of a typical microwave-assisted pyrolysis (MAP) set-up. A basic bench-scale set-up is generally composed of a microwave oven, a quartz reactor, some thermocouple probes, some collection flasks, some cooling lines, and some connection tubes. Before the pyrolysis, a vacuum pump is usually used or some inert gases (e.g., N_2) are purged in order to keep an inert atmosphere in the quartz reactor. When the biomass feedstock in the reactor is heated by microwave, its temperature increases to a high temperature

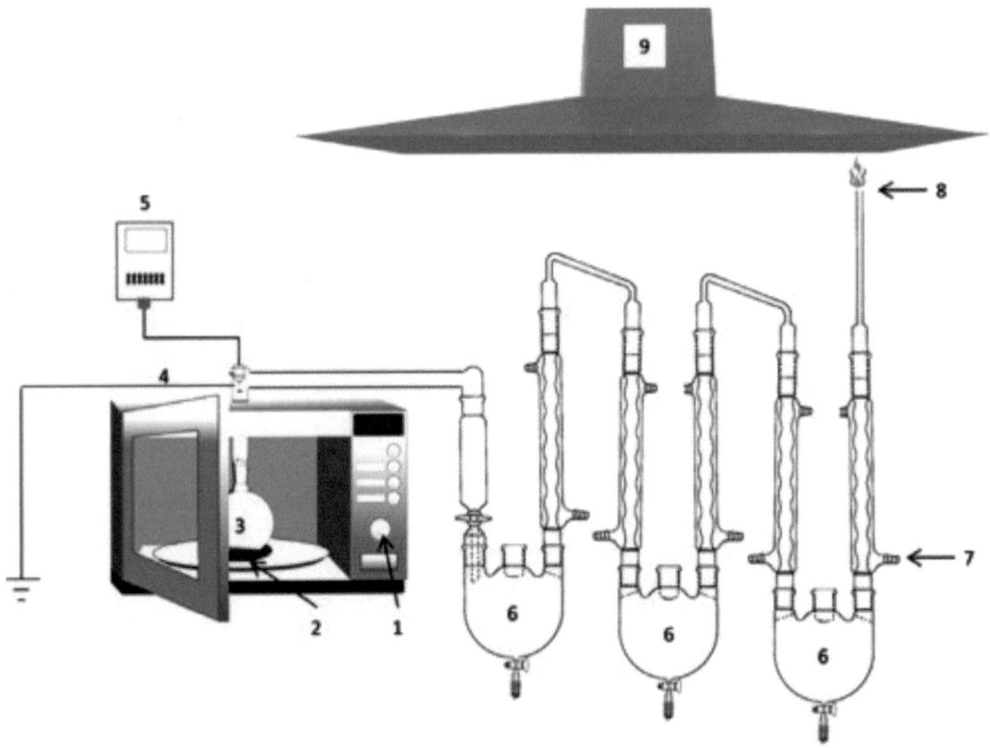

Figure 2. Schematic diagram of a typical MAP set-up [37]. (1) microwave control panel; (2) insulation plate; (3) quartz reactor with thermocouple probe; (4) grounded wire; (5) thermometer; (6) collection flasks; (7) cooling lines; (8) bio-gas outlet; (9) ventilation hood.

within a very short time. During this process, the biomass feedstock is dried initially, and it is then decomposed to form vapors and char. When the vapors go through the cooling lines (typically cycled by cooling water of 0–7°C), the condensables condense in the collection flasks to form liquid (usually called bio-oil), whereas the noncondensables exit the tube to form gas (usually called syngas). As the biomass feedstock in the reactor would be heated by microwave to reach a high temperature (300°C or even higher), some insulation materials, e.g. asbestos and bricks, are usually used to protect the microwave oven from high temperature damage and at the same time to avoid the loss of heat. Usually, the connection tubes between the reactor and the first collection flask are also insulated to prevent the emitted volatiles from condensing and adhering along the connection tubes. However, the condensing and adhering cannot be completely avoided even with very good insulation. The condensed and adhered liquids are therefore usually measured and counted into the bio-oil yields. As some of the tiny ash particles would fly with vapors, they would also form parts of bio-oil and syngas.

The typical microwave-assisted pyrolysis of biomass feedstock in the quartz reactor can be generally divided into three basic processes, namely drying, pyrolysis I, and pyrolysis II, as shown in **Figure 3**. During the drying process when the heating time goes from t_0 to t_1, the temperature of feedstock in the quartz reactor increases from the ambient temperature T_0 to a high temperature T_1 which is about 110°C [27, 37]. When the heating time furthers from t_1 to t_2, the feedstock temperature also increases from T_1 to T_2, pyrolysis I process begins, in which the feedstock particles are decomposed to release vapors including condensables (bio-oil) and noncondensables (syngas). During pyrolysis II when the heating time furthers from t_2 to t_3, the feedstock temperature generally fluctuates between T_2 and T_3 depending on the heat transformed (from microwave), absorbed (by reactions), and scattered (from reactor). If the

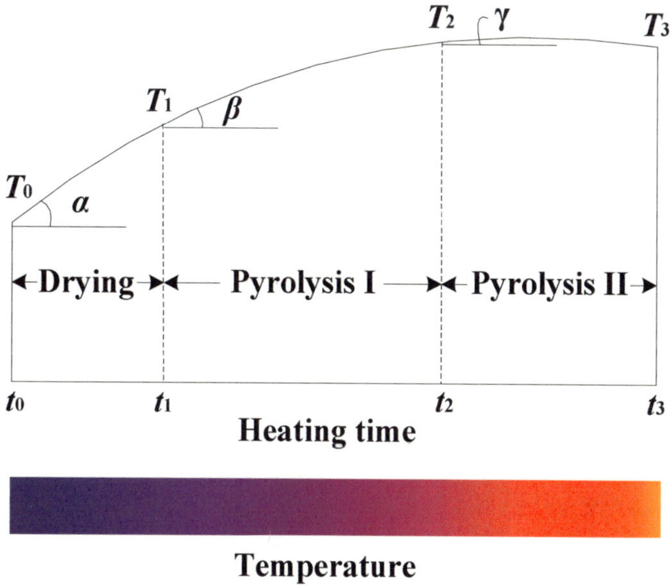

Figure 3. Typical microwave-assisted pyrolysis (MAP) process.

transformed heat is more (or less) than the sum of absorbed heat and scattered heat, T_3 is higher (or lower) than T_2. If the transformed heat is balanced by the absorbed heat and scattered heat, T_3 is equal to T_2. Regardless of the above relationships between T_2 and T_3, the feedstock particles continue to be decomposed to release vapors including condensables (bio-oil) and noncondensables (syngas) during this pyrolysis II process.

If the average temperature rising rates for the drying, pyrolysis I, and pyrolysis II processes are described by tgα, tgβ, and tgγ, respectively, an order of the tg values generally comes as tgα>tgβ>tgγ. For example, the average temperature rising rates for drying, pyrolysis I, and pyrolysis II processes are reported to be 100–150°C/min, 10–50°C/min, and ~0°C/min, respectively [27, 38]. The reason for tgα>tgβ is due to the evaporation of moisture content which has a very good microwave absorbance and this makes the feedstock have lower microwave absorbance as moisture evaporation goes on. However, tgβ>tgγ is mainly due to the fact that the microwave absorbance of the feedstock would decrease when pyrolysis temperature is increased. In some cases, tgγ may be lower than zero because the transformed heat does not compensate for the sum of absorbed heat and scattered heat.

The main operating parameters in pyrolysis are heating rate, pyrolysis temperature, and residence time [7]. In microwave-assisted pyrolysis, all these operating parameters are principally dependent on the microwave absorbance of biomass feedstocks. The microwave absorbance of a biomass feedstock is represented by the tanδ value of the biomass feedstock. **Table 7** shows the tanδ values of some biomass feedstocks at room temperature and 2.45 GHz. Compared with the conventional electrical heating pyrolysis, microwave-assisted pyrolysis generally has higher heating rates, pyrolysis temperatures, and conversion efficiencies. On the other hand, the particular heating mechanisms generally make the biomass feedstocks decompose at lower temperatures (e.g., 100–150°C lower [27]), whereas result in more bio-oil under microwave-assisted pyrolysis than under conventional electrical heating pyrolysis. However, the microwave-assisted pyrolysis process is significantly varied by many factors, e.g. pyrolysis temperature, microwave power, heating time, feedstock characteristics, etc.

3.2. Effect of pyrolysis temperature

Temperature has a strong effect on the pyrolysis of a biomass feedstock. Generally, microwave-assisted pyrolysis makes the biomass feedstock decompose at a lower temperature to result in a higher bio-oil yield [46] than the conventional electrical heating pyrolysis. Also, the microwave-assisted pyrolysis has its unique characteristics at different temperatures.

Bu et al. [47] studied the bio-oil yields (on wet basis) obtained from microwave-assisted pyrolysis of Douglas fir at different pyrolysis temperatures (316–484°C). When the pyrolysis temperature increased from 316 to 400°C, the bio-oil yield increased from 32.3 to 38.8 wt.%. The increase in the bio-oil yield was due to the devolatilization, depolymerisation, and decarboxylation, which were widely reported in some literature [48–50], and they were detailed as: (a) more energy was involved in the chemical reactions when there was an increase in the pyrolysis temperature [48, 51], and (b) more strong organic bonds in the biomass feedstock were broken and more volatiles were released to form condensable gases for bio-oil components [29, 48, 50].

http://dx.doi.org/10.5772/67442

Feedstock	tanδ	Source
Aspen Bark	0.22	[39]
Pine wood	0.19	[40]
Pine Bark	0.18	[39]
Oil palm shell	0.12	[41]
Water	0.12	[42]
Wood	0.11	[43]
Particle board	0.1–1.0	[39]
Oil palm fiber	0.08	[41]
Pulp mill sludge	0.08	[44]
Hemicelluloses	0.062	[44]
Lignin	0.052	[44]
Cellulose	0.035	[44]
Sludge	0.035	[44]
Wood polymer	0.03	[44]
Acid washed sludge	0.019	[44]
Fir plywood	0.01–0.05	[39]
Natural rubber	0.002–0.005	[45]
Polypropylene	0.0003–0.0004	[45]
Polystyrene	0.0002–0.0003	[45]
Polyethylene	0.0001–0.0002	[45]

Table 7. tan δ values of some biomass feedstocks at room temperature and 2.45 GHz.

However, when the pyrolysis temperature increased from 400 to 484°C, the bio-oil yield decreased from 38.8 to 36.2 wt.%. The decrease in the bio-oil yield was due to the fact that further high pyrolysis temperature favored the formation of noncondensable combustible gases rather than liquid products [52, 53]. This can be explained by (a) secondary reactions among CO_2, H_2O, and carbon were endothermic reactions [48], (b) high pyrolysis temperature would promote the conversions of condensable vapors to noncondensable gases [49, 50], and (c) the carbonization of volatiles for charcoal would also decrease the bio-oil yield [53].

Consequently, there is an optimal pyrolysis temperature for the bio-oil yield obtained from microwave-assisted pyrolysis of a biomass feedstock. Optimal pyrolysis temperatures in a wide range (400–800°C) have been reported and found to depend on the feedstocks and parameters used [29, 47–49, 54]. According to the optimal pyrolysis temperature, the bio-oil yield would exhibit different changes at different pyrolysis temperatures. If the pyrolysis temperatures are lower than the optimal pyrolysis temperature, there would be a monotonous increase in the bio-oil yield. If the pyrolysis temperatures are higher than the optimal pyrolysis temperature, there would be a monotonous decrease in the bio-oil yield. If pyrolysis temperature starts at a lower value and ends at a higher value than the optimal pyrolysis temperature, the bio-oil yield would increase initially and decrease finally. However, these changes would be varied because the microwave-assisted pyrolysis process is very complex and sensitive and would be affected by many other factors.

Pyrolysis temperature affects not only the bio-oil yield obtained from microwave-assisted pyrolysis of a biomass feedstock but also the chemical, physical, and energy properties of the bio-oil. **Table 8** shows the chemical compositions of the bio-oil obtained from microwave-assisted pyrolysis of corn stover at different pyrolysis temperatures [24]. It was observed that

Compound (% area)	Formula	Pyrolysis temperature (°C)		
		450	609	790
Phenol	C_6H_6O	4.29	17.21	17.75
Phenol, 3-methyl-	C_7H_8O	2.31	3.35	7.94
Phenol, 2-methoxy-	$C_7H_8O_2$	15.90	9.97	3.74
Styrene	C_8H_8	2.14	2.68	1.95
Benzofuran	C_8H_6O	0.18	0.28	2.67
Benzeneacetaldehyde	C_8H_8O	6.04	0	0.71
Phenol, 4-ethyl-	$C_8H_{10}O$	20.51	16.95	19.80
Phenol, 3,4-dimethyl	$C_8H_{10}O$	0.59	0.26	0.10
Phenol, 3,4-dimethoxy-	$C_8H_{10}O_3$	4.80	3.68	0.32
Benzenamine, 2,4-dimethyl-	$C_8H_{11}N$	0.82	0.25	0.69
Benzene, 1-propynyl-	C_9H_8	2.44	4.54	3.93
Furan, 3-pentyl-	$C_9H_{14}O$	2.59	1.09	1.70
2-Propanone, 1-phenoxy-	$C_9H_{10}O_2$	0	1.59	0.68
Acetic acid, 4-methylphenyl ester	$C_9H_{10}O_2$	0.80	0.53	0.49
Phenol, 4-ethyl-2-methoxy-	$C_9H_{12}O_2$	7.04	6.48	4.25
2-Propenoic acid, 3-(2-hydroxyphenyl)-, (E) -	$C_9H_8O_3$	12.29	1.77	1.60
Naphthalene	$C_{10}H_8$	10.97	16.19	21.40
Benzofuran, 4,7-dimethyl-	$C_{10}H_{10}O$	0	2.94	3.80
6-Nonynoic acid, methyl ester	$C_{10}H_{16}O_2$	0.12	2.26	0
1H-Indene, 1-ethylidene-	$C_{11}H_{10}$	0.69	0.92	2.05
Acenaphthylene	$C_{12}H_8$	0	2.49	1.61
1-Dodecanol, 3,7,11-trimethyl-	$C_{15}H_{32}O$	0.57	0.86	0.81

Table 8. The measured compounds of the bio-oil obtained from microwave-assisted pyrolysis of corn stover at different pyrolysis temperatures [24].

phenol and benzofuran were increased by about 3–14 times, whereas benzeneacetaldehyde and 3, 4-dimethoxy-phenol were reduced by about 88–93% when the pyrolysis temperature was increased from 450 to 790°C.

3.3. Effect of microwave power

Microwave power is an important parameter for microwave-assisted pyrolysis of biomass. Basically, it changes the pyrolysis temperatures for the microwave-assisted pyrolysis of a biomass feedstock. When microwave power increases, the microwave density of the cavity increases and the microwave energy absorption of the biomass feedstock becomes greater, making the interaction between the microwave field and biomass feedstock become more intensive [55]. A higher microwave power therefore leads to a higher heating rate and a higher pyrolysis temperature [46]. **Figure 4** shows the temperature profiles of corn stover during microwave-assisted pyrolysis at different microwave powers [24]. When 50 g of corn stover was heated by different microwave powers, the balanced pyrolysis temperature increased from 210°C at 200 W to 790°C at 900 W.

As the increased pyrolysis temperature may increase the bio-oil yield for the microwave-assisted pyrolysis of a biomass feedstock (Section 3.2), the increased microwave power would also increase the bio-oil yield for microwave-assisted pyrolysis of the biomass feedstock. **Figure 5** shows the bio-oil yields obtained from microwave-assisted pyrolysis of corn stover at different microwave powers [24]. When the microwave power was 200 W, no bio-oil was obtained because the low pyrolysis temperature (210°C) was not enough to start the pyrolysis process. When the microwave power was increased from 300 to 900 W (200% in increase), the bio-oil yiled was nearly increased monotonously from 16.3 to 30.2 wt.% (85% in increase).

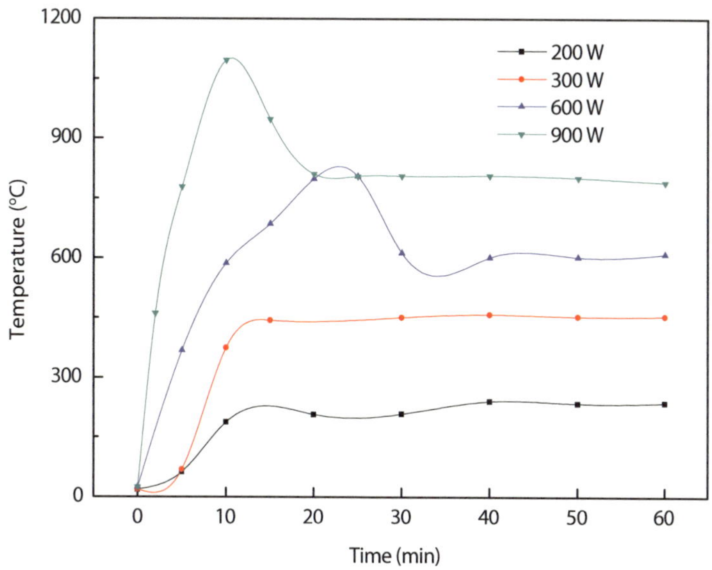

Figure 4. Temperature profiles of corn stover during microwave-assisted pyrolysis at different microwave powers [24].

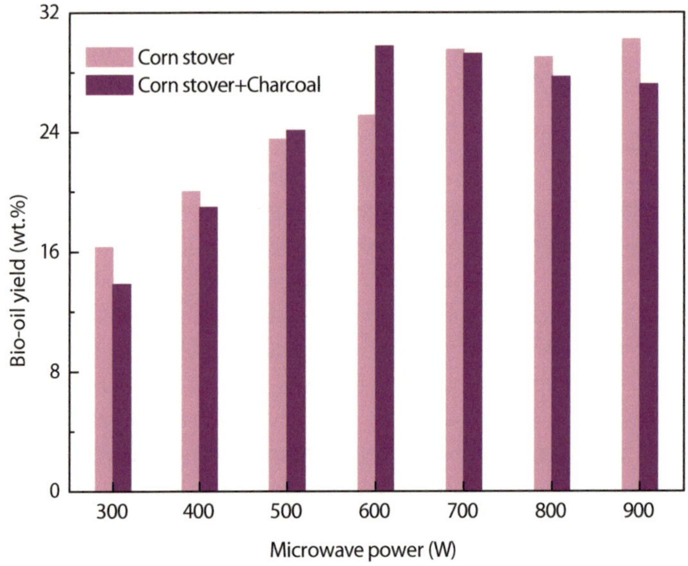

Figure 5. Bio-oil yields obtained from microwave-assisted pyrolysis of corn stover at different microwave powers [24].

However, this influence would also be varied by the other factors. When the corn stover was mixed with 1 wt.% charcoal, the bio-oil yiled increased initially and then reduced in the range of 13.8–29.7 wt.% when the microwave power was increased in the same range (300–900 W).

3.4. Effect of pyrolysis time

Pyrolysis time is also an important parameter for microwave-assisted pyrolysis of biomass. It controls the time for the interaction between microwave field and biomass feedstock. If the pyrolysis time is not long enough, the pyrolysis process would not be completed due to the short pyrolysis time or the pyrolysis process would even not occur due to the low pyrolysis temperature (which is also resulted from the short pyrolysis time). A pyrolysis time is therefore generally required to be long enough for the maximum or desirable pyrolysis temperature. In this section, required pyrolysis time is defined as the time the pyrolysis process takes to achieve the desirable pyrolysis temperature from its initial state when the biomass feedstock is heated by microwave. The desirable pyrolysis temperature is acutally sometimes different from the maximum pyrolysis temperature. As it is shown in **Figure 4**, the desirable pyrolysis temperatures are nearly also the maximum pyrolysis temperatures for the microwave powers of 200 and 300 W, whereas the desirable pyrolysis temperatures are lower than the maximum pyrolysis temperatures for the microwave powers of 600 and 900 W. It is also shown in **Figure 4** that the required pyrolysis time varied with microwave powers and many other factors, e.g., initial state conditions, feedstock charateristics, and so on.

When the pyrolysis time is longer than the required pyrolysis time, the difference is usually treated as the residence time. Residence time is generally more meaningful than pyrolysis time because it gives a more exact pyrolysis time for the microwave-assisted pyrolysis of a biomass feedstock. Bu et al. [47] studied the bio-oil yields obtained from microwave-assisted pyrolysis of corn stover at different residence times. When the residence time was increased from 1.27 to 12 min, the bio-oil yield increased from 31.48 to 37.04 wt.%. This increase was due to the fact that the longer residence time ensured a complete pyrolysis process for the biomass feedstock, and more volatiles were released from the feedstock or char to form bio-oil components [29, 47]. However, when the residence time was furthered from 12 to 14.73 min, the bio-oil yield decreased from 37.04 to 24.29 wt.%. This decrease was due to the fact that the condensable vapors (bio-oil components) were decomposed by secondary reactions to form noncondensable combustible gases (syngas components) in the high pyrolysis temperatures [48, 49]. Consequently, there is an optimal residence time for microwave-assisted pyrolysis of a biomass feedstock, and this optimal residence time is also varied by the other factors. The reported optimal residence times are mainly in the range of 6–12 min [20, 47, 56].

3.5. Effect of feedstock characteristics

All the operating parameters in microwave-assisted pyrolysis, e.g., heating rate, pyrolysis temperature, and residence time are principally dependent on the microwave absorbance of biomass feedstocks. The microwave absorbance of a biomass feedstock is evaluated by its tanδ value, which is significantly affected by the moisture content of the biomass feedstock because water has a very high tanδ value (0.12 [42]). Omar et al. [57] studied the tanδ values of an empty fruit bunch at different moisture contents. When the moisture content increased from 18 to 45% (150% increase), the tanδ value increased from 0.30 to 0.54 (80% increase), significantly increasing the microwave absorbance of the biomass feedstock.

However, as the microwave heating goes on, the temperature of the biomass feedstock rises and the moisture in the biomass feedstock evaporates, these collectively changing the microwave absorbance of the biomass feedstock. **Figure 6** shows the tanδ values of a switch-grass biomass during microwave-assisted heating process. When the reaction temperature increased from ~23 to ~440°C (1813% in increase), the tanδ value decreased from 0.064 to 0.003 (95.31% in decrease). Because of the decreases in the tanδ value, the biomass feedstock becomes less microwave absorptive, and the temperature rising rate then slows down (**Figure 7**).

The moisture content in a biomass feedstock not only increases the microwave absorbance of the biomass feedstock during the pyrolysis process, but it also generally contributes to the bio-oil yield. Menéndez et al. [59] studied the microwave-assisted pyrolysis of sewage sludge with different moisture contents. The results showed that the sewage sludge with a higher moisture content resulted in a higher bio-oil yield.

Although water in a biomass feedstock would help the microwave-assisted pyrolysis process and improve the bio-oil production, the bio-oil yield would contain a significantly higher aqueous fraction. **Table 9** shows the bio-oil yields and aqueous fractions obtained from microwave-assisted pyrolysis of different biomass feedstocks. The moisture content in the biomass feedstock can significantly contribute to the aqueous fractions, and the aqueous fractions can

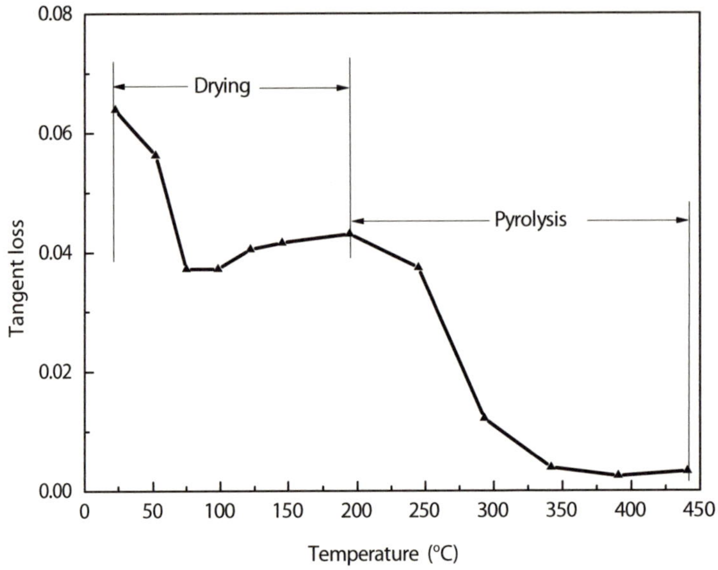

Figure 6. tanδ values of biomass feedstock at different temperatures [58].

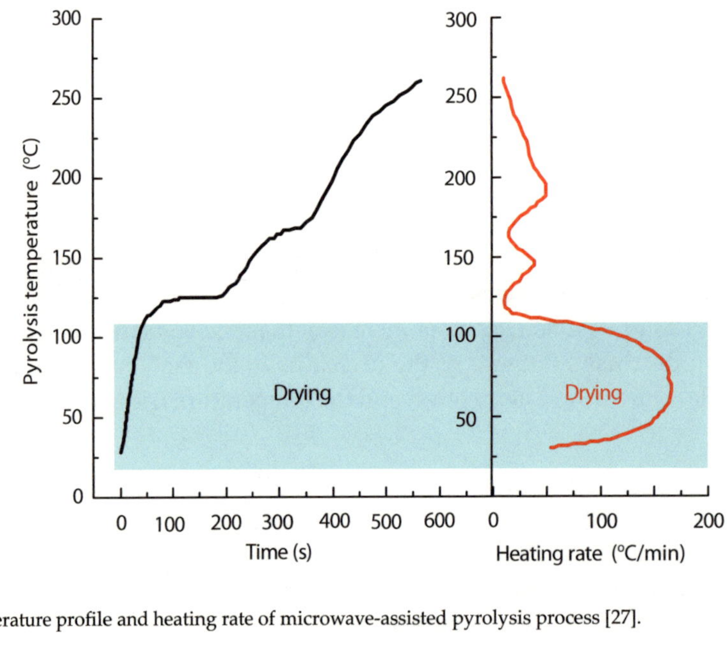

Figure 7. Temperature profile and heating rate of microwave-assisted pyrolysis process [27].

be 7.9–92.6 wt.% of the bio-oil yield [12, 14, 29] or 15.3–76.0 wt.% of the biomass feedstock [27, 66]. Even if the biomass feedstock itself contains no moisture content, a high aqueous fraction (e.g., 17–22 wt.% of feedstock) would also be obtained because chemical reactions may generate water molecules during the pyrolysis process [60]. As the aqueous fractions

Feedstock	Moisture content (%)	Bio-oil yield (wt.%)	Aqueous fraction (wt.%)	Source
Lignin	0	15–19[a]	17–22[d]	[60]
Nannochloropsis	3.21	42.5–59.4[b]	36–55[c]	[23]
Wood sawdust	5.15	48.08–64.25[b]	35–54[c]	[61]
Corn stover	5.27	46.67–65.00[b]	48–64[c]	[61]
Poplar clone	5.80	17.4[b]	17.5[c]	[62]
Aspen Pellet	5.90	18.95–22.97[a]	15.76–23.98 [d]	[63]
Poplar clone	6.40	26.6–29.6[b]	33.5–47.2[c]	[62]
Fir sawdust	7.00	31.4–53.9[b]	17.2–24.7[d]	[64]
Chlorella sp.	7.28	41.98–56.79[b]	51–69[c]	[23]
Poplar clone	7.80	12.2–32.0[b]	24.3–47.5[c]	[62]
Peanut shell	8.03	11.13–24.26[b]	7.86–26.70[c]	[29]
Pine sawdust	9.47	16.00–35.38[b]	11.15–28.92[c]	[29]
Wheat straw	10.00	7.3–22.1[a]	22.3–36.4[d]	[21]
Chlorella sp.	13.70	17.84–28.62[a]	20–22 [d]	[10]
Oil shale	14.50	5.4–6.2[a]	16.8–17.0[d]	[65]
Sewage sludge	71.00	2.2–4.0[a]	60.7–68.8[d]	[59]
Sewage sludge	76.80	4.9–6.0[a]	74.1–76.0[d]	[66]
Sewage sludge	84.30	3.1[a]	70[d]	[59]

[a] Dry bio-oil.
[b] Wet bio-oil.
[c] Of bio-oil.
[d] Of biomass.

Table 9. Bio-oil yields and aqueous fractions from microwave-assisted pyrolysis.

have no heating values or low heating values, they would dilute the bio-oil and degrade the bio-oil quality. On the other hand, the positive effects of moisture content on the pyrolysis temperature and pyrolysis process is limited because of its evaporation during the pyrolysis process. A biomass feedstock with lower moisture content is therefore preferred for microwave-assisted pyrolysis.

The ash content of a biomass feedstock also affects the microwave absorbance of the biomass feedstock, because ash components (Al_2O_3, CaO, Fe_2O_3, K_2O, MgO, MnO, MnO_2, Na_2O, TiO_2, etc.) are generally good microwave absorbents, and thus high ash content in biomass feedstock leads to increase in heating rates and maximum reaction temperatures accompanied by high bio-oil yield [67, 68]. However, as the ash components themselves cannot be changed to bio-oil components, the high ash content may reduce the overall bio-oil yields [69, 70].

4. Fast microwave-assisted pyrolysis

4.1. Microwave absorbent

Generally, biomass feedstocks are poor absorbents or receptors of microwave energy and therefore cannot be directly heated up to the high temperatures required for the complete pyrolysis [71]. To improve heating, exogenous microwave absorbents are therefore used in fast microwave-assisted pyrolysis (fMAP) of biomass.

Microwave absorbents used in microwave-assisted pyrolysis of biomass mainly include SiC, activated carbon (AC), carbon, char, and graphite, and sometimes may include the others e.g., fly ash, waste tire, etc. [12, 72]. **Table 10** shows the tanδ values of some microwave absorbents used in microwave-assisted pyrolysis of biomass. Generally, microwave absorbents have higher tanδ values (0.08–1.05) than biomass feedstocks (0.0001–0.22).

If microwave absorbents are mixed with biomass feedstocks and heated by microwave, the heating rates and pyrolysis temperatures are usually increased and the residence times shortened. **Figure 8** shows the temperature profiles of a biomass feedstock under microwave heating with and without microwave absorbents. When the biomass feedstock was heated by a 450 W microwave, its temperature reached about 325°C at 34 min with an average heating rate of 8.3°C/min. However, its temperature reached about 605°C at 7.5 min with an average heating rate of 75.9°C/min when the microwave absorbent of activated carbon was added [68].

Although the initial high heating rate and increased pyrolysis temperature as a result of using microwave absorbent would increase the bio-oil yield for microwave-assisted pyrolysis of biomass [47], too high a temperature especially in the later stage of pyrolysis may cause secondary reactions that break down vapors to noncondensable gases, reducing the bio-oil yield [10, 47, 77].

Microwave absorbent	tanδ	Source
SiC	0.02–1.05	[73]
Activated carbon	0.62	[74]
Kernel shell-activated carbon	0.40	[57]
Wet fruit bunch char	0.30	[57]
Carbon	0.28	[52]
Biochar	0.20	[75]
Graphite	0.10–0.17	[76]
Dried fruit bunch char	0.13	[57]
Palm shell char	0.08	[41]

Table 10. tan δ values of some microwave absorbents at room temperature and 2.45 GHz.

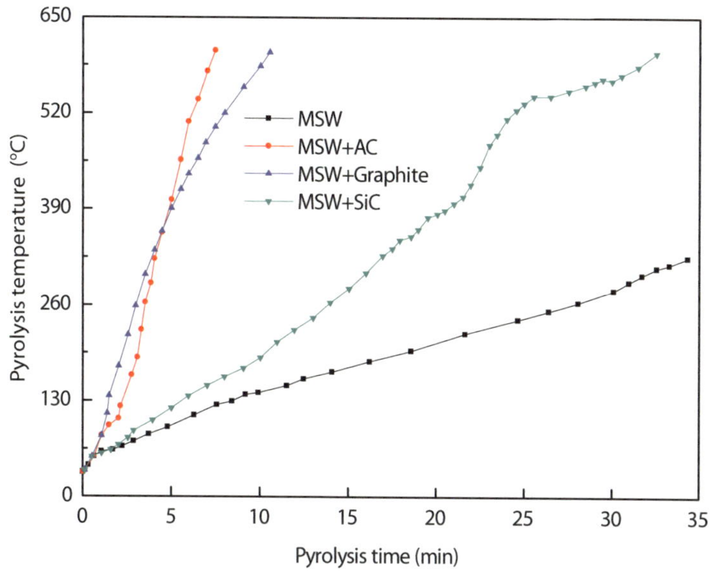

Figure 8. Temperature profiles of biomass feedstock with and without microwave absorbents [68].

4.2. The process

Generally, there are two ways of mixing the biomass feedstock with a microwave absorbent for the microwave-assisted pyrolysis process. The first way is to mix the biomass feedstock with the microwave absorbent before the microwave-assisted pyrolysis. Because the biomass feedstock usually has different bulk density with that of microwave absorbent and some feedstock particles may be smaller than that of microwave absorbent, the smaller feedstock particles may pass though the microwave absorbent layer and stay on the bottom of the mixing plate during the mixing process, causing a nonuniform mixing for the pyrolysis process. Measures should therefore be taken to ensure uniform mixing.

The other way is to mix the biomass feedstock with the microwave absorbent during the microwave-assisted pyrolysis. This can be realized by dropping biomass directly onto heated microwave absorbents which may or may not be agitated mechanically throughout the pyrolysis. The biomass is heated by the absorbents as well as microwave. Since the microwave absorbents are preheated to high temperature, pyrolytic reactions can take place instantly when biomass is dropped onto the absorbent bed, resulting in fast pyrolysis. As the heating rate of biomass feedstock in this mixing is much higher than that in the first-way mixing, this pyrolysis can be considered as fast microwave-assisted pyrolysis (fMAP). Because the biomass feedstock is poured onto the microwave absorbent layer, without agitation, the biomass feedstock would remain on top of the microwave absorbent layer, making the conduction heat transfer from the microwave absorbent particles to the biomass feedstock particles significantly retarded. Consequently, mechanical agitation is usually needed to improve heat transfer.

The fast microwave-assisted pyrolysis of biomass can be conducted using a typical fast microwave-assisted pyrolysis set-up as shown in **Figure 9**. Compared with the typical moderate microwave-assisted pyrolysis set-up in **Figure 2**, a bench-scale fast microwave-assisted pyrolysis set-up generally has two more features: (a) the feeder (1) is used to feed the biomass feedstock into the quartz reactor during the pyrolysis process and (b) the quartz reactor (4) generally has one more neck for the feeder.

Fast microwave-assisted pyrolysis (fMAP) has different pyrolysis process as compared with the microwave-assisted pyrolysis (MAP) process. During the fast microwave-assisted pyrolysis process, the pyrolysis temperature is generally a constant or it fluctuates slightly around a constant value, which is also the set or required pyrolysis temperature. For the fast microwave-assisted pyrolysis, both the drying and pyrolysis I processes (discussed in Section 3.1) are fulfilled within a much shorter time with no obvious differences in the pyrolysis temperatures, whereas the pyrolysis II dominates the whole pyrolysis process also with no obvious differences in the pyrolysis temperatures.

4.3. Research results: fMAP vs. MAP

Fast microwave-assisted pyrolysis (fMAP) is a relatively new technology to obtain bio-oil from biomass, and the related characteristics are not well investigated. Wang et al. [50] compared fast microwave-assisted pyrolysis (fMAP) and microwave-assisted pyrolysis (MAP) of

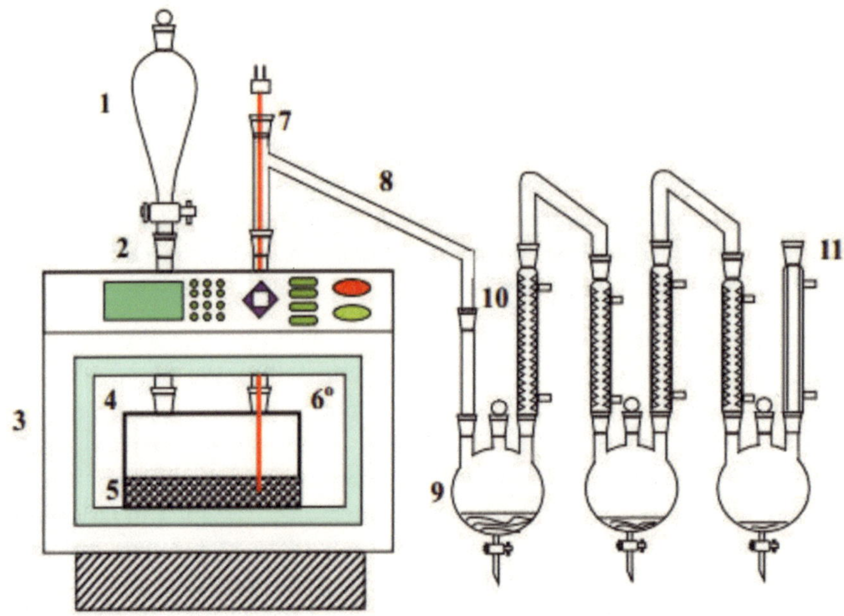

Figure 9. Schematic diagram of a typical fMAP set-up [23, 78, 79]. (1) biomass feeder; (2) quartz connector; (3) microwave oven; (4) quartz reactor; (5) bed with absorbent particles; (6) K-type thermocouple to measure cavity temperature; (7) K-type thermocouple to measure bed temperature; (8) quartz connector; (9) collection flasks; (10) cooling lines; (11) bio-gas outlet.

soapstock for bio-oil production. **Table 11** shows the bio-oil yields obtained from fast microwave-assisted pyrolysis and microwave-assisted pyrolysis of soapstock at different pyrolysis temperatures. Generally, fast microwave-assisted pyrolysis resulted in higher bio-oil yields (50.18–64.74 wt.%) than microwave-assisted pyrolysis (48.16–62.91 wt.%). The results also showed that the bio-oil yield increased initially and decreased finally when the pyrolysis temperature was increased, and the optimal pyrolysis temperature was 550°C. **Table 12** shows the main compounds of the bio-oils obtained from fast microwave-assisted pyrolysis and microwave-assisted pyrolysis of soapstock at the optimal pyrolysis temperature (550°C). Generally, fast microwave-assisted pyrolysis resulted in higher alkanes, aromatics, and oxygenates, whereas lower alkenes, alkadienes, cycloalkenes, and cycloalkanes than microwave-assisted pyrolysis. They also studied fast microwave-assisted pyrolysis of soapstock at different feed rates, and the results showed that the bio-oil yield increased (44.18–69.37 wt.%) when the feed rate was increased (2–10 g/min).

Borges et al. [61] compared fast microwave-assisted pyrolysis (fMAP) of wood sawdust and corn stover at pyrolysis temperatures of 450–550°C with SiC as microwave absorbent. **Tables 13–15** show the yields, properties, and compounds of the bio-oils for wood sawdust and corn stover,

Pyrolysis temperature (°C)	Bio-oil yield (wt.%)	
	fMAP	MAP
400	50.18	48.16
450	51.00	52.07
500	58.37	57.75
550	64.74	62.91
600	62.52	58.63

Table 11. Bio-oil yields obtained from fMAP and MAP of soapstock at different pyrolysis temperatures [50].

Compound	Percentages (% area)	
	fMAP	MAP
Alkenes	17.21	24.83
Cycloalkenes	3.67	9.03
Alkadienes	13.25	17.48
Alkynes	1.85	1.88
Alkanes	18.23	0.00
Cycloalkanes	0.00	12.58
Aromatics	34.72	25.53
Oxygenates	7.46	4.06

Table 12. Main compounds of the bio-oil obtained from fMAP and MAP of soapstock at 550°C [50].

Item	Wood sawdust	Corn stover
Bio-oil yield (wt.%)	46.7–65.0	48.1–64.3
Operation conditions	450–550°C, Feedstock 1–5 g/min, SiC 8–70 Grit	450–550°C, Feedstock 0.5–2.0 mm, Vacuum 0–270 mmHg
Optimal condition	500°C, 3 g/min, 36 Grit	500°C, 1 mm, 170 mmHg

Table 13. Yields of the bio-oils obtained from fMAP of wood sawdust and corn stover [61].

Item	Unit	Wood sawdust	Corn stover
Moisture	wt.%	34.6–53.6	48.2–63.6
Density (at 25°C)	kg/L	1.06	1.02
pH		2.07	2.64
Viscosity (at 40°C)	cP	14	13
Elemental analysis			
C	wt.%	24.86	13.00
H	wt.%	7.17	8.08
N	wt.%	0.35	0.53
O	wt.%	67.61	78.39
HHV	MJ/kg	20.38	20.39

Table 14. Properties of the bio-oils obtained from fMAP of wood sawdust and corn stover [61].

Compound (% area)	Formula	Wood sawdust	Corn stover
Phenol	C_6H_6O	4.32	17.75
Phenol, 3-methyl-	C_7H_8O		7.94
Phenol, 2-methoxy-	$C_7H_8O_2$	3.09	3.74
Styrene	C_8H_8		1.95
Benzofuran	C_8H_6O		2.67
Benzofuran, 2,3-dihydro-	C_8H_8O	11.54	
Phenol, 4-ethyl-	$C_8H_{10}O$	6.91	19.80
Benzene, 1-propynyl-	C_9H_8		3.93
Furan, 3-penty-	$C_9H_{14}O$		1.70
2-Methoxy-4-vinylphenol	$C_9H_{10}O_2$	8.06	
Phenol, 4-ethyl-2-methoxy-	$C_9H_{12}O_2$	2.89	4.25
2-Propenoic acid, 3-(2-hydroxyphenyl)-, (E)-	$C_9H_8O_3$		1.60
Naphthalene	$C_{10}H_8$	3.09	21.40

http://dx.doi.org/10.5772/67442

Compound (% area)	Formula	Wood sawdust	Corn stover
Benzofuran, 4,7-dimethyl-	$C_{10}H_{10}O$		3.80
1H-Indene, 1-ethylidene-	$C_{11}H_{10}$		2.05
Acenaphthylene	$C_{12}H_8$		1.61

Table 15. Main compounds of the bio-oils obtained from fMAP of wood sawdust and corn stover [61].

respectively. The results show that the bio-oils have similar yields, densities, viscosities, and HHVs (higher heating values). These may be due to the facts that the two biomass feedstocks had similar characteristics including moisture content, C content, H content, N content, and O content, and they were pyrolyzed at similar pyrolysis conditions, including microwave power, microwave absorbent, pyrolysis temperature, and feedstock loading. However, the bio-oils have different moisture contents, pH values, elemental contents, and chemical compositions. These may be due to the fact that fast microwave-assisted pyrolysis (fMAP) of biomass is a complex process, and the pyrolysis process and results are varied by many factors, e.g., pyrolysis temperature, microwave absorbent, biomass feedstock, etc.

5. Catalytic fast microwave-assisted pyrolysis

5.1. Reaction catalyst

To improve microwave-assisted pyrolysis of biomass for bio-oil production, another approach is to use reaction catalyst. The reaction catalysts used in microwave-assisted pyrolysis of biomass for bio-oil production mainly include metals (Al, Ni, Fe, Cu, etc.), metal oxides (CaO, CuO, MgO, Fe_2O_3, Fe_3O_4, NiO, Ni_2O_3, MnO_2, TiO_2, ZrO_2, Al_2O_3, HZSM-5, etc.), metal salts (NaCl, KCl, $CaCl_2$, $MgCl_2$, $AlCl_3$, $ZnCl_2$, $FeCl_2$, $FeCl_3$, $CuCl_2$, NaOH, KOH, KAc, K_2CO_3, Na_2CO_3, $CaCO_3$, NaH_2PO_3, $K_2Cr_2O_7$, etc.), and acids (H_2SO_4, H_3PO_4, H_3BO_3, etc.). Basically, these reaction catalysts can: (a) improve the catalytic cracking of biomass feedstock into products and (b) improve the bio-oil quality through catalyst selectivity [77], thus improving the microwave-assisted pyrolysis of biomass for bio-oil production.

Table 16 shows the tan δ values of some reaction catalysts at room temperature and 2.45 GHz. Because some of the reaction catalysts such as metal salts (KOH and $MgCO_3$), metal oxides (ZnO), and metals (Al) may have higher tanδ values (0.001–0.05) than biomass feedstocks (0.0001–0.22), they therefore also have good microwave absorbance and sometimes are used as microwave absorbents in the microwave-assisted pyrolysis of biomass for bio-oil production [44, 68].

When the reaction catalysts are used in microwave-assisted pyrolysis, the heating rates and pyrolysis temperatures would be increased due to the microwave absorbance of the reaction catalysts. **Figure 10** shows the temperature profiles of a biomass feedstock under microwave heating with and without reaction catalysts. When the biomass feedstock was heated by a 450

Reaction catalyst	tanδ	Source
KOH	0.04	[80]
ZnO	0.03	[81]
Al	0.001	[73]
$MgCO_3$	0.01–0.05	[45]
TiO_2	0.002–0.005	[45]
Al_2O_3	0.0006–0.0007	[45]
MgO	0.0004–0.0005	[45]

Table 16. tan δ values of some reaction catalysts at room temperature and 2.45 GHz.

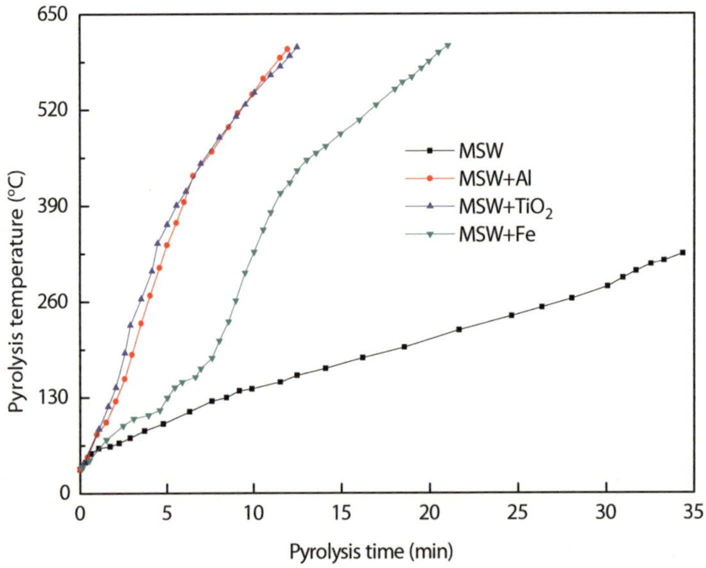

Figure 10. Temperature profiles of biomass feedstock with and without reaction catalysts [68].

W microwave, its temperature reached about 325°C at 34 min with an average heating rate of 8.3°C/min. However, its temperature reached more than 600°C within a much shorter time when the reaction catalysts were added, with an average heating rate of 47.5°C/min for aluminum, 45.5°C/min for TiO_2, and 27.2°C/min for iron [68]. Similar to the case where microwave absorbents are used, reaction catalysts can lead to reduction in bio-oil yields by causing too high a temperature that induces secondary reactions or over-cracking [47, 49, 55].

5.2. Introduction of reaction catalyst to fast microwave-assisted pyrolysis

There are generally four ways of including a reaction catalyst in the fast microwave-assisted pyrolysis (fMAP) as shown in **Figure 9**.

The first way is to mix the reaction catalyst with the microwave absorbent and the mixture is then placed into the quartz reactor to form a cold bed before the pyrolysis. Mixture is then heated with microwave to become a hot bed. When the temperature of the hot bed achieves the temperature required or designed for the pyrolysis, biomass feedstock is then dropped onto the hot bed from the feeder, and the catalytic fast microwave-assisted pyrolysis (cfMAP) is started. However, this method is seldom used because there is an inadequate contact between the reaction catalyst and biomass feedstock. In addition, this method would suffer from an uneven mixing of the reaction catalyst and microwave absorbent due to the differences in the properties (bulk density and particle size, etc.) of the two materials.

The second way is to mix the reaction catalyst with the biomass feedstock. The microwave absorbent is poured into the quartz reactor and heated by microwave to form a hot bed initially. When the temperature of the hot bed achieves the temperature required or designed for the pyrolysis, the mixed reaction catalyst and biomass feedstock are then poured onto the hot bed, and the catalytic fast microwave-assisted pyrolysis (cfMAP) is started. This method is widely used in practice and the details for this method can be found in the literature [23, 48, 78, 79, 82, 83].

The third way is to place the reaction catalyst at the outlet of the connection tube in the quartz reactor, where the vapors exit the quartz reactor. **Figure 11** shows a typical experimental setup for this catalytic fast microwave-assisted pyrolysis (cfMAP). As regard to this set-up, reaction

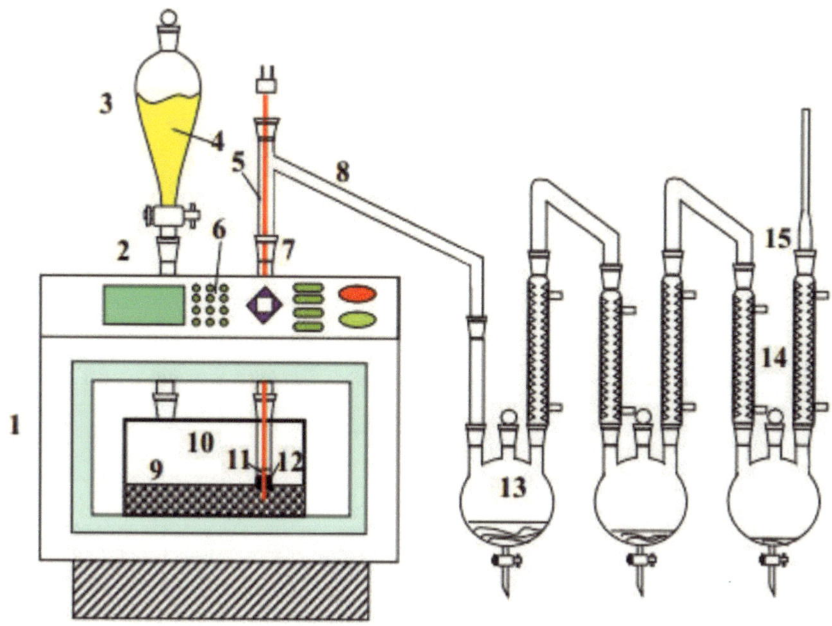

Figure 11. Schematic diagram of an "*in situ*" cfMAP experimental setup [51, 84]. (1) microwave oven; (2) quartz connector; (3) biomass feeder; (4) biomass feedstock; (5) K-type thermocouple; (6) microwave control panel; (7) quartz connector; (8) quartz connector; (9) bed with absorbent particles; (10) quartz reactor; (11) quartz wool; (12) catalyst; (13) collection flasks; (14) cooling lines; (15) bio-gas outlet.

catalyst can be put in the outlet connection tube (7) for the catalytic reactions between reaction catalyst and released vapors. Because the inlet of the outlet connection tube (7) is closer to the hot bed, it has the highest temperature along the whole outlet connection tube and the reaction catalyst can be located at this point. To improve the heating of the reaction catalyst, microwave absorbent is sometimes mixed with the reaction catalyst, and the mixture is then located at the inlet of the outlet connection tube [51, 84]. However, the temperature of the mixture would also be lower than the temperature of the hot bed because the heat absorbed by the mixture would be diffused to the colder environment through convection and radiation heat transfer mechanisms. This method is therefore also seldom used.

As the catalytic reactions in the three ways mentioned above are occurred in the quartz reactor, the three ways are therefore also called *"in situ"* catalytic fast microwave-assisted pyrolysis (cfMAP). As regard to this pyrolysis, the temperature of the reaction catalyst (catalytic temperature) cannot be easily varied or controlled, the fourth way is to place the reaction catalyst outside the reaction volume and thus comes the term *"ex situ"* catalytic fast microwave-assisted pyrolysis. **Figure 12** shows a typical experimental setup of the *"ex situ"* catalytic fast microwave-assisted pyrolysis (cfMAP). Compared with the *"in situ"* catalytic fast microwave-assisted pyrolysis, an extra electrical heater (11) (also means more energy) is required to vary and control the catalytic temperatures for more precise upgrading and refining of the bio-oil.

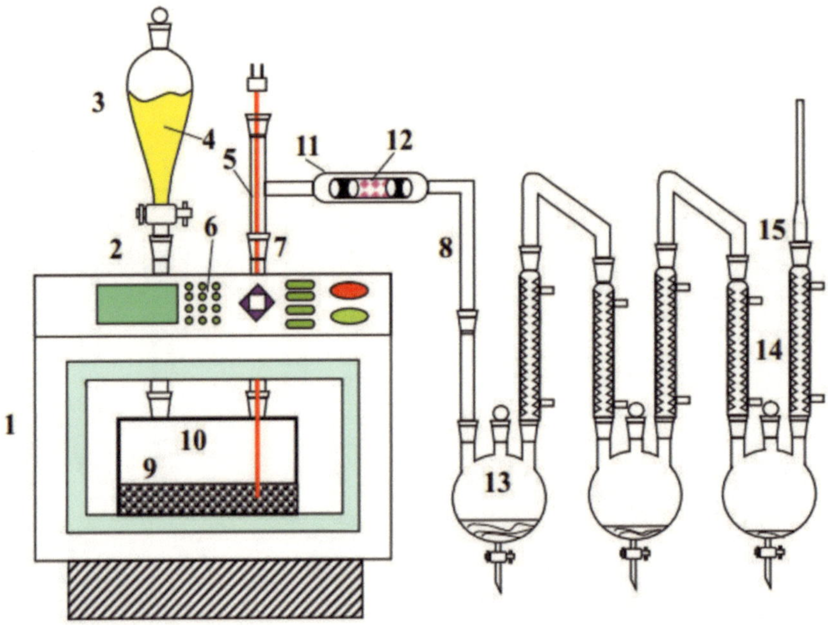

Figure 12. Schematic diagram of an *"ex situ"* cfMAP experimental setup. (1) microwave oven; (2) quartz connector; (3) biomass feeder; (4) biomass feedstock; (5) K-type thermocouple; (6) microwave control panel; (7) quartz connector; (8) quartz connector; (9) bed with absorbent particles; (10) quartz reactor; (11) heater; (12) catalyst; (13) collection flasks; (14) cooling lines; (15) bio-gas outlet.

5.3. Research results: cfMAP vs. fMAP

Borges et al. [23] compared catalytic fast microwave-assisted pyrolysis (cfMAP) and fast microwave-assisted pyrolysis (fMAP) of microalgae for bio-oil production. **Table 17** shows the bio-oil yields obtained from catalytic fast microwave-assisted pyrolysis and fast microwave-assisted pyrolysis of microalgae. For *Nannochloropsis*, catalytic fast microwave-assisted pyrolysis generally resulted in higher bio-oil yields (42.5–59.4 wt.%) than the fast microwave-assisted pyrolysis (45.9–49.0 wt.%). This may be due to the fact that the HZSM-5 catalyst improved the catalytic cracking of Nannochloropsis into bio-oil at a lower temperature. However, for Chlorella sp., catalytic fast microwave-assisted pyrolysis resulted in lower bio-oil yields (42.0–50.4 wt.%) than the fast microwave-assisted pyrolysis (53.3–56.8 wt.%). This may be due to the fact that the HZSM-5 catalyst improved the secondary cracking of bio-oil vapors into incondensable gases [23]. The results that catalytic fast microwave-assisted pyrolysis resulted in lower bio-oil yields than fast microwave-assisted pyrolysis were also reported in the literature [48, 82, 85].

Catalytic fast microwave-assisted pyrolysis (cfMAP) would also result in different bio-oil properties as compared with the fast microwave-assisted pyrolysis (fMAP). **Table 18** shows the physical-chemical properties of the bio-oils obtained from catalytic fast microwave-assisted pyrolysis and fast microwave-assisted pyrolysis of microalgae. It seems that the bio-oils obtained from catalytic fast microwave-assisted pyrolysis have slight higher moisture content and O content, whereas lower C content and H content, thereby having lower HHV (higher heating value) than those obtained from fast microwave-assisted pyrolysis. However, more important industrial chemicals such as phenolics and aromatics can usually be selectively produced when reaction catalysts are introduced in microwave-assisted pyrolysis of biomass [17, 47, 49, 86]. **Table 19** shows some main compounds of the bio-oils obtained from catalytic fast microwave-assisted pyrolysis (cfMAP) and fast microwave-assisted pyrolysis (fMAP) of microalgae. The results show that phenol content was doubled in the bio-oil (9.49 vs. 4.00%) when catalytic fast microwave-assisted pyrolysis was used instead of fast microwave-assisted pyrolysis. Similar results were also reported in the literature [82].

Although catalytic fast microwave-assisted pyrolysis (cfMAP) generally results in lower bio-oil yields than fast microwave-assisted pyrolysis (fMAP) [23, 48, 82, 85], it may selectively reduce

Item	***Chlorella* sp.**		***Nannochloropsis***	
	cfMAP	**fMAP**	**cfMAP**	**fMAP**
Bio-oil yield (wt.%)	42.0–50.4	53.3–56.8	42.5–59.4	45.9–49.0
Operation conditions	450–550°C	450–550°C	450–550°C	450–550°C
	Catalyst/feed: 0.5–1.0		Catalyst/feed: 0.5–1.0	
Optimal condition	500°C	550°C	500°C	550°C
	Catalyst/feed: 0.5		Catalyst/feed: 0.5	

Table 17. Bio-oil yields obtained from cfMAP and fMAP of microalgae [23].

Property		*Chlorella* sp.		*Nannochloropsis*	
		cfMAP	fMAP	cfMAP	fMAP
Moisture	wt.%	62.1–69.2	51.4–64.3	47.4–54.8	36.3–48.7
Density	g/mL	1.01	1.00	1.18	1.18
pH	–	9.54	9.33	9.62	9.93
Elemental analysis					
C	wt.%	59.27	65.70	59.75	81.64
H	wt.%	7.75	9.34	6.75	8.20
N	wt.%	9.46	8.45	16.34	5.24
O	wt.%	23.52	15.78	17.16	4.90
HHV	MJ/kg	26.80	32.37	27.15	42.00

Table 18. Physical-chemical properties of the bio-oils from cfMAP and fMAP of microalgae [23].

Compound (% area)	Formula	*Chlorella* sp.		*Nannochloropsis*	
		cfMAP	fMAP	cfMAP	fMAP
Acetamide	C_2H_5NO	0.89	1.45	8.15	1.07
Propanoic acid	$C_3H_6O_2$	0.55			
Acetic acid	$C_4H_4O_2$	2.94	2.73		
Pyridine	C_5H_5N	0.66	0.31		
Phenol	C_6H_6O	0.49	0.87	9.49	4.00
Aniline	C_6H_7N	0.05	0.04	0.85	
Pyridine, 3-methyl-	C_6H_7N	0.18		2.46	1.01
Phenol, 4-methyl-	C_7H_8O	0.42	1.41		
Styrene	C_8H_8				5.15
Indole	C_8H_7N	0.26	1.12	2.76	4.12
Benzyl nitrile	C_8H_7N	0.14	0.26	3.03	3.00
n-Hexadecanoic acid	$C_{16}H_{32}O_2$		2.83		

Table 19. Some main compounds of the bio-oils from cfMAP and fMAP of microalgae [23].

the number of compounds in the bio-oil [82], selectively produce more amount of important industrial chemicals such as phenolics and aromatics [23, 82], and also enable the process to achieve higher bio-oil yield at a lower pyrolysis temperature [23]. Consequently, catalytic fast microwave-assisted pyrolysis (cfMAP) of biomass for bio-oil production is widely studied.

Catalytic fast microwave-assisted pyrolysis (cfMAP) is also a relatively new technology to obtain bio-oil from biomass. Although the various effects on the pyrolysis processes and results were studied and reported, it is still difficult to accurately predict the detailed pyrolysis processes and results for a particular biomass feedstock treated by catalytic fast microwave-assisted pyrolysis (cfMAP).

6. Conclusions and future outlook

Pyrolysis of biomass using microwave-induced heat is a relatively new thermochemical conversion process. Bio-oils from microwave-assisted pyrolysis of biomass have lower oxygen content and higher HHV than those from conventional biomass pyrolysis. The dielectric properties of biomass play a large role in microwave-assisted pyrolysis as they affect the microwave absorbance of biomass. Utilizing the superior microwave absorbance of exogenous microwave absorbents and reaction catalysts, we can achieve very high heating rates, high pyrolysis temperatures, and catalytic cracking of large molecules, resulting in higher yield and better quality of bio-oils. However, too high a pyrolysis temperature can adversely affect the bio-oil yield by inducing secondary reactions that break down the volatiles to incondensable gases. Nonetheless, microwave-assisted pyrolysis has a bright future as it offers many advantages over conventional pyrolysis processes. Particularly, microwave heating is a mature technology which can be easily implemented and precisely controlled. Future work in the areas of catalyst screening and applications, co-pyrolysis of feedstocks with complementary microwave heating characteristics, pilot study, and equipment development is necessary in order to cost-effectively produce high quality bio-oils and commercialize the technology.

Acknowledgements

This study was supported in part by grants from China Scholarship Council (No. 201506125122), Minnesota's Environment and Natural Resources Trust Fund (ENRTF) through the processes of the Legislative-Citizen Commission on Minnesota Resources (LCCMR), and University of Minnesota Center for Biorefining.

Author details

Yaning Zhang[1,2], Paul Chen[2], Shiyu Liu[2], Liangliang Fan[2,3], Nan Zhou[2], Min Min[2], Yanling Cheng[2], Peng Peng[2], Erik Anderson[2], Yunpu Wang[3], Yiqin Wan[3], Yuhuan Liu[3], Bingxi Li[1] and Roger Ruan[2,3]*

*Address all correspondence to: ruanx001@umn.edu

1 School of Energy Science and Engineering, Harbin Institute of Technology, Harbin, China

2 Department of Bioproducts and Biosystems Engineering, Center for Biorefining, University of Minnesota, St Paul, MN, USA

3 State Key Laboratory of Food Science and Technology, Nanchang University, Nanchang, China

References

[1] Bridgwater AV, Peacocke GVC. Fast pyrolysis processes for biomass. Renewable and Sustainable Energy Reviews. 2000; **4**: 1-73.

[2] Czernik S, Bridgwater AV. Overview of applications of biomass fast pyrolysis oil. Energy Fuel. 2004; **18**: 590-598.

[3] Mohan D, Pittman CU, Steele JPH. Pyrolysis of wood/biomass for bio-oil: A critical review. Energy Fuel. 2006; **20**: 848-889.

[4] Kan T, Strezov V, Evans TJ. Lignocellulosic biomass pyrolysis: A review of product properties and effects of pyrolysis parameters. Renewable and Sustainable Energy Reviews. 2016; **57**: 1126-1140.

[5] Bridgwater AV. Review of fast pyrolysis of biomass and product upgrading. Biomass and Bioenergy. 2012; **38**: 68-94.

[6] Motasemi F, Ani FN. A review on microwave-assisted production of biodiesel. Renewable and Sustainable Energy Reviews. 2012; **16**: 4719-4733.

[7] Saber M, Nakhshiniev B, Yoshikawa K. A review of production and upgrading of algal bio-oil. Renewable and Sustainable Energy Reviews. 2016; **58**: 918-930.

[8] Motasemi F, Afzal MT. A review on the microwave-assisted pyrolysis technique. Renewable and Sustainable Energy Reviews. 2013; **28**: 317-330.

[9] Chen P, Xie Q, Addy M, Zhou W, Liu Y, Wang Y, Cheng Y, Li K, Ruan R. Utilization of municipal solid and liquid wastes for bioenergy and bioproducts production. Bioresource Technology. 2016; **215**: 163-172.

[10] Du Z, Li Y, Wang X, Wan Y, Chen Q, Wang C, Lin X, Liu Y, Chen P, Ruan R. Microwave-assisted pyrolysis of microalgae for biofuel production. Bioresource Technology. 2011; **102**: 4890-4896.

[11] Fourier Transform-Infrared Spectroscopy (FTIR) [Internet]. 2016. Available from: http://www.mee-inc.com/hamm/fourier-transform-infrared-spectroscopy-ftir/ [Accessed: 2016-11-20]

[12] Undri A, Rosi L, Frediani M, Frediani P. Fuel from microwave assisted pyrolysis of waste multilayer packaging beverage. Fuel. 2014; **133**: 7-16.

[13] Undri A, Abou-Zaid M, Briens C, Berruti F, Rosi L, Bartoli M, Frediani M, Frediani P. Bio-oil from pyrolysis of wood pellets using a microwave multimode oven and different microwave absorbers. Fuel. 2015; **153**: 464-482.

[14] Undri A, Rosi L, Frediani M, Frediani P. Efficient disposal of waste polyolefins through microwave assisted pyrolysis. Fuel. 2014; **116**: 662-671.

[15] Undri A, Rosi L, Frediani M, Frediani P. Conversion of poly(lactic acid) to lactide via microwave assisted pyrolysis. Journal of Analytical and Applied Pyrolysis. 2014; **110**: 55-65.

[16] Wan Y, Chen P, Zhang B, Yang C, Liu Y, Lin X, Ruan R, Microwave-assisted pyrolysis of biomass: Catalysts to improve product selectivity. Journal of Analytical and Applied Pyrolysis. 2009; **86**: 161-167.

[17] Bu Q, Lei H, Wang L, Wei Y, Zhu L, Liu Y, Liang J, Tang J. Renewable phenols production by catalytic microwave pyrolysis of Douglas fir sawdust pellets with activated carbon catalysts. Bioresource Technology. 2013; **142**: 546-552.

[18] Wang L, Lei H, Bu Q, Ren S, Wei Y, Zhu L, Zhang X, Liu Y, Yadavalli G, Lee J, Chen S, Tang J. Aromatic hydrocarbons production from ex situ catalysis of pyrolysis vapor over Zinc modified ZSM-5 in a packed-bed catalysis coupled with microwave pyrolysis reactor. Fuel. 2014; **129**: 78-85.

[19] Wang X, Chen H, Luo K, Shao J, Yang H. The influence of microwave drying on biomass pyrolysis. Energy Fuel. 2008; **22**: 67-74.

[20] Aziz SMA, Wahi R, Ngaini Z, Hamdan S. Bio-oils from microwave pyrolysis of agricultural wastes. Fuel Processing Technology. 2013; **106**: 744-750.

[21] Budarin VL, Clark JH, Lanigan BA, Shuttleworth P, Breeden SW, Wilson AJ, Macquarrie DJ, Milkowski K, Jones J, Bridgeman T, Ross A. The preparation of high-grade bio-oils through the controlled, low temperature microwave activation of wheat straw. Bioresource Technology. 2009; **100**: 6064-6068.

[22] Ferrera-Lorenzo N, Fuente E, Bermúdez JM, Suárez-Ruiz I, Ruiz B. Conventional and microwave pyrolysis of a macroalgae waste from the Agar-Agar industry. Prospects for bio-fuel production. Bioresource Technology. 2014; **151**: 199-206.

[23] Borges FC, Xie Q, Min M, Muniz LA, Farenzena M, Trierweiler JO, Chen P, Ruan R. Fast microwave-assisted pyrolysis of microalgae using microwave absorbent and HZSM-5 catalyst. Bioresource Technology. 2014; **166**: 518-526.

[24] Yu F, Ruan R, Steele P. Microwave pyrolysis of corn stover. Transactions of the ASABE. 2009; **52**: 1595-1601.

[25] Lei H, Ren S, Julson J. The effects of reaction temperature and time and particle size of corn stover on microwave pyrolysis. Energy Fuel. 2009; **23**: 3254-3261.

[26] Benerosoa D, Bermúdeza JM, Arenillasa A, de la Peñab F, Garcíab JL, Prietob MA, Menéndeza JA. Oil fractions from the pyrolysis of diverse organic wastes: The different effects of conventional and microwave induced pyrolysis. Journal of Analytical and Applied Pyrolysis. 2015, **114**: 256-264.

[27] Budarin VL, Shuttleworth PS, De bruyn M, Farmer TJ, Gronnow MJ, Pfaltzgraff L, Macquarrie DJ, Clark JH. The potential of microwave technology for the recovery, synthesis and manufacturing of chemicals from bio-wastes. Catalysis Today. 2015; **239**: 80-89.

[28] Yu F, Deng S, Chen P, Liu Y, Wan Y, Olson A, Kittelson D, Ruan R. Physical and chemical properties of bio-oils from microwave pyrolysis of corn stover. Applied Biochemistry and Biotechnology. 2007; **137-140**: 957-970.

[29] Mamaeva A, Tahmasebi A, Tian L, Yu J. Microwave-assisted catalytic pyrolysis of lignocellulosic biomass for production of phenolic-rich bio-oil. Bioresource Technology. 2016; **211**: 382-389.

[30] Diebold J P, Czernik S. Additives to lower and stabilize the viscosity of pyrolysis oils during storage. Energy Fuel. 1997; **11**: 1081-1091.

[31] Oasmaa A, Czernik S. Fuel oil quality of biomass pyrolysis oils-State of the art for the end user. Energy Fuel. 1999; **13**: 914-921.

[32] Tat ME, van Gerpen JH. The kinematic viscosity of biodiesel and its blends with diesel fuel. Journal of the American Oil Chemists' Society. 1999; **76**: 1511-1513.

[33] Channiwala SA, Parikh PP. A unified correlation for estimating HHV of solid, liquid and gaseous fuels. Fuel. 2002; **81**:1051-1063.

[34] Ates F, Isikdag MA. Evaluation of the role of the pyrolysis temperature in straw biomass samples and characterization of the oils by GC-MS. Energy Fuel. 2008; **22**: 1936-1943.

[35] Szargut J, Morris DR, Stewart FR. Exergy analysis of thermal, chemical, and metallurgical processes. Ann Arbor: Edwards Brothers; 1988. p. 103.

[36] Omoriyekomwan JE, Tahmasebi A, Yu J. Production of phenol-rich bio-oil during catalytic fixed-bed and microwave pyrolysis of palm kernel shell. Bioresource Technology. 2016; **207**: 188-196.

[37] Wang X, Morrison W, Du Z, Wan Y, Lin X, Chen P, Ruan R. Biomass temperature profile development and its implications under the microwave-assisted pyrolysis condition. Applied Energy. 2012; **99**: 386-392.

[38] Zhang S, Dong Q, Zhang L, Xiong Y. High quality syngas production from microwave pyrolysis of rice husk with char-supported metallic catalysts. Bioresource Technology. 2015; **191**: 17-23.

[39] Mushtaq F, Mat R, Ani FN. A review on microwave assisted pyrolysis of coal and biomass for fuel production. Renewable and Sustainable Energy Reviews. 2014; **39**: 555-574.

[40] Vos B, Mosman J, Zhang Y, Poels E, Bliek A. Impregnated carbon as a susceptor material for low loss oxides in dielectric heating. Journal of Materials Science. 2003; **38**: 173-182.

[41] Salemaa AA, Yeowb YK, Ishaquec K, Anid FN, Afzala MT, Hassane A. Dielectric properties and microwave heating of oil palm biomass and biochar. Industrial Crops and Products. 2013; **50**: 366-374.

[42] Larhed M, Moberg C, Hallberg A. Microwave-accelerated homogeneous catalysis in organic chemistry. Accounts of Chemical Research. 2002; **35**: 717-727.

[43] Torgovnikov GI. Dielectric properties of wood and wood-based materials. New York: Springer-Verlag; 1993. p. 72.

[44] Namazi AB, Allen DG, Jia CQ. Probing microwave heating of lignocellulosic biomasses. Journal of Analytical and Applied Pyrolysis. 2015; **112**: 121-128.

[45] Dielectric Chart [Internet]. 2016. Available from: www.eccosorb.com/file/1138/dielectric-chart.pdf [Accessed: 2016-11-20]

[46] Tian Y, Zuo W, Ren Z, Chen D. Estimation of a novel method to produce bio-oil from sewage sludge by microwave pyrolysis with the consideration of efficiency and safety. Bioresource Technology. 2011; **102**: 2053-2061.

[47] Bu Q, Lei H, Zacher AH, Wang L, Ren S, Liang J, Wei Y, Liu Y, Tang J, Zhang Q, Ruan R. A review of catalytic hydrodeoxygenation of lignin-derived phenols from biomass pyrolysis. Bioresource Technology. 2012; **124**: 470-477.

[48] Xie Q, Peng P, Liu S, Min M, Cheng Y, Wan Y, Li Y, Lin X, Liu Y, Chen P, Ruan R. Fast microwave-assisted catalytic pyrolysis of sewage sludge for bio-oil production. Bioresource Technology. 2014; **172**: 162-168.

[49] Wang L, Lei H, Ren S, Bu Q, Liang J, Wei Y, Liu Y, Lee GSJ, Chen S, Tang J, Zhang Q, Ruan R. Aromatics and phenols from catalytic pyrolysis of Douglas fir pellets in microwave with ZSM-5 as a catalyst. Journal of Analytical and Applied Pyrolysis. 2012; **98**: 194-200.

[50] Wang Y, Dai L, Wang R, Fan L, Liu Y, Xie Q, Ruan R. Hydrocarbon fuel production from soapstone through fast microwave-assisted pyrolysis using microwave absorbent. Journal of Analytical and Applied Pyrolysis. 2016; **119**: 251-258.

[51] Zhang B, Zhong Z, Xie Q, Liu S, Ruan R. Two-step fast microwave-assisted pyrolysis of biomass for bio-oil production using microwave absorbent and HZSM-5 catalyst. Journal of Environmental Sciences. 2016; **45**: 240-247.

[52] Salema AA, Ani FN. Microwave-assisted pyrolysis of oil palm shell biomass using an overhead stirrer. Journal of Analytical and Applied Pyrolysis. 2012; **96**: 162-172.

[53] Zhang B, Yang C, Moen J, Le Z, Hennessy K, Wan Y, Liu Y, Lei H, Chen P, Ruan R. Catalytic conversion of microwave-assisted pyrolysis vapors. Energy Sources, Part A: Recovery, Utilization, and Environmental Effects. 2010; **32**: 1756-1762.

[54] Wahi R, Idris A, Mohd MA, Khalid K. Low-temperature microwave pyrolysis of sewage sludge. International Journal of Engineering and Technology. 2006; **3**: 132-138.

[55] Hu Z, Ma X, Chen C. A study on experimental characteristic of microwave-assisted pyrolysis of microalgae. Bioresource Technology. 2012; **107**: 487-493.

[56] Wang N, Tahmasebi A, Yu J, Xu J, Huang F, Mamaeva A. A comparative study of microwave-induced pyrolysis of lignocellulosic and algal biomass. Bioresource Technology. 2015; **190**: 89-96.

[57] Omar R, Idris A, Yunus R, Khalid K, Isma MIA. Characterization of empty fruit bunch for microwave-assisted pyrolysis. Fuel. 2011; **90**: 1536-1544.

[58] Motasemi F, Afzal MT, Salema AA, Mouris J, Hutcheon RM. Microwave dielectric characterization of switchgrass for bioenergy and biofuel. Fuel. 2014; **124**: 151-157.

[59] Menéndez JA, Domínguez A, Inguanzo M, Pis JJ. Microwave pyrolysis of sewage sludge: analysis of the gas fraction. Journal of Analytical and Applied Pyrolysis. 2004; **71**: 657-667.

[60] Farag S, Fu D, Jessop PG, Chaouki J. Detailed compositional analysis and structural investigation of a bio-oil from microwave pyrolysis of kraft lignin. Journal of Analytical and Applied Pyrolysis. 2014; **109**: 249-257.

[61] Borges FC, Du Z, Xie Q, Trierweiler JO, Cheng Y, Wan Y, Liu Y, Zhu R, Lin X, Chen P, Ruan R. Fast microwave assisted pyrolysis of biomass using microwave absorbent. Bioresource Technology. 2014; **156**: 267-274.

[62] Bartoli M, Rosi L, Giovannelli A, Frediani P, Frediani M. Bio-oil from residues of short rotation coppice of poplar using a microwave assisted pyrolysis. Journal of Analytical and Applied Pyrolysis. 2016; **119**: 224-232.

[63] Moen J, Yang C, Zhang B, Lei H, Hennessy K, Wan Y, Le Z, Liu Y, Chen P, Ruan R. Catalytic microwave assisted pyrolysis of aspen. International Journal of Agricultural and Biological Engineering. 2009; **2**: 70-75.

[64] Ren S, Lei H, Wang L, Bu Q, Chen S, Wu J, Julson J, Ruan R. Biofuel production and kinetics analysis for microwave pyrolysis of Douglas fir sawdust pellet. Journal of Analytical and Applied Pyrolysis. 2012; **94**: 163-169.

[65] El harfi K, Mokhlisse A, Chanâa MB, Outzourhit A. Pyrolysis of the Moroccan (Tarfaya) oil shales under microwave irradiation. Fuel. 2000; **79**: 733-742.

[66] Yu Y, Yu J, Sun B, Yan Z. Influence of catalyst types on the microwave-induced pyrolysis of sewage sludge. Journal of Analytical and Applied Pyrolysis. 2014; **106**: 86-91.

[67] Suriapparao DV, Pradeep N, Vinu R. Bio-oil production from *Prosopis juliflora* via microwave pyrolysis. Energy Fuel. 2015; **29**: 2571-2581.

[68] Suriapparao DV, Vinu R. Bio-oil production via catalytic microwave pyrolysis of model municipal solid waste component mixtures. RSC Advances. 2015; **5**: 57619-57631.

[69] Domínguez A, Fernández Y, Fidalgo B, Pis JJ, Menéndez JA. Bio-syngas production with low concentrations of CO_2 and CH_4 from microwave-induced pyrolysis of wet and dried sewage sludge. Chemosphere. 2008; **70**: 397-403.

[70] Hascakir B, Akin S. Recovery of Turkish oil shales by electromagnetic heating and determination of the dielectric properties of oil shales by an analytical method. Energy Fuel. 2010; **24**: 503-509.

[71] Menéndez JA, Arenillas A, Fidalgo B, Fernández Y, Zubizarreta L, Calvo EG, Bermúdez JM. Microwave heating processes involving carbon materials. Fuel Processing Technology. 2010; **91**: 1-8.

[72] Suriapparao DV, Vinu R. Resource recovery from synthetic polymers via microwave pyrolysis using different susceptors. Journal of Analytical and Applied Pyrolysis. 2015; **113**: 701-712.

[73] GubbT,BaranovaI,AllanS,FallM,ShulmanH,KrivenWM.Microwaveprocessingofgeopolymers and evolved glass-ceramic composites [Internet]. 2016. Available from: http://www.ceralink.com/sites/default/files/MicrowaveProcessingofGeopolymersandEvolvedGlass-CeramicCompositesPresentation.pdf [Accessed: 2016-11-20]

[74] Ramasamy S, Moghtaderi B. Dielectric properties of typical Australian wood-based biomass materials at microwave frequency. Energy Fuel. 2010; **24**: 4534-4548.

[75] Peng Z, Hwang J, Bell W, Andriese M, Xie S. Microwave dielectric properties of pyrolyzed carbon. 2nd International Symposium on High-Temperature Metallurgical Processing; February 27–March 3 2011; California, USA; 2011. pp. 77-83.

[76] Sanchez-Coronado J, Chung DDL. Thermomechanical behavior of a graphite foam. Carbon. 2003; **41**: 1175-1180.

[77] Mohamed BA, Kim CS, Ellis N, Bi X. Microwave-assisted catalytic pyrolysis of switchgrass for improving bio-oil and biochar properties. Bioresource Technology. 2016; **201**: 121-132.

[78] Xie Q, Addy M, Liu S, Zhang B, Cheng Y, Wan Y, Li Y, Liu Y, Lin X, Chen P, Ruan R. Fast microwave-assisted catalytic co-pyrolysis of microalgae and scum for bio-oil production. Fuel. 2015; **160**: 577-582.

[79] Liu S, Xie Q, Zhang B, Cheng Y, Liu Y, Chen P, Ruan R. Fast microwave-assisted catalytic co-pyrolysis of corn stover and scum for bio-oil production with CaO and HZSM-5 as the catalyst. Bioresource Technology. 2016; **204**: 164-170.

[80] Zaini MAA, Mohd N, Aini N, Kamaruddin MJ, Yeow YK, Setapar SHM. Dielectric properties of potassium hydroxide-treated palm kernel shell for microwave-assisted adsorbent preparation. Jurnal Teknologi (Sciences & Engineering). 2015; **74**: 13-18.

[81] Westphal UWB, Sils A. Dielectric constant and loss data [Internet]. 1972. Available from: http://www.dtic.mil/dtic/tr/fulltext/u2/746686.pdf [Accessed: 2016-11-20]

[82] Xie Q, Borges FC, Cheng Y, Wan Y, Li Y, Lin X, Liu Y, Hussain F, Chen P, Ruan R. Fast microwave-assisted catalytic gasification of biomass for syngas production and tar removal. Bioresource Technology. 2014; **156**: 291-296.

[83] Fan L, Chen P, Zhang Y, Liu S, Liu Y, Wang Y, Dai L, Ruan R. Fast microwave-assisted catalytic co-pyrolysis of lignin and low-density polyethylene with HZSM-5 and MgO for improved bio-oil yield and quality. Bioresource Technology. 2017; **225**: 199-205.

[84] Zhang B, Zhong Z, Chen P, Ruan R. Microwave-assisted catalytic fast pyrolysis of biomass for bio-oil production using chemical vapor deposition modified HZSM-5 catalyst. Bioresource Technology. 2015; **197**: 79-84.

[85] Zhang B, Zhong Z, Min M, Ding K, Xie Q, Ruan R. Catalytic fast co-pyrolysis of biomass and food waste to produce aromatics: Analytical Py-GC/MS study. Bioresource Technology. 2015; **189**: 30-35.

[86] Bu Q, Lei H, Ren S, Wang L, Holladay J, Zhang Q, Tang J, Ruan R. Phenol and phenolics from lignocellulosic biomass by catalytic microwave pyrolysis. Bioresource Technology. 2011; **102**: 7004-7007.

Chapter 7

Catalytic Pyrolysis of Biomass

Sibel Başakçılardan Kabakcı and

Şeyma Hacıbektaşoğlu

Additional information is available at the end of the chapter

http://dx.doi.org/10.5772/67569

Abstract

Biomass pyrolysis has been a popular research topic due to versatility in products: char, bio-oil, and syngas. What makes the biomass pyrolysis so important is that it provides product options such as fuels or green chemicals. In the case of biomass pyrolysis, lignocellulosic materials undergo thermal degradation in the absence of oxidative environment at atmospheric pressure during a definite residence time, which produces solid char, bio-oil, and syngas. If the bio-oil is going to be used as a fuel source or to be processed for producing chemicals, it requires an upgrading. Catalytic pyrolysis is the most promising method to improve the quality of bio-oil. The present study presents an updated review on catalytic pyrolysis of biomass for the production of upgraded bio-oil. In this context, this review considers recent advances in catalysts and catalytic pyrolysis process.

Keywords: pyrolysis, catalysts, bio-oil upgrading

1. Introduction

The continuous growth in economies over the world requires demand for energy at a high rate. For meeting the energy demand, high energy production via fossil fuels leads to an increase in CO_2 emissions. In addition, considering the issue in terms of the effects of global climate change and environmental responsibility, the need for renewable energy resources has become inevitable. Also, diminishing reserves of fossil fuels and the necessity of reducing CO_2 emissions that is emitted by fossil fuel burning processes force us to seek environmentally and economically beneficial ways to produce energy from renewable energy resources [1–5].

Biomass is a general term for nonfossilized and biodegradable organic material originating from plants, animals, and microorganisms. Given the fact that biomass is abundant, and it accounts for 38–43% of primary energy consumption in developing countries, the use of waste-based biomass sources and feedstock is of utmost importance for research in terms of clean energy generation via thermochemical or biochemical conversion processes [6–8]. Agricultural, aquaculture and forestry products (also by-products, residues, and wastes), dedicated energy crops, biodegradable organic fraction of industrial wastes, biodegradable organic fraction of municipal solid waste, and food processing wastes can be included in primary biomass resources [8–12].

Pyrolysis is a thermochemical conversion process in the absence of an oxidizing agent and can be regarded as the initial stage of gasification and combustion. Solid char, liquid pyrolysis oil, and gas are the main products of biomass pyrolysis. The amount of products and their fractions are influenced by many factors such as heating rate, pyrolysis temperature, biomass composition, and catalyst effect [8, 13, 14]. Since biomass pyrolysis is feedstock composition-dependent process, finding suitable catalysts to regulate pyrolysis processes is an alternative way to reduce overall energy consumption [15]. Also, catalytic pyrolysis of biomass provides, with optimum catalyst/biomass ratio, chemically more homogenous fractions of pyrolysis products. In recent years, different biomass feedstock or biomass originated waste materials have been studied using several catalysts in order to understand the effects of catalyst to pyrolysis process [16–22].

Bio-oil possesses many undesirable components, such as high content of volatile acids, water, and highly oxygenated compounds, which in turn lowers the heating value. In order to use bio-oil in fuel applications, upgrading of bio-oil is necessary. Catalytic pyrolysis is a promising way to improve bio-oil quality by removing of oxygenated compounds, increasing calorific value, lowering viscosity, and increasing stability. Of the various catalysts studied for biomass catalytic pyrolysis, zeolites have been shown to be effective for reducing oxygenated compounds of pyrolysis oil. One of the most common zeolites have been studied is ZSM-5, which can be synthesized with different Si/Al ratios, thus having different acidity characteristics [23]. Studies have shown that using ZSM-5 as catalyst in biomass pyrolysis increased aromatic hydrocarbons content and decreased oxygenated aromatic compounds content in bio-oil [1, 19, 23–25].

2. Fundamentals of biomass pyrolysis

Pyrolysis is the heating of the feedstock in the absence of oxygen at a specified heating rate to a definite temperature and holding it there for a certain time. During pyrolysis, large hydrocarbon molecules are broken into relatively smaller ones via reactions such as depolymerization, dehydration, decarbonylation, decarboxylation, deoxygenation, oligomerization, and aromatization [23, 26]. The amount and the ratio of these fractions are influenced by many factors such as heating rate, pyrolysis temperature, biomass composition, and catalyst effect, which can been seen in **Table 1** [8, 13, 14].

http://dx.doi.org/10.5772/67569

Reactor type	T (°C)	Sweeping gas flow rate	Residence time (s)	Catalyst type	Catalyst to biomass ratio	Bio-oil (wt.%)			Gas (wt.%)	Char (wt.%)	Ref.
						Aqueous	Organic	Total (aqueous + organic)			
Fixed bed	600	30 cm^3 min^{-1}	10	La/HZSM-5	3	14.4	7.7	22.4	34.2	39.9	[27]
Fluidized bed	450	3.75 l min^{-1}	*	H-Beta	0.4	13.1	15.8	28.9	49.4	21.7	[28]
Auger reactor	500	120 l h^{-1}	2	ZSM-5	5	*	*	50.3	26.2	13.6	[29]
Fixed bed	500	30 cm^3 min^{-1}	4.5	Al-MCM-41	0.46	37.5	10.91	48.41	7.49	37.43	[30]
Fixed bed	500	50 cm^3 min^{-1}	0.03	CoO	0.46	22.37	34.10	56.47	21.29	22.24	[31]
Fixed bed	500	50 cm^3 min^{-1}	0.03	Co_3O_4	0.46	26.30	29.18	55.48	33.78	22.55	[31]
Fixed bed	500	50 cm^3 min^{-1}	0.03	NiO	0.46	24.81	22.65	47.46	27.73	24.84	[31]
Fixed bed	550	150 ml min^{-1}	*	ZnO	0.05	*	*	47.02	21.38	31.6	[17]
Fixed bed	490–540	50 cm^3 min^{-1}	*	ZSM-5	0.01	*	*	38.29	19.45	42.27	[32]
Fixed bed	490–540	50 cm^3 min^{-1}	*	Al-MCM-41	0.01	*	*	39.98	18.80	43.15	[32]
Fixed bed	490–540	50 cm^3 min^{-1}	*	Al-MSU-F	0.01	*	*	39.59	19.18	43.31	[32]
Fluidized bed	425-450	*	3	HZSM-5	*	20.9	11.9	32.8	46.8	20.3	[33]
Fluidized bed	400	27 l min^{-1}	*	ZnO	*	*	*	57	20	12	[34]
Fixed bed	450	50 ml min^{-1}	*	α-Al_2O_3	0.5	19.1	42.4	61.5	10.5	17.6	[20]
Fixed bed	500	100 cm^3/min^{-1}	*	ZnO	0.15	*	*	45.22	30.46	24.32	[35]
Fixed bed	550	100 cm^3/min^{-1}	*	Al_2O_3	0.1	*	*	40.95	37.69	21.36	[35]
Fixed bed	500	70 ml min^{-1}	4	Na_2CO_3/γ-Al_2O_3	0.5	28	9	37	23	19	[36]

* Data is not available

Table 1. Pyrolysis conditions and pyrolysis products of several research studies.

2.1. Pyrolysis operating parameters

2.1.1. Temperature

The effect of reaction temperature on amount of each pyrolysis product is regarded as one of the most important and significant parameters. Studies have shown that increasing pyrolysis temperature causes a reduction in char yields, suggesting that secondary reactions of the liquid fraction and further char decomposition reactions are promoted with increase in temperature, resulting in enhanced gas yields at temperatures over 600°C [17, 27, 37, 38]. Previous studies have confirmed that the maximum liquid yield is obtained at temperatures in the range of 500–550°C [17, 37–39]. In a study of Pütün [37], bio-oil yields of 41, 46, and 43% were obtained at temperatures 400, 550, and 700°C, respectively. At temperatures between 350 and 400°C, char yields were close to maximum within the range of 29–38.48% (wt.) for most cases, suggesting that biomass could not be decomposed completely [17, 37, 39]. Similarly, solid char formation is favored at low temperatures (<350°C) mainly due to reason of cross-linking reaction of cellulose and lignin. However, volatiles are released at temperatures higher than 350°C as a result of depolymerization reactions [40].

2.1.2. Particle size

Biomass, being poor conductor of heat, often possesses heat transfer difficulties during pyrolysis. The size of particle has a direct effect on heat transfer, which in turn influences the yield and properties of bio-oil. Understanding the effect of particle size distribution on pyrolysis product yields also help to optimize residence times of the pyrolysis reactions. It is a known fact that the particle size of the feed influences the heat and mass transfer rate and release rate of volatile matter during pyrolysis process [35]. Studies have revealed that for larger particles of biomass, the yield of gaseous fraction was higher than that obtained from smaller particles. This situation can be explained by the fact that larger biomass particles increase the residence time of volatile matter, favoring secondary cracking reactions of tar, thus increasing the gas yield. Enhanced bio-oil yield is obtained using smaller particle size due to the reason that pyrolysis process is kinetically controlled with smaller particle size, whereas it mainly happens on the surface of the biomass in the case of larger particle size. Due to low heat transfer rate, the temperature inside the larger particle is lower than expected, thus causing pyrolysis process to be incomplete; resulting higher char yields and lower bio-oil yields [41, 42]. Also, Asadullah et al. suggested that mass transfer resistance inside the biomass particle was higher in larger particles than smaller ones; resulting increased char yield [40]. Unexpected differences between the product fractions might be due to differences in type of biomasses (for example, oxygen content of the biomass has a known effect on heat transfer mechanism) [40].

2.1.3. Heating rate

As affecting heat transfer, adjusting heating rate to optimum level is of high importance regarding product yields of pyrolysis process. Increasing heating rate has an effect of enhancing liquid yields and decreasing char yields, which suggests that optimum heating rate prevents secondary reactions resulting maximum liquid yields [39, 40]. Higher heating rate favors rapid formation of volatiles, while lower heating rate causes longer residence time for volatiles, thus enabling the repolymerization reactions forming char [40]. A study on olive residue and sugar cane bagasse pyrolysis has shown that, between the heating rates of 2 and 50 $K.min^{-1}$, the pyrolysis rate is increased with heating rate. It was suggested that the overlapping differential thermogravimetry (DTG) peaks with high heating rates was a consequence of the fact that some of the constituents of the biomass samples decomposed simultaneously while at low heating rate, those peaks were clear. This can be explained by the fact that the temperature inside the biomass particle increases with an increase in heating rate, and rate of decomposition is higher than the rate of formation and rate of volatile release [4]. This is the reason why optimum heating rate is necessary for an effective pyrolysis process.

As stated above, higher heating rates minimize char formation. In case of lower heating rates, which run for several days, the main product is char. This process is called carbonization. Carbonization enables the conversion of condensable vapor into char and noncondensable gases [26].

2.1.4. *Residence time and sweeping gas flow rate*

Sweeping gas flow rate is another important parameter, which influences the residence time of volatiles and the yield of gaseous fraction of pyrolysis products. Numerous studies have shown that higher sweeping gas flow rate favors rapid removal of vapors from the reaction medium, thereby reducing secondary reactions such as thermal cracking, repolymerization, recondensation, and char formation. Hence, char yields are decreased, whereas the yields of gaseous products are increased [17, 37, 39]. Additionally, in a study of catalytic and conventional pyrolysis of several biomass types, Huang et al. [27] showed that increasing gas residence time affected the yield of gaseous products of biomass pyrolysis so that secondary cracking reactions are attributable for enhanced gas yield and decreased liquid yield. Moreover, optimum sweeping gas flow is required in order to obtain maximum pyrolysis product yield from the reaction system. A study of Pütün [37] revealed that nitrogen flow rate of 200 mL min^{-1} was adequate for 48.30% maximum yield of bio-oil for the current system. In fact, flow rate of nitrogen does not have a significant effect on liquid yield of pyrolysis systems as it much more depends on sufficient quenching of pyrolysis vapors and downstream cooling mechanism. The study of Uzun and Sarıoğlu [39] has shown that gas yields reached its maximum value of 30.08% with a nitrogen flow rate of 800 cm^3 min^{-1}.

Other than the sweeping gas flow rate, the type of sweeping gas plays an important role in terms of product composition and quality. Melligan et al. revealed that bio-oil obtained by using H_2 as sweeping gas had higher heating value of 24.4 MJ kg^{-1} relatively higher than that of obtained under N_2 atmosphere with a higher heating value of 17.8 MJ kg^{-1} [43]. Pütün et al. [18] compared the outcome of biomass pyrolysis using N_2 and steam atmospheres as sweeping gases. The effect of nitrogen as a carrier gas was consistent with other studies in literature [39], while steam lowered the yield of char by diffusing into biomass particles, facilitating desorption, and removing the volatiles.

2.1.5. Biomass composition

Biomass is composed of lignin, hemicellulose, cellulose, extractives, and inorganic elements. The composition and amount of pyrolysis products vary depending on the content of biomass constituents as well as the distribution and percentages of these constituents, which vary with biomass species [44].

Cellulose, being linear-structured polymer, consists of b-1,4 linked glucose units, whereas hemicellulose is a branched-structured polymer composed of sugars such as pentoses and hexoses. Being formed by cross-linked phenylpropane units, lignin is a more complex aromatic compound and more resistant to thermal decomposition than cellulose and hemicellulose [3, 45, 46]. Cellulose is decomposed via two types of reactions: depolymerization and ring scission, which will result in the formation of different compounds. The ring scission reaction mainly results in the formation of hydroxyacetaldehyde, acetol, linear carbonyls, linear alcohols, and esters. However, anhydro oligosaccharides, monomeric anhydrosugars, furans, cyclopentanones, and pyrans are obtained with the depolymerization reactions [43]. Melligan et al. [43] revealed that during conventional pyrolysis of biomass, cellulose decomposition follows depolymerization reaction pathway. Lignin conversion is mainly due to depolymerization reactions and fragmentation reactions, which lead to formation of aromatic compounds of the main precursors of hazardous materials present in tar [46].

During the pyrolysis of biomass, the decomposition of biomass constituents takes place after moisture evaporation. It is a known fact that hemicellulose is the first component to decompose within the temperature range of 160–240°C, and cellulose is generally degraded at temperatures between 240 and 372°C. The decomposition of lignin, however, occurs within wider range of temperature (160–625°C) and at a low rate due to its resistant nature compared to hemicellulose and cellulose [4, 47].

Biomass constituents possess different characteristics and reactivity due to their heterogeneous nature, the effect of main constituents to pyrolysis products is an important parameter to take into consideration. In a study of Haykiri-Acma and Yaman [45], the behavior of sunflower shell and olive refuse was investigated during pyrolysis. It was revealed that sunflower shell with high content of volatile matter (76.0%) favored the formation of gaseous products, whereas high lignin and high ash content in olive refuse samples (34.7 and 13.8%, respectively) provided enhanced yield of solid char. Huang et al. studied various types of biomasses with addition of catalyst at temperature of 600°C and compared them with main biomass components. Cellulose, as one of the main biomass components that gave the highest yield of gaseous products (60.7%), was followed by sugar cane bagasse with yield of 57.0%. It was also found that samples with high lignin content gave more char yield, which can be explained by the presence of stable aromatic compounds present in the lignin matrix. For the major components of biomass, cellulose has the highest olefin yield, whereas lignin has the lowest, which is attributable to the fact that levoglucosan is the main product of cellulose pyrolysis [27, 45]. Collard et al. [46] compared the pyrolysis behavior of beechwood with main biomass constituents. It was noted that cellulose produced highest amount of volatiles followed by the gas and char fractions. Also, beechwood gave high yield of tar (36.5%) due to its high content of cellulose. However, in a study of Bertero et al. [48], bio-oil obtained from pine sawdust, which had lower cellulose content than that of mesquite sawdust had lower yield.

2.1.6. Catalyst effect

The presence of catalyst in pyrolysis process has also been investigated with different biomass sources [34]. Using catalysts in biomass, pyrolysis influences the decomposition behavior of biomass and composition and quantity of pyrolysis products. ZSM-5 zeolites were found to be effective in improving deoxygenated aromatic content of pyrolysis products while reducing liquid yield [19, 31, 49]. Using silica-alumina with or without deposition of alkali metals favors water formation, increases gas, and char yield, consequently reducing total organic yield comparing to conventional pyrolysis [20].

Catalysts also affect the distribution and the content of chemical compounds of pyrolysis products. Zhou et al. investigated the effect of catalyst to biomass ratio by using zinc oxide as catalyst. With increasing the catalyst amount, approximately 6% (wt.) decrease in bio-oil yield was observed, and the gas yield increased in the range of 21.38–28.74% (wt.) [17]. In a study by Huang et al., HZSM-5 zeolite impregnated with 6% wt. Lanthanum showed that increasing catalyst to biomass ratio improved the chemical composition of pyrolysis liquid in terms of olefin content [27].

2.2. Pyrolysis mechanism

Understanding the pyrolytic behavior of cellulose is crucial due to the fact that it accounts for approximately 50% of biomass [44, 50]. Levoglucosan is the main component of cellulose pyrolysis product [44, 51, 52]. Bai et al. studied the effect of levoglucosan in cellulose pyrolysis. It was suggested that during pyrolysis, cellulose conversion into levoglucosan occurred faster than its vaporization rate [51]. It was suggested that vaporized levoglucosan leaves reaction environment, whereas its polymerized form leads to yield low-molecular-weight volatiles including 5-hydroxymethyl furfural, furan, furfural, H_2O, CO_2, and acetic acid [51].

Stefanidis et al. investigated the pyrolysis of lignocellulosic biomass to see the effect of pyrolysis of each constituent on the overall yield [44]. Similarly, the main product of cellulose pyrolysis was found to be levoglucosan. Also, small amount of phenolic compounds, which were formed as a result of secondary reactions were detected. In this study, formation of acetic acid was considered to be due to dehydration of hydroxyacetaldehyde, resulting formation of ketene, which produces acetic acid via hydration [44].

In order to understand the mechanism of thermal decomposition of hemicellulose and to reveal reaction pathways for the formation of main pyrolysis products, xylan is considered to be a model compound as a replacement of hemicellulose. In a study of Stefanidis et al., the main products of xylan pyrolysis were reported to be phenols and cyclic ketones, which were derived from the cleavage of the ferulic acid ester branch of xylan and cleavage of o-glucosidic bonds followed by removal of hydroxyl groups of xylose rings, respectively [44].

Lignin decomposition mainly leads to the formation of phenolic compounds, which are of more complicated structure than those derived from cellulose and xylan, due to their methoxy and poly-substituted structure [44, 52]. Xin et al. proposed that the primary reaction of lignin pyrolysis was depolymerization and dealkylation due to increased formation of guaiacol at 350°C. As temperature increases to 450 or 550°C, the decomposition of guaiacol leads to formation of phenols and CH_4. Dehydroxylation of phenolic compounds followed by polymerization of ring-containing monomers leads to the formation of Polycyclic Aromatic Hydrocarbons (PAHs) [52].

2.3. Pyrolysis products

2.3.1. Gas

The gaseous fraction of biomass pyrolysis products contains mostly hydrogen, carbon monoxide, carbon dioxide, methane, ethane, and ethylene [5, 27, 29, 33, 53]. Pyrolysis gas can be used to provide heat for pyrolysis reactor or for heat and electricity generation in a gas turbine–combined cycle system [38]. Depending on the biomass feedstock and reactor configuration, primary gases of biomass pyrolysis contain 86.7 wt.% of CO_2, 1.2 wt.%, and 6.5 wt.% H_2 [48]. Dependent on the process temperature, the high heating value of gaseous products varies between 6.28 and 14.77 MJ m^{-3} [38].

2.3.2. Char

As solid product of biomass pyrolysis, char, which is comprised of the condensed organic residues and the inorganic phases, with an average high heating value of 28.5–29 MJ kg^{-1} (depending on the biomass feedstock) can be used as a solid fuel. It is also a well-known material for activated carbon production [38, 45]. Char with >70 wt.% fixed carbon content can be used as a raw material for production.

2.3.3. Pyrolysis oil (bio-oil)

Bio-oil is a dark brown colored liquid mixture, which is composed of an aqueous phase and an organic phase having a large number of different chemical compounds derived from depolymerization and fragmentation reactions of biomass main components: cellulose, hemicellulose, and lignin [17, 20, 30, 32, 48, 54]. Aqueous phase of bio-oil contains a wide variety of oxygenated organic compounds such as acetic acid, methanol, and acetone. Organic phase of bio-oil contains single-ring aromatic hydrocarbons such as benzene, toluene, indene, alkylated derivatives, and polycyclic aromatic hydrocarbons including naphthalene, fluorene, phenanthrene, and oxygenated organic compounds such as aliphatic alcohols, carbonyls, acids, phenols, cresols, benzenediols, and guaiacol [17, 30, 32, 48]. Due to its relatively higher organic compound content, organic phase of bio-oil has higher carbon concentration and higher calorific value than aqueous phase [21, 48].

The chemical composition of bio-oil varies depending on biomass feedstock characteristics, mainly the percentages of main biomass constituents. The decomposition of cellulose leads to formation of levoglucosan and furfural which are considered valuable organic compounds in terms of fuel quality of bio-oil [32]. Phenolic compounds are the products of lignin depolymerization and cracking reactions. Among the phenolic compounds present in bio-oil, guaiacol and its alkylated compounds are the most important ones due to being thermally unstable and being able to transform through secondary reactions to alkylated phenols and aromatic compounds, which are desirable for bio-oil quality [54]. Acetic acid, generally the main compound in the group of acids in bio-oils, is formed by the deacetylation of hemicellulose [54].

2.3.3.1. Bio-oil characteristics

Due to heterogeneous structure of biomass, bio-oil contains different types of acidic compounds as well as high amount of water and highly oxygenated hydrocarbons, which lead to poor combustion properties, instability, lower calorific value, and higher viscosity compared to fossil fuels [24, 55].

Water in bio-oil is primarily formed as a result of dehydration reactions and depolymerization reactions of hemicellulose, cellulose, and lignin [48]. The presence of water in bio-oil is a drawback in terms of utilization of bio-oil, because it lowers the heating value and delays the ignition [17]. However, high water content also leads to low viscosity so that the fluidity of bio-oil is enhanced, which is a good indicator of bio-oil quality for utilization in combustion engines [32, 48]. The water yield in bio-oil varies depending upon the biomass composition and process conditions, such as 15.0 wt.% for hybrid poplar [33], 15.32 wt.% for rice husk [17], 44.3 wt.% for chañar fruit [54], 16.0 wt.% for rice straw, and 18.0 wt.% for sawdust [56].

Traditionally obtained bio-oil shows high viscosity due to presence of large molecules. Levoglucosan, as a heavily oxygenated compound produced from cellulose pyrolysis, significantly affects the viscosity of bio-oil by causing crystallization in time at room temperature [20, 33, 43]. Also, high level of heavy phenols, which are derived from lignin depolymerization reactions, increases the viscosity of bio-oil [43]. Due to the fact that chemical reactions at higher temperatures between reactive components such as ketones and aldehydes resulting heavier compounds increase viscosity and cause instability [31, 43], the storage of bio-oil becomes a major issue [20, 57]. In a study of Duman et al., it was reported that addition of methanol improved the stability of bio-oil such that the viscosity of bio-oil increased at 29.82% in 168 h period instead of 46.63% [57]. Bio-oils with lower content of carbonyl compounds are considered thermally more stable [20].

High oxygen content (generally 35–40 wt.%), which means low H/O ratio, lowers the energy density of bio-oil. High oxygen content is also the reason of the immiscibility with petroleum products [20]. Oxygenated aromatic hydrocarbons decrease the heating value of bio-oil and stability but increase the viscosity due to their high molecular weight [17, 31, 54].

Bio-oil is considered highly acidic compared to conventional fuels and need to be upgraded before utilized commercially [17, 31]. Chemical composition, especially the amount of acidic compounds, determines the acidity of bio-oil. Acid content in bio-oil leads to corrosive characteristics towards metals such as copper and iron, which makes transportation, utilization, and storage of bio-oil a major issue [43]. Biomass pyrolysis oils typically contain 3–6 wt.%

volatile acids, with the main compounds being acetic and formic acids [58]. Carboxylic acids, mainly acetic acid, are formed from the cleavage of acetyl groups in the hemicellulose components of biomass [20, 58] and from ring scission of cellulose [43]. For noncatalytic bio-oil, acidity accounts for its pH in the range of 2.0–3.0 [32, 33, 54].

2.3.3.2. *Utilization of bio-oil*

Bio-oils can be used in many applications. It can be used as a substitute in chemical industry as food flavorings, fertilizers, emission control agents [39], and as an energy carrier in stationary applications for heat and electricity generation in boilers, furnaces, engines, and turbines [16, 39].

Furans are valuable chemical compounds present in bio-oil because they can be used for organic solvent production and for substituting fossil fuels [43]. Aromatic hydrocarbons in bio-oils with lower molecular weight are more favorable than the ones with higher molecular weight, in terms of utilization as fuel additives because they have lower boiling points [16]. Oxygenated compounds decrease the stability and energy density [54] and thus the quality of bio-oil [43]. However, low levels of oxygenated molecules such as (alkyl)-furans compounds are favorable due to the reason that high octane number of these compounds enhances energy density of bio-oil, thereby facilitating the utilization of bio-oil as fuel additive [20]. Phenolic compounds in bio-oil (particularly phenolic ethers such as vanillin, guaiacol, and syringol) are the precursors for the synthesis of pharmaceutical and polymeric compounds or adhesives [54].

Using the bio-oil as a fuel without upgrading results in several significant problems such as poor volatility, high viscosity, high water content, low polymerization temperature (<100°C), and corrosiveness for engine equipment [59]. Emulsification and blending are among the most preferred methods to upgrade the bio-oil when substituting conventional diesel [59–61]. Van de Beld et al. investigated the performance of bio-oil derived from pine wood in a modified diesel engine connected to a generator to convert mechanical power to electricity. It was noted that, at air inlet temperatures in the range of 100–120°C and at an engine compression ratio of 17.6, bio-oil/ethanol blends were found to reduce CO emissions and increase NO_x emissions. Adding up to 30 wt.% of ethanol to the bio-oil improved combustion performance due to better atomization of the fuel, resulting in lower CO emissions [59]. Similar results were obtained by Yang et al. [62]. In a study of wood pyrolysis oil, it was revealed that using pure bio-oil in diesel engines leads to the widening of spray channels, thus damaging the injector. The results of this study showed that at 200°C tip temperature, the needle of nozzle was stuck in a short period of operation time [63]. Sugarcane bio-oil was blended with gasoline in the range of 5–14 vol.%. The results showed that blending up to 10 vol.% of bio-oil with gasoline did not affect the operation of Otto engine of 4 kW capacity connected to a 2-kWe generator suggesting that power and fuel consumption were similar with that of gasoline-operated engine [64]. However, specific modifications of diesel engines are necessary for using the bio-oil or bio-oil–derived fuels in diesel engine systems. Dedicated fuel feeding system parallel to diesel feeding line, a pilot diesel fuel injection, use of cleaning fuels such as methanol and corrosive resistant or stainless steel fuel pump, and injector design are among the modifications adapted by researchers [60, 62, 63]. Furthermore, in order to reduce viscosity of the fuel, short preheating at temperatures <90°C is needed, and direct heating is not recommended [65].

Boucher et al. investigated softwood bark pyrolysis oil with the addition of methanol as a possible liquid fuel for gas turbines. The results of this study showed that the viscosity of bio-oil was close to the gas turbine requirements with the value of 5.3 cSt at 90°C, and the net heating value of 32 MJ kg^{-1} was recorded as relatively high. However, due to the fact that the ash and solid content present in the bio-oil will eventually cause degradation in gas turbines, filtering and upgrading of bio-oil are required [66]. Lopez Juste and Salva Monfort [67] conducted a preliminary study on combustion performance of a gas turbine operated with bio-oil/ethanol (80/20, wt.%) mixture. At high air flow rate, considerable increase in CO emission was observed suggesting that using bio-oil/ethanol mixture leads to inefficient operation of the combustor. In order to utilize bio-oil for gas turbine applications, fuel preheating at temperatures 70–90°C is necessary to keep viscosity lower than that of 10 cSt. Resistant materials to acidity of bio-oil should be selected in order to prevent wear corrosion damage [65]. Modifications must also include nozzle adaptations for bio-oil–specific properties because using standard nozzles does not allow for full load (thus causing power decrease). It is of high importance that start-ups and shut-downs must be done using standard fossil fuels in order to warm up combustion chamber facilitating bio-oil ignition [65].

3. Utilization of catalysts in biomass pyrolysis

Depending on the biomass type used, the low calorific value, high water content, high viscosity (due to large molecules), high oxygen content (due to oxygenated hydrocarbon content) as well as instability, immiscibility with other fossil fuels make bio-oil difficult to use directly as a fuel without upgrading [24, 55]. In order to upgrade bio-oil to use in fuel applications, two different methods have been utilized: hydrodeoxygenation and catalytic cracking. In hydrodeoxygenation, bio-oil compounds react with hydrogen under high pressure and moderate temperature to produce hydrocarbon compounds and water. Catalytic cracking is used to upgrade bio-oil through a catalytic medium, removing oxygen from bio-oil compounds in the form of H_2O and CO_2, involving the chemical reactions of rupturing the C–C bonds via dehydration, decarboxylation, and decarbonylation [1, 16, 24, 55]. Catalytic cracking has several advantages over hydrodeoxygenation including operating at atmospheric pressure and in an environment with no need of extra hydrogen supply [1].

3.1. System configurations for catalysts used in biomass pyrolysis

There are two methods for catalytic pyrolysis in use: catalytic bed and catalyst mixing. In catalytic bed method, which is also referred to as *"in-situ"* upgrading or ex-bed method, pyrolytic vapors coming from the first reactor pass through a catalytic reactor which is called catalytic bed, resulting bio-oil, char, and gaseous products. In catalyst mixing (in-bed) method, however, biomass and catalyst samples are mixed physically before being inserted in pyrolysis reactor [1, 30, 37].

3.1.1. Catalyst bed method (in situ)

Catalyst bed method, also called *in-situ* pyrolysis, involves catalytic upgrading subsequent to thermal conversion of biomass resulting pyrolysis vapors [1]. In general, *in-situ* catalytic pyrolysis of biomass can be performed using different reactor configurations: single stage and two stage reactor configurations. Single stage pyrolysis involves catalytic pyrolysis of biomass in the same reactor with catalyst, whereas two-stage configuration involves fixed bed/fluidized bed reactor followed by a fixed bed catalytic reactor. The former produces more coke than the latter [33]. A study by Mante and Agblevor [33] using two stage reactor configuration for catalytic pyrolysis of hybrid poplar wood with HZSM-5 reported low yield of coke with 3.8% of value which was relative to the weight of biomass. Advantage is that catalytic pyrolysis of evolved vapors from biomass can be operated at a different temperature than that of the main pyrolysis reactor in case of two stage reactor configuration is employed [18]. However, compared to catalyst mixing, the catalyst bed method also leads to high amount of char resulting clogging of catalyst pores, which prevents the diffusion of vapors through the pores [1].

In a study of Thangalazhy-Gopakumar et al., it was noted that some noncatalytic bio-oil compounds were detected when using catalyst bed method, suggesting that primary tar compounds were converted into secondary and tertiary tar compounds before reaching the catalyst bed to be cracking into aromatics [1]. Uzun and Sarıoğlu reported that using catalyst bed method with several types of catalysts decreased the liquid yield compared to catalyst mixing method [39]. Iliopoulou et al. studied catalytic pyrolysis of lignocellulosic biomass and explained the *in-situ* effect of metal modified ZSM-5 with various percentages. It was suggested that transition metals favored the formation of hydrogen, which leads to hydrocarbon reactions on the zeolite acid sites via catalyst bed mode [31].

3.1.2. Catalyst mixing method (in-bed)

Catalyst mixing method can be done by either addition of catalyst to biomass with defined amounts or by wet impregnation of biomass. Due to better physical surface contact between biomass and catalyst in pyrolysis reactor, mixing allows immediate interaction of evolved pyrolysis vapors with the catalyst suggesting that evolved vapors can be adsorbed on the catalyst surface to be diffused into the pores for catalytic cracking [1, 39]. Disadvantage is that the catalytic conditions are irreversible so that biomass and catalyst should be operated under the same conditions [18].

Aromatic content of bio-oil is of utmost importance, and utilizing catalysts is one alternative way to increase aromatization reactions (thus enhance bio-oil quality). Studies have shown that catalytic mixing method provides better mass transfer for cracking of bio-oil compounds in terms of aromatization and deoxygenation [1, 39].

Thangalazhy-Gopakumar et al. investigated the catalytic effect of ZSM-5 zeolite in pyrolysis of pine wood chips under helium environment. It was revealed that using catalyst mixing method with 1:9 biomass to catalyst ratio gave 41.5% of aromatic yield compared to that of 9.8% using catalyst bed method with 1:5 biomass to catalyst ratio. In this study, the absence of guaiacol compounds in bio-oil shows that catalyst mixing is an effective method for cracking of lignin-derived compounds to aromatics [1]. Pütün et al. studied pyrolysis of cotton seed with the addition of MgO with defined proportions to the biomass samples. Compared to conventional pyrolysis results, aromatic and aliphatic content was enhanced to the values of 35 and 23%, and the oxygen content was reduced from 9.56 to 4.90% [37]. The results of rice husk pyrolysis with ZnO studied by Zhou et al. showed that catalyst mixing with various amounts significantly improved bio-oil quality in terms of hydrogen content, H/C ratio, higher heating value, and reducing carboxylic acid content of bio-oil [17]. Thus, in order to design large-scale pyrolysis plants, effective and homogenous mixing systems are necessary.

3.2. Metal oxide catalysts

Metal oxide catalysts have been extensively studied in literature, using miscellaneous biomass species, with regard to their effect on the quality and quantity of pyrolysis products. It is a known fact that metal oxides, as any catalysts, influence the decomposition temperature. Accordingly, in a thermogravimetric study conducted by Chattopadhyay et al. using Cu/Al_2O_3 as a catalyst, it was noted that transition metal supported alumina had a strong effect on decreasing the devolatilization temperature of volatiles [15]. Zabeti et al. applied amorphous silica alumina (ASA) supported with alkali or alkaline earth metals in pyrolysis of pinewood. Maximum bio-oil was obtained with nonsupported amorphous silica alumina with the value of 42.4 wt.%. However, the fraction with Cs/ASA showed the best performance in terms of oxygen elimination in aromatic hydrocarbons thus increasing the heating value of bio-oil [20]. Wang et al. [68] investigated the catalytic effect on pyrolysis of lignocellulosic biomass using catalysts including $NiMo/Al_2O_3$, $CoMo/Al_2O_3$, $CoMo\text{-}S/Al_2O_3$, activated alumina, and porous silica. In order to enhance the production of pyrolysis intermediates (benzene, toluene, xylene, naphthalene), it was suggested that $CoMo\text{-}S/Al_2O_3$ was most favorable among all. However, $NiMo/Al_2O_3$ gave highest yield of CH_4 with the value of 51.82%. Shadangi and Mohanty [69] studied CaO and Al_2O_3 in pyrolysis of *Hyoscyamus niger L.* and confirmed that, consistent with other studies mentioned above, adding catalysts to pyrolysis process considerably decreased the bio-oil yield, however, eliminated the oxygenated groups present in bio-oil thus improving its fuel quality. Aysu and Küçük investigated the pyrolysis of eastern giant fennel (*Ferula orientalis L.*) comparing the effect of ZnO and Al_2O_3 catalysts [35]. Al_2O_3, with the value of 79.94%, was found to be more effective than ZnO in terms of conversion of biomass. However, the effect of the catalyst on the bio-oil yield was different from each other. ZnO increased the bio-oil yield with increasing catalyst to biomass ratio, whereas bio-oil yield decreased with increasing Al_2O_3 addition. This indicated that Al_2O_3 promoted gas formation. In a study of Yorgun and Şimşek, activated alumina was used in pyrolysis of *Miscanthus × giganteus*, and it was noted that at high heating rates, 60 wt.% of catalyst loading to biomass is effective for maximum liquid production with the value of 51 wt.%. The oxygen content of bio-oil was found to be higher than that of noncatalytic bio-oil [70]. Nguyen et al. investigated the pyrolysis vapors of pine wood chips over 20 wt.% $Na_2CO_3/\gamma\text{-}Al_2O_3$. It was reported that liquid yield was lowered but decarboxylation of carboxylic acids was favored with the catalyst, resulting a pH value of 6.5 suggesting that the sodium-based alumina catalyst is effective on improving acidity of bio-oil. Hydrocarbon concentration was increased from 0.5 to 17.5% indicating a higher energy density of bio-oil [36]. Chen et al. presented the gaseous product distribution of biomass (rice straw and saw dust) pyrolysis, at a temperature of 800°C, utilizing different metal oxides such as Cr_2O_3, MnO, FeO, Al_2O_3, CaO, and CuO. It was noted that except for CuO and Al_2O_3, all the catalysts noticeably improved gas production [56]. Zhou et al. [17] studied the pyrolysis of rice husk with the addition of ZnO. The results of this study indicated that ZnO showed a trend to decrease bio-oil yield with increasing biomass to catalyst ratio. However, ZnO improved the bio-oil compositional quality in terms of low-molecular-weight compounds including alkanes, alkenes, styrene, and alkyl phenols (thus increasing bio-oil stability). Nokkosmaki et al. also studied the same catalyst for the conversion of pyrolysis vapors of pine sawdust and the viscosity of catalytic bio-oil was shown to be reduced by 40% compared to noncatalytic pyrolysis results [34]. Pütün et al. [37] revealed that utilizing MgO as a catalyst in cotton seed pyrolysis in a fixed bed reactor improved the bio-oil quality by removing oxygenated compounds, enhancing higher heating value, and increasing aromatic content of bio-oil.

3.3. Zeolite catalysts

Zeolites, having a tetrahedral structure and acidic nature, are three-dimensional aluminosilicates linked through oxygen atoms and supported with channels and cavities, resulting porous structure of exceptional catalytic activity. Each type of these tetrahedral zeolites with overall charge balance of minus one have Si or Al at the center and oxygen atoms at the corners of the structure [71].

Zeolites possess following characteristics: (1) cracking of deoxygenated compounds via shape selectivity, (2) high surface area, (3) varied dimensions of channels and pores, and (4) high adsorption capacity [22, 72]. The physical properties of zeolites depend on the synthesis conditions including temperature, gel precursors, structure-directing agent [73]. The pore size and framework of zeolites tend to affect the product composition via several reactions restricting formation of hydrocarbons larger than that of pore size of zeolites. This is called shape selectivity and is one of the most important factors differentiating zeolites from other type of catalysts. Selectivity is based on whether the aromatics derived from pyrolysis vapors can enter, form in, and diffuse out of the pores of the zeolite [22, 23]. Shape selectivity of zeolites is elaborately discussed in Section 3.3.1.1 demonstrating ZSM-5 as a zeolite type.

The main reason why zeolites are commonly used in biomass pyrolysis is that their varied acidity and shape-selectivity provide advantage over silica-alumina catalysts of amorphous structure, in terms of aromatization reactions. Acidity depends on the Si/Al ratio of zeolite structure and can be caused by Brønsted and Lewis acid sites [22]. Acidity affects the catalytic reactions by providing enhanced cracking activity with decreasing Si/Al ratio [23]. Hence, the distribution of acid sites in the pores of zeolites is of high importance in terms of preventing coke-forming reactions in internal pores of zeolite. Low of Si/Al ratio causes higher amounts of acid sites with close proximity. Due to this reason, coke-forming reactions, which convert aromatic hydrocarbons into coke compounds, will increase. Therefore, an optimum acidity is required for zeolites if it is used in biomass pyrolysis [19].

Aside from the distribution of acid sites, the pore structure of the individual zeolite plays an important role for selectivity of products and in terms of aromatic content of bio-oil. The pores of zeolites are characterized by the size of the ring defining the pore, that is, the n-number of the ring, which is referred to as the number of Si or Al atoms in the ring [74]. Hydrocarbon chain length of pyrolysis products, thus the size distribution of aromatic compounds, depends on the zeolite pore size and internal pore surface area. In general, larger pores and surface area lead to long chain–structured hydrocarbons. Microporous surface area of the catalyst determines the yield of gaseous products of pyrolysis, whereas macroporous area determines the liquid yield [24, 37]. A range of zeolites with different pore sizes has been studied in literature [39, 75, 76]. Y zeolite (faujasite), having a cubic structure with a pore system composed of 12-membered circular ring channels, has the largest average pore size (7.4 Å) and internal pore space (11.24 Å) [22, 23]. Such relatively large pore size affects the catalyzed reactions resulting less contact between pore surface and pyrolysis vapors, thereby leading to less cracking of biomass-derived oxygenates. Also, straight channels with larger pore size do not provide shape selectivity compared to other zeolites, which have smaller and sinusoidal channels that enable shape selectivity [23]. Despite having large pore size [73], beta zeolite is of tetragonal crystal structure and has 12-membered straight channels crossed with 10-membered ring channels, which makes it more effective for production of aromatic hydrocarbons than Y-zeolites [23]. Ferrierite, as a medium pore size zeolite, has orthorhombic structure with 8 and 10 membered channels with internal pore space 6.31 Å [23, 73].

Due to having two parallel channels connected with 12-membered rings and 8-membered rings, Mordenite is classified as a large pore-sized zeolite and have orthorhombic structure [22, 23, 73]. ZSM-5, consisting of an MFI orthorhombic structure, is composed of 10-membered straight channels connected by 10-membered sinusoidal channels [71]. General physicochemical properties of zeolites are presented in **Table 2**.

Zeolites with medium and large pore size facilitate faster reactant diffusion compared to zeolites with smaller pore size, thus yielding more aromatics in liquid fraction of pyrolysis product. However, large-pore size zeolites produce less aromatic than medium-pore size zeolites because large pores promote coke formation [22]. Accordingly, a recent study confirmed that ZSM-5 of medium pore size and medium internal pore surface area favors higher aromatics production and lower coke yield [19].

Catalyst	**ZSM-5**	**Mordenite**	**Beta zeolite**	**Y zeolite**	**Ferrierite**
IZA code	MFI	MOR	BEA	FAU	FER
Pore dimension	3	2	3	3	2
Channel system	10–10	12–8	12–12	12–12	8–10
Pore size (Å)	5.1 × 5.5	7.0 × 6.5	7.6 × 6.4	7.4 × 7.4	4.2 × 5.4
	5.3 × 5.6	5.7 × 2.6	5.6 × 5.6		3.5 × 4.8
Internal pore space (Å)	5.2–5.5	4.2–6.7	6.1–6.68	11.24	6.31
BET surface area (m^2/g)	395.5	558.7	643.1	809.1	*

*Information not available.

Table 2. Physicochemical properties of zeolites most commonly used [22, 72, 76, 77].

3.3.1. ZSM-5 zeolite

ZSM-5, as one of the most common used zeolites in biomass pyrolysis [23], consists of pentasil units and has orthorhombic structure [74]. Being composed of 10-membered straight channels connected by 10-membered sinusoidal channels enables ZSM-5 to have significantly more cracking activity than the other zeolites [23, 27]. ZSM-5 has been widely used as catalyst in petroleum industry due to its shape selectivity, exceptional pore size with steric hindrance, thermal stability, and solid acidity [78].

Compared to other zeolites, medium-scaled pore size of ZSM-5 makes it difficult for larger aromatic coke precursors to form inside the pores [79]. Studies have shown that, regardless of feedstock, utilizing ZSM-5 in biomass pyrolysis reduces the oxygenated compound content in bio-oil and simultaneously increases aromatic species [80–84]. Deoxygenation of oxygenated organic compounds occurs inside the ZSM-5 zeolite pores via reactions such as dehydration, decarboxylation, and decarbonylation [27, 31, 32]. At lower temperatures, oxygen is found to be removed in the form of H_2O, whereas in case of higher temperatures, CO and CO_2 are the main products of oxygen removal [31]. Oxygen removal, primarily in the form of CO and CO_2, is more preferable as it leads to less carbon deposition on the zeolite and more hydrogen formation and consequently less water content in bio-oil [31].

3.3.1.1. Shape selectivity of ZSM-5

The phenomenon of shape selectivity can be explained by the combined effect of molecular sieve and catalytic reaction that occurs at the external and internal acidic sites of zeolites [85]. ZSM-5, having porous structure, can be used for shape-selective catalysis providing that not only the pore size but also dimensions of reacting and diffusing molecules are similar to zeolite pores [19]. Thus, the effect of pore size and steric hindrance of ZSM-5 on the catalytic reactions must be investigated if zeolites with better performance are to be designed for biomass conversion [73].

The formation of pyrolysis products with shape selective catalysis depends on two types of selectivity: (1) reactant and product selectivity and (2) transition state selectivity, which are described with their mass transfer effects and intrinsic chemical effects, respectively [22, 73, 85]. The main idea behind reactant and product selectivity lies behind the fact of hindered diffusion of reactants and products inside zeolite pores. Specific pore size of ZSM-5 affects the diffusion of reactants inside the pores excluding the ones with dimensions larger than the pore openings of ZSM-5, thereby preventing them from reaching the catalytic active sites and consequently allowing catalytic decomposition only at the external active sites [73, 85]. Due to pore geometry of ZSM-5, formation of certain products is restricted affecting the chemical reaction and thus causing selectively homogenization of pyrolysis products [85].

Selectivity of ZSM-5 has been extensively studied, and it is more commonly found to cause selectivity over aromatic compounds [49]. Mihalcik et al. [23] studied different zeolites for the conversion of several types of biomass and biomass components. According the results of this study, ZSM-5 was found to promote the formation of *p*-xylene in abundance for every case of biomass pyrolysis. In a study of Foster et al. [19], HZSM-5 for furan conversion showed a tendency for aromatic selectivity giving higher yield of aromatics as naphthalene having the highest percentage of 30.4% of overall aromatic species. Fogassy et al. [86], investigating the shape selectivity of zeolites for lignin fragments, revealed that majority of the phenolic compounds derived from lignin decomposition are too large to enter through zeolite pores, therefore, the conversion of these compounds occurs at the external active sites. As Yu et al. [22] suggested, however, at higher temperatures effective pore size of ZSM-5 increases, enabling bigger molecules than that of ZSM-5 pore size to reach the internal catalytic active sites. Jae et al. investigated the role of pore size of several types of zeolites in glucose pyrolysis (using kinetic diameters for products and reactants as affecting parameters) to determine whether the catalytic reaction occurs inside the pores or at the external surface [73]. Kinetic diameter was estimated from the properties at the critical point. It was revealed that ZSM-5 allowed pyrolysis intermediates and products (such as benzene, toluene, indene, ethylbenzene, *p*-xylenes) to diffuse into pores due to their significantly small kinetic diameters than that of ZSM-5 pore size.

As temperature increases to 600°C, due to thermal distortion, compounds like naphthalene which gave the highest yield in aromatics are likely to be formed inside the pores as well as on the surface. According to this study, it was concluded that, in addition to pore size, internal pore space of ZSM-5 affects the catalytic reaction. This suggests that biomass conversion with ZSM-5 is affected from mass transfer limitations as well as transition state effects [73].

3.3.1.2. *Acidity of ZSM-5*

Aside from the shape selectivity, the acidity of ZSM-5 plays an important role for the conversion of oxygenates into aromatics. It is generally accepted that Brønsted acid sites are the active sites that converts oxygenated compounds into aromatics rather than Lewis acid sites. Cracking of large oxygenated occurs at the acid sites of external surface of ZSM-5, whereas conversion of smaller ones into aromatics takes place at the acid sites inside the pores [22, 87]. Therefore, the abundance of both external and internal acid sites must be investigated in order to design a better process for aromatics formation. As explained by Van Santen, Brønsted acid sites are generated as a result of replacement of silica, which has a valency of four, by a metal atom, most commonly aluminum with a valency of three [88]. Thus, this makes Brønsted acid sites proton donors. Si and Al are connected with a proton-attached oxygen atom, which leads to a chemically more stable structure [88, 89]. On the other hand, Lewis acid sites are electron pair acceptors, and the nature of these sites is related to aluminum atoms positioned in the framework [89, 90].

In addition to nature of the acid sites, the molar ratio between SiO_2 and Al_2O_3 in zeolite framework also influences the reactivity and performance of ZSM-5. Optimum Si/Al ratio is necessary to provide high availability of Brønsted acid sites for adequate acidity and to maintain the distance between acid sites in order to limit coke-forming reactions [19]. As Si/Al ratio decreases (increase the acidity of ZSM-5), the acid sites will be in close proximity to each other, resulting in secondary reactions for conversion of aromatic compounds to coke species [19]. Foster et al. [19] investigated the effect of ZSM-5 with varying Si/Al ratio on pyrolysis of glucose. This study showed that decreasing Si/Al ratio contributed formation of the additional acid sites for ZSM-5 with increasing coke yields. In a study of Carlson et al. [79], ZSM-5 (SiO_2/Al_2O_3 = 15) promoted coke formation mostly on the catalyst surface giving the highest yield of 33% for coke (where the catalyst to biomass ratio was 19). It was also indicated that coke formation on the external pores of ZSM-5 did not significantly decrease the yield of aromatics but affected the selectivity toward light hydrocarbons, resulting lower yields for benzene and toluene, higher yields for naphthalene and indane.

3.4. General effects of catalysts on bio-oil

It is a known fact that catalysts have strong influence on pyrolysis in terms of product distribution, chemically homogenization, and enhancing fractional product yield, upgrading the pyrolysis products to better quality. Among catalysts, zeolites and metal oxides have been mainly investigated for biomass conversion and found to be effective in changing the composition of bio-oil by reducing the oxygenated compounds via deoxygenation reactions and enhancing the aromatic yield, thus producing a more homogeneous and stable organic fraction that can be upgraded to diesel grade fuels [30, 39]. To consider pyrolysis products, particularly bio-oil, for stationary fuel applications or heat/electricity generation, properties including acidity, viscosity, stability, and aromatic content of bio-oil should be evaluated [20]. Therefore, the effects of catalysts to bio-oil must be addressed, as elaborately discussed below, in order for better understanding of biomass conversion.

3.4.1. Aromatic yield of bio-oil

Aromatic content of bio-oil is of high importance in terms of producing diesel grade fuels from biomass feedstock and biomass-originated waste materials. Among aromatics, the amount of benzene, toluene, ethyl-benzene, and xylenes (BTEX components) are the most significant feedstock materials to take into consideration for petroleum chemical industry [68, 87].

Utilizing catalysts has been shown to increase the yield of bio-oil as well as the content of aromatics in bio-oil, which is a good indicator fuel quality. Kim et al. [87] studied catalytic pyrolysis of mandarin residues with high lignin content and found that using HZSM-5 of 23 and 80 acidity increased yield of monoaromatics from 3.4 to 36.0 and 41.0%, respectively. From the study of Zheng et al. [76], changing crystal size of ZSM-5 affected the aromatic yield and BTX selectivity so that smaller crystal size gave maximum aromatic yield and lowest BTX selectivity with the values of 38.4 and 36.3%, respectively. However, larger crystal size exhibited lowest aromatic yield and highest BTX selectivity with the values of 31.1 and 42.6%, respectively. Thus, smaller crystal size (200 nm) was found to be optimum for high aromatic yield.

Zhang et al. [83] compared the behavior of pyrolysis of aspen lignin under the effect of H-Y and HZSM-5 catalyst. At catalyst to feed ratio of 3:1, production of aromatics exhibited a maximum value of 23% using HZSM-5 as catalyst, where the oxygen content of the aromatics decreased to about 4% and the HHV of the fraction was estimated to be approximately 46 MJ/kg, which is closer to that of gasoline and diesel. It was indicated that HZSM-5 was more effective than H-Y in converting phenolic compounds to aromatic hydrocarbons [83] due to its higher acidity and smaller pore size compared to HY [83]. Similarly, Pattiya et al. [75] studied ZSM-5 and two mesoporous materials including Al-MCM-41 and Al-MSU-F to investigate the fast pyrolysis of cassava rhizome. It was revealed that, of all the catalysts tested in the study, ZSM-5 gave the highest yield of aromatic hydrocarbons in order of toluene > benzene > 4,7-dimethylindane > *p*-ethyl-stryrene > 5-methylindane > xylenes.

Aside from zeolites, the effect of metal oxides toward aromatization has been studied by researchers [7, 49, 91, 92]. From the results of the study of Ateş and Işıkdağ [91], using alumina as catalyst on corncob pyrolysis exhibited a trend to promote the formation of 1,1,3,3-tetramethylindane, benzene, and 1-methyl-4-(penylmethyl) being most significant monoaromatic compounds. The formation of naphthalene, 1-(2-propenyl)-, a PAH compound, was found to be increased at moderate temperature using catalyst. Smets et al. [93] compared various catalysts including HZSM-5, γ-Al_2O_3, and Na_2CO_3. Sodium carbonate was the most effective catalyst to increase the yield of aromatics following HZSM-5. Wang et al. [92] also conducted a comparative study for catalytic conversion of herb residues over alumina, ZSM-5 and Al-SBA-15, where alumina was found to give the highest bio-oil yield. Thus, the researchers of this study investigated the effect of alumina to the aromatic yield in terms of toluene, ethylbenzene, and *p*-xylene compounds and revealed that the percentage of aromatic fractions increased from 8.02 to 10.93%.

3.4.2. *Acidity of bio-oil*

The acidity of bio-oil is due to volatile acids, mainly carboxylic acids, that is, formic acid and acetic acid [58]. Phenolic compounds also contribute the acidity of bio-oil [20]. Determination of acidity in bio-oil is performed either by measuring pH value or total acid number. pH value is an indicator for evaluating the corrosiveness of bio-oil, whereas total acid number is used as a quality indicator for bio-oil utilization in co-processing of petroleum-refining facilities, and it relates to the level of acidic components in the oil [58].

Studies have clearly shown that organic acids are reduced by catalysts [36], thereby, facilitating the utilization of bio-oil in fuel applications. The main question is to find the most suitable catalyst-biomass match, taking the process conditions into account for pyrolysis systems, in order to replace bio-oil with fossil fuel equivalents (such as diesel and gasoline).

Zabeti et al. [20] studied amorphous silica alumina modified with alkali or alkali earth metals such as Na, K, Cs, Mg, and Ca. It was concluded that among all the catalysts tested in the study, K/ASA was the most effective catalyst to reduce the carboxylic acids and carbonyl substituted-phenols content in bio-oil. Zhou et al. [17] investigated the effect of ZnO to physicochemical properties of rice husk bio-oil. pH value of catalytic bio-oil was recorded as 4.35, whereas noncatalytic bio-oil had 4.15 of pH value. Thus, this indicated the effect of ZnO catalyst toward reduction of acidic compounds in bio-oil. Abu Bakar and Titiloye [32] studied rice husk pyrolysis over various catalysts including ZSM-5, Al-MCM-41, Al-MSU-F, and (Brunei rice husk ash) BRHA. Catalysts have been shown to reduce the acid number from 55 mg/KOH to 39–47 mg/KOH, with ZSM-5and BRHA having the lowest value. Also, pH value of the catalytic rice husk bio-oil was recorded in the range of 2.7–3.0. Majority of acidic compounds were carboxylic acids, as acetic acid having the highest percentage. Mante and Agblevor [33] studied HZSM-5 as deoxygenating catalyst to convert hybrid poplar wood to biosyncrude oil. As indicated in the study, the pH value of light biosyncrude oil containing mainly aromatic hydrocarbons was increased from 2.60 to 4.05 due to HZSM-5.

3.4.3. *Viscosity of bio-oil*

High viscosity of bio-oil, compared to conventional fuels, is one of the drawbacks for its utilization in fuel applications. Most importantly, in case of bio-oil using combustion engines, high viscosity increases the droplet size from the injector spray thereby affecting the ignition of droplets [94]. Therefore, decreasing viscosity of bio-oil to improve fuel properties is essential. Studies revealed that use of catalysts improved fuel properties of bio-oil by decreasing viscosity [53, 95].

Azargohar et al. conducted noncatalytic pyrolysis experiments for several biomass waste materials, and it was observed that the viscosity of bio-oil, ranging between 63 and 418 cP, was much more higher than that of crude petroleum oil (~23 cP), requiring a further upgrading process. It was also revealed that the reason of high viscosity was mainly due to lignin-derived hydrocarbons of large molecular weight [96]. However, in a study of Fan et al. on rape straw pyrolysis over nanocrystalline HZSM-5, the dynamic viscosity was reported to be 5.12 $mm^2 s^{-1}$ which was between the accepted limits for diesel fuel as indicated in the study [95]. Mante et al. investigated the hybrid poplar wood pyrolysis with the additive effect of Y-zeolite-based FCC catalyst to ZSM-5. FCC/ZSM-5 catalyst was found to be more effective than pure ZSM-5 in decreasing viscosity of bio-oil samples, indicating synergistic effect of hybrid catalyst, also suggesting that lower weight hydrocarbons was attributed to be formed by the presence of catalyst [53]. Mante and Agblevor [33] studied hybrid poplar wood pyrolysis with the addition of ZSM-5. They classified the liquid fraction of pyrolysis product as LBS (low biosyncrude) oil containing mainly aromatic hydrocarbons and HBS (high biosyncrude) oil which consists of mainly phenols, methyl-substituted phenols, naphthalenes, benzenediols, and naphtalenol. The viscosity of LBS oil, which was significantly lower than that of non-catalytic bio-oil (285 cSt), was reported to be 4.90 cSt. It was suggested that lower viscosity was attributed to the catalytic cracking of levoglucosan and depolymerization of lignin-derived products [33]. Shadangy and Mohanty conducted several studies using various biomass species over CaO, kaolin, and Al_2O_3 [69, 97, 98]. Regardless of the biomass type, CaO was found to produce bio-oil of lower viscosity than that of noncatalytic bio-oil compared to other catalysts used for their studies. The viscosities of bio-oil obtained by using CaO were 0.019629 Pas [98] and 9.007 cP [69], indicating that utilizing catalyst favored a decrease in viscosity about 62 and 74.5%, respectively. Abu Bakar and Titiloye studied ZSM-5, Al-MCM-41, Al-MSU-F, and BRHA (Brunei rice husk ash) for the conversion of rice husk to bio-oil, and the viscosities of bio-oils were, as indicated, 1.55, 1.65, 1.49, and 1.57 cSt, respectively. All the catalysts used in the study decreased viscosity about 1.7–11.3% and slightly increased water content, which indicates that the catalysts favored dehydration reaction [32].

3.4.4. *Stability of bio-oil*

Bio-oil is not as chemically or thermally stable as fossil fuels due to its high content of oxygenated compounds [94]. At temperatures above 40°C or during long-term storage situations, the viscosity of bio-oil is reported to increase due to chemical reactions between components such as ketones and aldehydes, leading to formation of compounds of heavy molecular weight [20]. Thus, it is expected that bio-oil with lower content of carbonyl groups would be more stable. Utilizing catalysts, in order to facilitate transportation and storage of bio-oil, leads to enhanced cracking reactions of heavy molecules as well as removal of oxygenated compounds, thus leading to production of bio-oil with high stability [99]. There is no standard method for determination of stability of bio-oil; however, several methods have been developed by researchers [100–102].

In a study of Zabeti et al., where Cs/ASA was found to be most effective catalyst to eliminate oxygenated compounds and increase aromatic yield compared all the catalysts tested in the study. The results of size exclusion chromatography (SEC) showed that bio-oil molecular weight shifted to higher weight regions after aging [20]. Mante and Agblevor conducted a stability test for the catalytic bio-oils (low and high biosyncrude oil) produced from hybrid poplar wood. The stability and aging tests were performed in a gravity oven at 90°C for 24 hours. Also, the viscosities of the bio-oil samples stored at 40°C for over 10 months were also measured, and the change in viscosity was found to be 5% for low biosyncrude oil and 27.9% for high bio syncrude oil. It was concluded that catalytic bio-oils were thermally stable and could be stored in room temperature for over 10 months without any significant increase in viscosity [33]. Nokkosmaki et al. studied pine sawdust pyrolysis with the addition of ZnO as catalyst. The stability test was performed at 80°C for 24 h and showed that viscosity was changed with the use of ZnO. The change in viscosity was 55%, which was significantly lower than that of noncatalytic bio-oil (129%) [34]. Duman et al. investigated the effect of methanol addition to the stability of bio-oil produced from safflower oil cake using FCC as catalyst. Addition of methanol reduced the viscosity. The viscosity was much lower at higher temperatures, thus indicating a more stable bio-oil. After aging test at 40°C for 168 h, the viscosity increased by 46.63 and 21.08% in case of raw bio-oil and methanol amended bio-oil, respectively [57].

4. Deactivation and regeneration of catalysts

Catalyst deactivation, the loss of catalytic activity and selectivity over time, is one of the major problems concerning industrial application of catalyst in pyrolysis process [25, 53].

The causes of deactivation are mainly due to chemical, mechanical, and thermal mechanisms of catalyst delay [103], but in this section only physical deactivation is addressed. Physical deactivation is done either by coke deposition on the catalyst pores or by covering active catalytic sites preventing pyrolysis vapors to enter through the pores for catalytic reactions such as aromatization, depolymerization, and isomerization [53, 103, 104]. As suggested by Forzatti and Lietti [104], catalyst deactivation reactions proceeding via carbonium ion intermediates involve series of chemical reactions, which vary by the variety of components of reaction mixture, catalyst type, reactions conditions.

Coke is deposited on catalyst pores, in the form of carbon oxides via oxidation. In case of deactivation, if irreversible, regeneration of catalysts under more severe conditions than that of main process is necessary to remove coke [105]. Mante et al. compared the catalytic activity of different zeolites in terms of coke deposition and reported that Y-zeolite was more prone to produce coke than ZSM-5 because of its faujasite structure, larger pore size, and acidity [53]. In a study of Lisa et al., ZSM-5 zeolite was regenerated via oxidation and after five regeneration cycles; no reduction in catalytic activity was recorded [25]. Nokkosmaki et al. [34] observed catalyst deactivation at 400°C. The catalyst affected the degradation of polysaccharides, so that between the first and the fifth pyrolysis cycles, the ratio of levoglucosan was raised from 1.0 to 2.0%. Zabeti et al. studied regeneration of Cs/ASA catalyst by calcinating under air atmosphere at 600°C for 5 h operation [20]. The regenerability of catalyst was evaluated measuring the BET surface area and bio-oil yield obtained after regeneration. It was reported that BET surface area of regenerated catalyst decreased by 10.6%, whereas bio-oil yield remained the same with 45.0 wt.%, suggesting that Cs/ASA catalyst is regenerable. Similar results were obtained by Aho et al. investigating the effect of regenerated catalyst to pyrolysis of biomass with H-Beta zeolite with Si/Al ratio of 25. After regeneration, surface area of catalysts decreased, though, the bio-oil yield increased [28].

Author details

Sibel Başakçılardan Kabakcı* and Şeyma Hacıbektaşoğlu

*Address all correspondence to: sibel.kabakci@yalova.edu.tr

Energy Systems Engineering Department, Faculty of Engineering, Yalova University, Yalova, Turkey

References

[1] Thangalazhy-Gopakumar S., Adhikari S., Gupta R.B., Tu M., Taylor S. Production of hydrocarbon fuels from biomass using catalytic pyrolysis under helium and hydrogen environment. Bioresource Technology. 2011;**102**(12):6742-6749. DOI: 10.1016/j.biortech.2011.03.104

[2] Chouchene A., Jeguirim M., Khiari B., Zagrouba F., Trouvé G. Thermal degradation of olive solid waste: influence of particle size and oxygen concentration. In: Klemeš J.J., Stehlík P., editors. Selected Papers from the 11th Conference Process Integration, Modelling and Optimisation for Energy Saving and Pollution Reduction; 24-28 August 2008; Prague, Czech Republic Oxford: Elsevier; 2010. pp. 271-288. DOI: 10.1016/j.resconrec.2009.04.010

[3] Garcia-Maraver A., Salvachúa D., Martínez M.J., Diaz L.F., Zamorano M. Analysis of the relation between the cellulose, hemicellulose and lignin content and the thermal behavior of residual biomass from olive trees. Waste Management. 2013;**33**(11):2245-2249. DOI: 10.1016/j.wasman.2013.07.010

[4] Ounas A., Aboulkas A., El harfi K., Bacaoui A., Yaacoubi A. Pyrolysis of olive residue and sugar cane bagasse: non-isothermal thermogravimetric kinetic analysis. Bioresource Technology. 2011;**102**(24):11234-11238. DOI: 10.1016/j.biortech.2011.09.010

[5] Encinar J.M., González J.F., Martínez G., González J.M. Two stages catalytic pyrolysis of olive oil waste. Fuel Processing Technology. 2008;**89**:1448-1455.. DOI: 10.1016/j.fuproc.2008.07.005

[6] Seo D.K., Park S.S., Hwang J., Yu T.U. Study of the pyrolysis of biomass using thermogravimetric analysis (TGA) and concentration measurements of the evolved species. Journal of Analytical and Applied Pyrolysis. 2010;**89**(1):66-73. DOI: 10.1016/j.jaap.2010.05.008

[7] Demiral İ., Şensöz S. The effects of different catalysts on the pyrolysis of industrial wastes (olive and hazelnut bagasse). Bioresource Technology. 2008;**99**(17):8002-8007. DOI: 10.1016/j.biortech.2008.03.053

[8] Balat M., Balat H., Balat M., Kırtay E. Main routes for the thermo-conversion of biomass into fuels and chemicals. Part 1: pyrolysis systems. Energy Conversion and Management. 2009;**50**(12):3147-3157. DOI: 10.1016/j.enconman.2009.08.014

[9] García-Ibañez P., Sánchez M., Cabanillas A. Thermogravimetric analysis of olive-oil residue in air atmosphere. Fuel Processing Technology. 2006;**87**(2):103-107. DOI: 10.1016/j.fuproc.2005.08.005

[10] Jauhiainen J., Conesa J.A., Font R., Martín-Gullón I. Kinetics of the pyrolysis and combustion of olive oil solid waste. Journal of Analytical and Applied Pyrolysis. 2004;**72**(1):9-15. DOI: 10.1016/j.jaap.2004.01.003

[11] Özveren U., Özdoğan Z.S. Investigation of the slow pyrolysis kinetics of olive oil pomace using thermo-gravimetric analysis coupled with mass spectrometry. Biomass and Bioenergy. 2013;**58**:168-179. DOI: 10.1016/j.biombioe.2013.08.011

[12] Başakçılardan-Kabakcı S., Aydemir H. Pyrolysis of olive pomace and copyrolysis of olive pomace with refuse derived fuel. Environmental Progress & Sustainable Energy. 2014;**33**(2):649-656. DOI: 10.1002/ep.11827

[13] Bulushev D.A., Ross J.R.H. Catalysis for conversion of biomass via pyrolysis and gasification: A review. Catalysis Today. 2011;**171**:1-13. DOI: 10.1016/j.cattod.2011.02.005

[14] Bu Q., Lei H., Ren S., Wang L., Zhang Q., Tang J., Ruan R. . Production of phenols and biofuels by catalytic microwave pyrolysis of lignocellulosic biomass. Bioresource Technology. 2012;**108**:274-279. DOI: 10.1016/j.biortech.2011.12.125

[15] Chattopadhyay J., Kim C., Kim R., Pak D. Thermogravimetric study on pyrolysis of biomass with Cu/Al_2O_3 catalysts. Journal of Industrial and Engineering Chemistry. 2009;**15**(1):72-76. DOI: 10.1016/j.jiec.2008.08.022

[16] Thangalazhy-Gopakumar S., Adhikari S., Gupta R.B. Catalytic pyrolysis of biomass over H+ZSM 5 under hydrogen pressure. Energy & Fuels. 2012;**26**(8):5300-5306. DOI: 10.1021/ef3008213

[17] Zhou L., Yang H., Wu H., Wang M., Cheng D. Catalytic pyrolysis of rice husk by mixing with zinc oxide: characterization of bio-oil and its rheological behavior. Fuel Processing Technology. 2013;**106**:385-391. DOI: 10.1016/j.fuproc.2012.09.003

[18] Pütün E., Ateş F., Pütün A.E. Catalytic pyrolysis of biomass in inert and steam atmospheres. Fuel. 2008;**87**(6):815-824. DOI: 10.1016/j.fuel.2007.05.042

[19] Foster A.J., Jae J., Cheng Y.T., Huber G.W., Lobo R.F. Optimizing the aromatic yield and distribution from catalytic fast pyrolysis of biomass over ZSM-5. Applied Catalysis A: General. 2012;**423-424**:154-161. DOI: 10.1016/j.apcata.2012.02.030

[20] Zabeti M., Nguyen T.S., Heeres H.J., Seshan K. In situ catalytic pyrolysis of lignocellulose using alkali-modified amorphous silica alumina. Bioresource Technology. 2012;**118**:374-381. DOI: 10.1016/j.biortech.2012.05.034

[21] Mansur D., Yoshikawa T., Norinaga K., Hayashi J., Tago T., Masuda T. Production of ketones from pyroligneous acid of woody biomass pyrolysis over an iron-oxide catalyst. Fuel. 2013;**103**:130-134. DOI: 10.1016/j.fuel.2011.04.003

[22] Yu Y., Li X., Su L., Zhang Y., Wang Y., Zhang H. The role of shape selectivity in catalytic fast pyrolysis of lignin with zeolite catalysts. Applied Catalysis A: General. 2012;**447-448**:115-123. DOI: 10.1016/j.apcata.2012.09.012

[23] Mihalcik D.J., Mullen C.A., Boateng A.A. Screening acidic zeolites for catalytic fast pyrolysis of biomass and its components. Journal of Analytical and Applied Pyrolysis. 2011;**92**(1):224-232. DOI: 10.1016/j.jaap.2011.06.001

[24] French R., Czernik S. Catalytic pyrolysis of biomass for biofuels production. Fuel Processing Technology. 2010;**91**(1):25-32. DOI: 10.1016/j.fuproc.2009.08.011

[25] Iisa K., Stanton A.R., Czernik S.. Production of Hydrocarbon Fuels from Biomass by Catalytic Fast Pyrolysis. Golden, CO: National Renewable Energy Laboratory (NREL); 2012

[26] Basu P. Pyrolysis and torrefaction. In: Biomass Gasification and Pyrolysis: Practical Design and Theory Burlington: Academic Press; 2010. pp. 65-96. DOI: 10.1016/B978-0-12-374988-8.00003-9

[27] Huang W., Gong F., Fan M., Zhai Q., Hong C., Li Q. Production of light olefins by catalytic conversion of lignocellulosic biomass with HZSM-5 zeolite impregnated with 6 wt.% lanthanum. Bioresource Technology. 2012;**121**:248-255. DOI: 10.1016/j.biortech.2012.05.141

[28] Aho A., Kumar N., Eranen K., Salmi T., Hupa M., Murzin D.Y. Catalytic pyrolysis of biomass in a fluidized bed reactor: influence of acidity of H-Beta Zeolite. Process Safety and Environmental Protection. 2007;**85**(B5):473-480. DOI: 10.1205/psep07012

[29] Yildiz G., Pronk M., Djokic M., van Geem K.M., Ronsse F., van Duren R., Prins W. Validation of a new set-up for continuous catalytic fast pyrolysis of biomass coupled with vapour phase upgrading. Journal of Analytical and Applied Pyrolysis. 2013;**103**:343-351. DOI: 10.1016/j.jaap.2013.02.001

[30] Iliopoulou E.F., Antonakou E.V., Karakoulia S.A., Vasalos I.A., Lappas A.A., Triantafyllidis K.S. Catalytic conversion of biomass pyrolysis products by mesoporous materials: effect of steam stability and acidity of Al-MCM-41 catalysts. The Chemical Engineering Journal. 2007;**134**(1):51-57. DOI: 10.1016/j.cej.2007.03.066

[31] Iliopoulou E.F., Stefanidis S.D., Kalogiannis K.G., Delimitis A., Lappas A.A., Triantafyllidis K.S. Catalytic upgrading of biomass pyrolysis vapors using transition metal-modified ZSM-5 zeolite. Applied Catalysis B: Environmental. 2012;**127**:281-290. DOI: 10.1016/j.apcatb.2012.08.030

[32] Abu Bakar M.S., Titiloye J.O. Catalytic pyrolysis of rice husk for bio-oil production Journal of Analytical and Applied Pyrolysis. 2012;**103**:362-368. DOI: 10.1016/j.jaap.2012.09.005

[33] Mante O.D., Agblevor F.A. Catalytic conversion of biomass to bio-syncrude oil. Biomass Conversion and Biorefinery. 2011;**1**:203-215. DOI: 10.1007/s13399-011-0020-4

[34] Nokkosmaki M.I., Kuoppala E.T., Leppamaki E.A., Krause A.O.I. Catalytic conversion of biomass pyrolysis vapours with zinc oxide. Journal of Analytical and Applied Pyrolysis. 2000;**55**(1):119-131.

[35] Aysu T., Küçük M.M. Biomass pyrolysis in a fixed-bed reactor: effects of pyrolysis parameters on product yields and characterization of products. Energy. 2014;**64**:1002-1025. DOI: 10.1016/j.energy.2013.11.053

[36] Nguyen T.S., Zabeti M., Lefferts L., Brem G., Seshan K. Conversion of lignocellulosic biomass to green fuel oil over sodium based catalysts. Bioresource Technology. 2013;**142**:353-360. DOI: 10.1016/j.biortech.2013.05.023

[37] Pütün E. Catalytic pyrolysis of biomass: effect of pyrolysis temperature, sweeping gas flow rate and MgO catalyst. Energy. 2010;**35**(7):2761-2766. DOI: 10.1016/j.energy.2010.02.024

[38] Encinar J.M., Gonzalez J.F., Martinez G., Roman S. Catalytic pyrolysis of exhausted olive oil waste. Journal of Analytical and Applied Pyrolysis. 2009;**85**(1-2):197-203. DOI: 10.1016/j.jaap.2008.11.018

[39] Uzun B.B., Sarıoğlu N. Rapid and catalytic pyrolysis of corn stalks. Fuel Processing Technology. 2009;**90**(5):705-716. DOI: 10.1016/j.fuproc.2009.01.012

[40] Asadullah M., Zhang S., Li C.Z. Evaluation of structural features of chars from pyrolysis of biomass of different particle sizes. Fuel Processing Technology. 2010;**91**(8):877-881. DOI: 10.1016/j.fuproc.2009.08.008

[41] Luo S., Yi C., Zhou Y. Bio-oil production by pyrolysis of biomass using hot blast furnace slag. Renewable Energy. 2013;**50**:373-377. DOI: 10.1016/j.renene.2012.07.008

[42] Choi H.S., Choi Y.S., Park H.C. Fast pyrolysis characteristics of lignocellulosic biomass with varying reaction conditions. Renewable Energy. 2012;**42**:131-135. DOI: 10.1016/j.renene.2011.08.049

[43] Melligan F., Hayes M.H.B., Kwapinski W., Leahy J.J. Hydro-pyrolysis of biomass and online catalytic vapor upgrading with Ni-ZSM 5 and Ni-MCM-41. Energy & Fuels. 2012;**26**(10):6080-6090. DOI: 10.1021/ef301244h

[44] Stefanidis S.D., Kalogiannis K.G., Iliopoulou E.F., Michailof C.M., Pilavachi P.A., Lappas A.A.. A study of lignocellulosic biomass pyrolysis via the pyrolysis of cellulose, hemicellulose and lignin. Journal of Analytical and Applied Pyrolysis. 2014;**105**:143-150. DOI: 10.1016/j.jaap.2013.10.013

[45] Haykiri-Acma H., Yaman S. Thermogravimetric investigation on the thermal reactivity of biomass during slow pyrolysis. International Journal of Green Energy. 2009;**6**(4):333-342. DOI: 10.1080/15435070903106959

[46] Collard F.X., Blin J., Bensakhira A., Valette J. Influence of impregnated metal on the pyrolysis conversion of biomass constituents. Journal of Analytical and Applied Pyrolysis. 2012;**95**:213-226. DOI: 10.1016/j.jaap.2012.02.009

[47] Stenseng M., Jensen A., Dam-Johansen K. Investigation of biomass pyrolysis by thermogravimetric analysis and differential scanning calorimetry. Journal of Analytical and Applied Pyrolysis. 2001;**58-59**:765-780.

[48] Bertero M., de la Puente G., Sedran U. Fuels from bio-oil: bio-oil production from different residual sources, characterization and thermal conditioning. Fuel. 2012;**95**:263-271. DOI: 10.1016/j.fuel.2011.08.041

[49] Zhang H., Xiao R., Jin B., Shen D., Chen R., Xiao G. Catalytic fast pyrolysis of straw biomass in an internally interconnected fluidized bed to produce aromatics and olefins: effect of different catalyst. Bioresource Technology. 2013;**137**:82-87. DOI: 10.1016/j.biortech.2013.03.031

[50] Wang S., Guo X., Liang T., Zhou Y., Luo Z. Mechanism research on cellulose pyrolysis by Py-GC/MS and subsequent density functional theory studies. Bioresource Technology. 2012;**104**:722-728. DOI: 10.1016/j.biortech.2011.10.078

[51] Bai X., Johnston P., Sadula S., Brown R.C. Role of levoglucosan physiochemistry in cellulose pyrolysis. Journal of Analytical and Applied Pyrolysis. 2013;**99**:58-65. DOI: 10.1016/j.jaap.2012.10.028

[52] Xin S., Yang H., Chen Y., Wang X., Chen H. Assessment of pyrolysis polygeneration of biomass based on major components: product characterization and elucidation of degradation pathways. Fuel. 2013;**113**:266-273

[53] Mante O.D., Agblevor F.A., Oyama S.T., McClung R. Catalytic pyrolysis with ZSM-5 based additive as co-catalyst to Y-zeolite in two reactor configurations. Fuel. 2014;**117**:649-659. DOI: 10.1016/j.fuel.2013.09.034

[54] Bertero M., Gorostegui H.A., Orrabalis C.J., Guzmán C.A., Calandri E.L., Sedran U. Characterization of the liquid products in the pyrolysis of residual chañar and palm fruit biomasses. Fuel. 2014;**116**:409-414

[55] Wang D., Xiao R., Zhang H., He G. Comparison of catalytic pyrolysis of biomass with MCM-41 and CaO catalysts by using TGA-FTIR analysis Journal of Analytical and Applied Pyrolysis. 2010;**89**(2):171-177. DOI: 10.1016/j.jaap.2010.07.008

[56] Chen G. Catalytic application to biomass pyrolysis in a fixed bed reactor. Energy Sources. 2003;**25**(3):223-228. DOI: 10.1080/00908310390142271

[57] Duman G., Pala M., Ucar S., Yanik J. Two-step pyrolysis of safflower oil cake. Journal of Analytical and Applied Pyrolysis. 2013;**103**:352-361. DOI: 10.1016/j.jaap.2012.11.023

[58] Oasmaa A., Elliott D.C., Korhonen J. Acidity of biomass fast pyrolysis bio-oils. Energy Fuels. 2010;**24**:6548-6554. DOI: 10.1021/ef100935r

[59] Van de Beld B., Holle E., Florjin J. The use of pyrolysis oil and pyrolysis oil derived fuels in diesel engines for CHP applications. Applied Energy. 2013;**102**:190-197. DOI: 10.1016/j.apenergy.2012.05.047

[60] Chiaramonti D., Bonini M., Fratini E., Tondi G., Gartner K., Bridgwaterd A.V., et al Development of emulsions from biomass pyrolysis liquid and diesel and their use in engines—Part 1: emulsion production. Biomass and Bioenergy. 2003;**25**(1):85-99. DOI: 10.1016/S0961-9534(02)00183-6

[61] Calabria R., Chiariello F., Massoli P. Combustion fundamentals of pyrolysis oil based fuels. Experimental Thermal and Fluid Science. 2007;**31**:413-420.

[62] Yang S.I., Hsu T.C., Wu C.Y., Chen K.H., Hsu Y.L., Li Y.H. Application of biomass fast pyrolysis part II: the effects that bio-pyrolysis oil has on the performance of diesel engines. Energy. 2014;**66**:172-180. DOI: 10.1016/j.energy.2013.12.057

[63] Chiaramonti D., Bonini M., Fratini E., Tondi G., Gartner K., Bridgwater A.V., et al Development of emulsions from biomass pyrolysis liquid and diesel and their use in engines—Part 2: tests in diesel engines. Biomass and Bioenergy. 2003;**25**(1):101-111. DOI: 10.1016/S0961-9534(02)00184-8

[64] Pelaez-Samaniego M.R., Mesa-Pérez J., Cortez L.A.B., Rocha J.D., Sanchez C.G., Marín H. Use of blends of gasoline with biomass pyrolysis-oil derived fractions as fuels in an Otto engine. Energy for Sustainable Development. 2011;**15**(4):376-381. DOI: 10.1016/j.esd.2011.06.001

[65] Oasmaa A., Peacocke C. Properties and Fuel Use of Biomass-Derived Fast Pyrolysis Liquids—A Guide Espoo: VTT Publications 731; 2010. 134 p.

[66] Boucher M.E., Chaala A., Roy C. Bio-oils obtained by vacuum pyrolysis of softwood bark as a liquid fuel for gas turbines. Part I: properties of bio-oil and its blends with methanol and a pyrolytic aqueous phase. Biomass and Bioenergy. 2000;**19**:337-350.

[67] Lopez Juste G., Salva Monfort J.J. Preliminary test on combustion of wood derived fast pyrolysis oils in a gas turbine combustor. Biomass and Bioenergy. 2000;**19**(2):119-128. DOI: 10.1016/S0961-9534(00)00023-4

[68] Wang C., Hao Q., Lu D., Jia Q., Li G., Xu B. Production of light aromatic hyrdocarbons from biomass by catalytic pyrolysis. Chinese Journal of Catalysis. 2008;**29**(9):907-912. DOI: 10.1016/S1872-2067(08)60073-X

[69] Shadangi K.P., Mohanty K. Production and characterization of pyrolytic oil by catalytic pyrolysis of Niger seed. Fuel. 2014;**126**:109-115. DOI: 10.1016/j.fuel.2014.02.035

[70] Yorgun S., Şimşek Y.E. Catalytic pyrolysis of Miscanthus x giganteus over activated alumina. Bioresource Technology. 2008;**99**(1):8095-8100. DOI: 10.1016/j.biortech.2008.03.036

[71] Flanigen E.M. Zeolites and molecular sieves: a historical perspective. In: van Bekkum H., Flanigen E.M., Jacobs P.A., Jansen J.C.. Introduction to Zeolite Science and Practice. 2nd ed. Amsterdam: Elsevier; 2001. pp. 11-12. DOI: 10.1016/S0167-2991(01)80243-3

[72] Shen Y., Yoshikawa K. Recent progresses in catalytic tar elimination during biomass gasification or pyrolysis—a review. Renewable and Sustainable Energy Reviews. 2013;**21**:371-392. DOI: 10.1016/j.rser.2012.12.062

[73] Jae J., Tompsett G.A., Foster A.J., Hammond K.D., Auerbach S.M., Lobo R.F., et al Investigation into the shape selectivity of zeolite catalysts for biomass conversion. Journal of Catalysis. 2011;**297**:257-268. DOI: 10.1016/j.cat.2011.01.019

[74] McCusker L.B., Baerlocher C. Zeolite structures. In: Jacobs P.A., Flanigen E.M., Jansen J.C., van Bekkum H, editors. Introduction to Zeolite Science and Practice. 2nd ed. Amsterdam: Elsevier; 2001. pp. 38-40.

[75] Pattiya A., Titiloye J.O., Bridgwater A.V. Fast pyrolysis of cassava rhizome in the presence of catalysts. Journal of Analytical and Applied Pyrolysis. 2008;**81**:72-79.

[76] Zheng A., Zhao Z., Chang S., Huang Z., Wu H., Wang X., et al Effect of crystal size of ZSM-5 on the aromatic yield and selectivity from catalytic fast pyrolysis of biomass. Journal of Molecular Catalysis A: Chemical. 2014;**383-384**:23-30. DOI: 10.1016/j.molcata.2013.11.005

[77] Payra P., Dutta P.K. Zeolites: a primer. In: Auerbach S.M., Carrado K.A., Dutta P.K., editors. Handbook of Zeolite Science and Technology New York: Marcel Dekker; 2003. pp. 1-21.

[78] Shirazi L., Jamshidi E., Ghasemi M.R. The effect of Si/Al ratio of ZSM–5 zeolite on its morphology, acidity and crystal size. Crystal Research Technology. 2008;**43**(12):1300-1306. DOI: 10.1002/crat.200800149

[79] Carlson T.R., Jae J., Lin Y.C., Tompsett G.A., Huber G.W. Catalytic fast pyrolysis of glucose with HZSM-5: the combined homogeneous and heterogeneous reactions. Journal of Catalysis. 2010;**270**(1):110-124. DOI: 10.1016/j.jcat.2009.12.013

[80] Williams P.T., Horne P.A. The influence of catalyst type on the composition of upgraded biomass pyrolysis oils. Journal of Analytical and Applied Pyrolysis. 1995;**31**:39-61. DOI: 10.1016/0165-2370(94)00847-T

[81] Jae J., Coolman R., Mountziaris T.J., Huber G.W. Catalytic fast pyrolysis of lignocellulosic biomass in a process development unit with continual catalyst addition and removal. Chemical Engineering Science. 2014;**108**:33-46. DOI: 10.1016/j.ces.2013.12.023

[82] Naqvi S.R., Uemura Y., Bt Yusup S. Catalytic pyrolysis of paddy husk in a drop type pyrolyzer for bio-oil production: the role of temperature and catalyst. Journal of Analytical and Applied Pyrolysis. 2014;**104**:57-62. DOI: 10.1016/j.jaap.2013.12.009

[83] Zhang M., Resende F.L.P., Moutsoglou A. Catalytic fast pyrolysis of aspen lignin via Py-GC/MS. Fuel. 2014;**116**:358-369. DOI: 10.1016/j.fuel.2013.07.128

[84] Murata K., Liu Y., Inaba M., Takahara I. Catalytic fast pyrolysis of jatropha wastes. Journal of Analytical and Applied Pyrolysis. 2012;**94**:75-82. DOI: 10.1016/j.jaap.2011.11.008

[85] Weitkamp J., Ernst S., Puppe L. Shape-selective catalysis in zeolites. In: Weitkamp J., Puppe L., editors. Catalysis and Zeolites: Fundamentals and Applications Berlin: Springer-Verlag; 1999. pp. 327-370.

[86] Fogassy G., Thegarid N., Schuurman Y., Mirodatosa C. From biomass to bio-gasoline by FCC co-processing: effect of feed composition and catalyst structure on product quality. Energy & Environmental Science. 2011;**4**(12):5068-5076. DOI: 10.1039/C1EE02012A

[87] Kim J.W., Park S.H., Jinho J., Jeon J.K., Ko C.H., Jeong K.E., et al Catalytic pyrolysis of mandarin residue from the mandarin juice processing industry. Bioresource Technology. 2013;**136**:431-436. DOI: 10.1016/j.biortech.2013.03.062

[88] van Santen R.A. Theory of Bronsted acidity in zeolites. In: Jansen J.C., Stöcker M., Karge H.G., Weitkamp J, editors. Advanced Zeolite Science and Applications, Studies in Surface Science and Catalysis Amsterdam: Elsevier Science; 1994. pp. 273-294.

[89] Kerssens M.M., Sprung C., Whiting G.T., Weckhuysen B.M. Selective staining of zeolite acidity: recent progress and future perspectives on fluorescence microscopy. Microporous and Mesoporous Materials. 2014;**189**:136-143. DOI: 10.1016/j.micromeso.2013.10.015

[90] Laredo G.C., Quintana-Solórzano R., Castillo J.J., Armendáriz-Herrera H., Garcia-Gutier J.L. Benzene reduction in gasoline by alkylation with propylene over MCM-22 zeolite with a different Brønsted/Lewis acidity ratios. Applied Catalysis A: General. 2013;**454**:37-45. DOI: 10.1016/j.apcata.2013.01.001

[91] Ateş F., Işıkdağ M.A. Influence of temperature and alumina catalyst on pyrolysis of corncob. Fuel. 2009;**88**(10):1991-1997. DOI: 10.1016/j.fuel.2009.03.008

[92] Wang P., Zhan S., Yu H., Xue X., Hong N. The effects of temperature and catalysts on the pyrolysis of industrial wastes(herb residue). Bioresource Technology. 2010;**101**(9):3236-3241. DOI: 10.1016/j.biortech.2009.12.082

[93] Smets K., Roukaerts A., Czech J., Reggers G., Schreurs S., Carleer R., et al Slow catalytic pyrolysis of rapeseed cake: product yield and characterization of the pyrolysis liquid. Biomass and Bioenergy. 2013;**57**:180-190. DOI: 10.1016/j.biombioe.2013.07.001

[94] Lehto J., Oasmaa A., Solantausta Y., Kytö M., Chiaramonti D. Fuel Oil Quality and Combustion of Fast Pyrolysis of Bio-oils Espoo: VTT Technical Research Centre of Finland; 2013

[95] Fan Y., Cai Y., Li X., Yu N., Yin H. Catalytic upgrading of pyrolytic vapors from the vacuum pyrolysis of rape straw over nanocrystalline HZSM-5 zeolite in a two-stage-fixed-bed reactor. Journal of Analytical and Applied Pyrolysis. 2014;**108**:185-195. DOI: 10.1016/j.jaap.2014.05.001

[96] Azargohar R., Jacobson K.L., Powell E.E., Dalai A.K. Evaluation of properties of fast pyrolysis products obtained from Canadian waste biomass. Journal of Analytical and Applied Pyrolysis. 2013;**104**:330-340. DOI: 10.1016/j.jaap.2013.06.016

[97] Shadangi K.P., Mohanty K. Comparison of yield and fuel properties of thermal and catalytic Mahua seed pyrolytic oil. Fuel. 2014;**117 (Part A)**:372-380. DOI: 10.1016/j.fuel.2013.09.001

[98] Shadangi K.P., Mohanty K. Thermal and catalytic pyrolysis of Karanja seed to produce liquid fuel. Fuel. 2014;**114**:434-442. DOI: 10.1016/j.fuel.2013.07.053

[99] Yu F., Gao L., Wang W., Zhang G., Ji J. Bio-fuel production from the catalytic pyrolysis of soybean oil over Me-Al-MCM-41 (Me = La, Ni or Fe) mesoporous materials. Journal of Analytical and Applied Pyrolysis. 2013;**104**:325-329. DOI: 10.1016/j.jaap.2013.06.017

[100] Hoekstra E., Kersten S.R.A., Tudos A., Meier D., Hogendoorn K.J.A. Possibilities and pitfalls in analyzing (upgraded) pyrolysis oil by size exclusion chromatography (SEC). Journal fo Analytical and Applied Pyrolysis. 2011;**91**(1):76-88. DOI: 10.1016/j.jaap.2011.01.006

[101] Diebold J.P., Czernik S. Additives to lower and stabilize the viscosity of pyrolysis oils during storage. Energy & Fuels. 1997;**11**(5):1081-1091. DOI: 10.1021/ef9700339

[102] Oasmaa A., Leppämäki E., Koponen P., Levander J., Tapola E.. Physcial Characterisation of Biomass-based Pyrolysis Liquids. Application of Standard Fuel Oil Analyses Espoo: VTT Technical Centre of Finland: VTT Publications; 1997

[103] Bartholomew C.H. Mechanisms of catalyst deactivation. Applied Catalysis A: General. 2001;**212**:17-60. DOI: 10.1016/S0926-860X(00)00843-7

[104] Forzatti P., Lietti L. Catalyst deactivation. Catalysis Today. 1999;**52**(2-3):165-181. DOI: 10.1016/S0920-5861(99)00074-7

[105] Sie S.T. Consequences of catalyst deactivation for process design and operation. Applied Catalysis A: General. 2001;**212**(1-2):129-151. DOI: 10.1016/S0926-860X(00)00851-6

Chapter 8

Spray Pyrolysis Processing for Optoelectronic Applications

Oleksandr Malik,
Francisco Javier De La Hidalga-Wade and
Raquel Ramírez Amador

Additional information is available at the end of the chapter

http://dx.doi.org/10.5772/67431

Abstract

Spray pyrolysis is a low-cost and simple technique for the fabrication of high-quality transparent and conducting oxide thin films for different optoelectronic applications. The fabrication method, structural, morphological, and electro-optical properties of fluorine-doped tin oxide (FTO) and tin-doped indium oxide (ITO) films have been investigated. The deposited films have low resistivity and high transparency. Applications of such films are shown in high-efficiency surface-barrier photodetectors and solar cells, where the films serve as an active and antireflection electrode. A short description of other undoped and doped oxide films such as ZnO and TiO_2 fabricated by spray pyrolysis is presented.

Keywords: spray pyrolysis, thin film, fluorine-doped tin oxide, tin-doped indium oxide, surface-barrier photodetectors, solar cells

1. Introduction

It is well known that some thin oxide films that are heavily doped n-type semiconductors present both high conductivity and high transparency. Materials exhibiting simultaneously both high conductivity and optical transmittance are named transparent conducting oxides (TCOs). The most representative of such materials are tin-doped indium oxide (In_2O_3) and fluorine-doped tin oxide (SnO_2), known as ITO and FTO, respectively. These materials have been thoroughly studied due to their innumerable optoelectronic applications for devices, such as in solar cells, liquid crystal displays, organic light-emitting diodes, and heat mirrors [1]. A number of techniques, such as oxidation of metal films, sputtering, chemical vapor deposition, and growth from chemical solutions, have been investigated in the search for the most reliable and cheapest fabrication method of the TCO thin films [1]. Chemical fabrication techniques have been studied extensively due to their simplicity, low-cost, and the flexibility for the doping process. The spray pyrolysis processing is one of the most simple fabrication method that has been known for more than three decades since the first published work in 1966 when it was used for the spray-deposited CdS films [2]. Because of the simplicity of the technical apparatus used as well as the inherent suitability for large-scale production, the spray pyrolysis is the most attractive method for the TCO films fabrication. This method presents a numbers of advantages [3] such as the extremely easy way of doping the films by adding certain elements to the spray solution, the process is conducted in air ambient (vacuum conditions are not required) and operated at moderate temperatures, resulting into a compact and simple fabrication process. The aim of this chapter is to describe the fabrication method and properties of some useful TCO films and their optoelectronic properties suitable for applications as transparent ohmic contacts in thin film solar cells as well as an active and antireflection electrode in the design of efficient surface-barrier semiconductor photodetectors. Recently, the quite important optoelectronic application of this method for the fabrication of efficient silicon solar cells and modules has been reported [4]. Readers can find many such applications of the TCO films in the literature.

2. Spray pyrolysis processing: general remarks

Spray pyrolysis is a process in which a thin film is deposited by spraying a solution on a heated surface, where the constituents react to form a chemical compound [5]. The chemical spray deposition process, according to the type of reaction, can be divided into three groups [6]: In the first group, the droplets of the solution reside on the heated surface as the solvent evaporates and components may further react in the dry state. The second group represents a process in which the solvent evaporates before the drops reach the heated surface and the dry solid impinges on the surface by decomposition. In the third group, there are processes where the solvent vaporizes as the droplets approach the substrate with the consequent heterogeneous reaction of the solution components. The most important parameters to be controlled in all of these processes are the substrate temperature, carrier gas flow rate, nozzle-to-substrate distance, and the solution content and concentration. Among these variables, the substrate temperature has been considered as the most important factor in producing thin film from spray pyrolysis processing; this is because the droplets drying, decomposition, crystallization, and grain growth depend strongly on this parameter [7]. The main part of the apparatus used for the spray pyrolysis deposition is the atomizer for obtaining aerosol from the precursor solution. The design of this equipment can be variable from inexpensive cosmetics or perfume purpose atomizers [8, 9] through a Pyrex glass or metallic individual or commercial design atomizers to much more complicated ultrasonic equipments [10]. Despite the apparent simplicity of the spray pyrolysis technique, a tight correlation of the deposition parameters called for a theoretical modeling spray pyrolysis deposition [11], the aim of which was the understanding of this correlation with the optimization of the deposition process. Such modeling examines the changing of the films topography, where the droplets can be seen as

http://dx.doi.org/10.5772/67431

a flux and not as individual drops, and when they evaporate near the surface prior to fully contacting the substrate in liquid form. A typical spray deposition system includes the spray atomizer containing the precursor solution, a heater for the substrate, pressurized air, liquid flow, and temperature controllers. Inherently, to the spray process, the thickness of the film can be nonuniform, thus a random motion of the spray nozzle or the substrate is useful for an uniform deposition [5]. The following sections of this chapter are dedicated to the description of the experimental method for the deposition of thin TCO films, as well to show their structural, electric, and optical properties for their applications in designing different optoelectronic devices. Our research activity in this field started by 1979 [12, 13] when the spray pyrolysis was used for the fabrication of silicon solar cells.

3. Spray-deposited fluorine-doped tin oxide films

Figure 1 shows schematical representation of our spraying system [4]. It produces films presenting uniform thicknesses for spraying areas of few square centimeters by a mechanical moving of a Pyrex glass spray head.

The substrate holder is a graphite block mounted on the electric heater, the temperature of which is controlled by a thermocouple. To prevent a rapid cooling of the substrate by

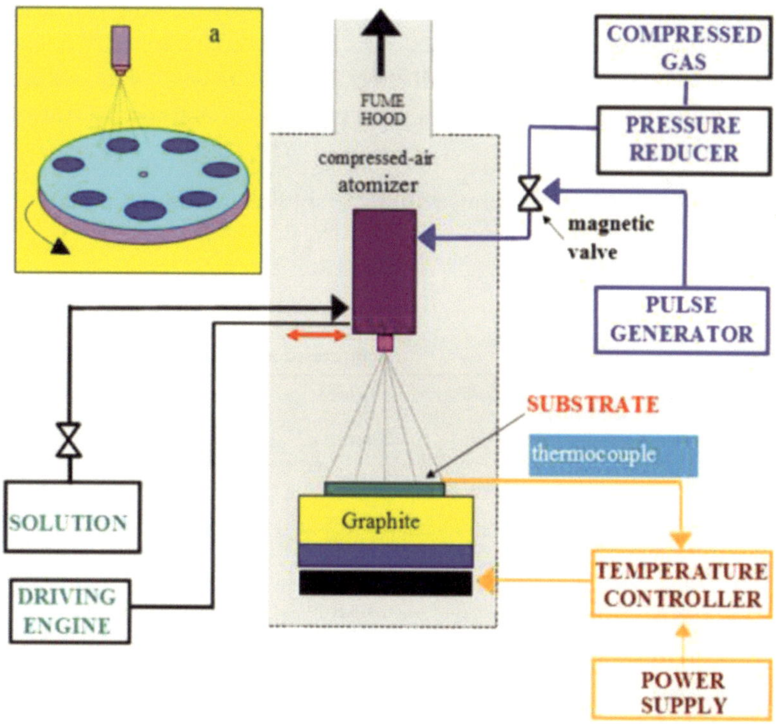

Figure 1. Schematical representation of our spraying system.

the spraying solution, the compressed air is injected into the atomizer during a short time (around 1–2 s) with a long pause (10–30 s); this period of time is controlled by the magnetic valve. Thus, the pulsed spraying leads to the deposition of uniform films. The fixed temperature (480°C) is controlled and supported by a microprocessor. The inset of **Figure 1** shows the holder modification for deposition of TCO films on semiconductor wafers with large areas. The film thickness depends on several parameters: the distance between the spray nozzle and the substrate, the substrate temperature, the concentration of the precursor solution, and the deposition time.

The deposition method used in this work was conducted, using compressed air at a pressure of 0.5 atm and a Pyrex atomizer located at 30 cm from either the Corning glass or the sapphire substrate [14]. Additional technological parameters are shown in **Table 1**. A 0.2 M solution of tin (IV) chloride pentahydrate ($SnCl_4{\cdot}5H_2O$) dissolved in methanol was used as the starting solution. A small amount of ammonium fluoride (NH_4F) dissolved in water was added in the starting solution for the preparation of the precursor with a 0–1 F/Sn molar ratio (**Table 2**). Since the presence of water can lead to the hydrolysis of the tin chloride, we added a few drops of hydrochloric acid (HCl) into the precursor solution; a 10 ml/min precursor flow rate was used for this deposition process [14].

The films' thickness was measured using an Alpha Step 200 profilometer. The X-ray diffraction (XRD) measurements were carried out with an X-ray diffractometer Brucker AXS D8 Advance, with a Cu cathode (1.54059 Å) operating in the Bragg-Brentano Two-Theta geometry. An atomic force microscope JEOL JSPM-4310 was used to study the films' surface. The electrical resistivity (ρ) and carrier density (n_e) were measured at room temperature using the standard van der Pauw method, where the Hall effect parameters were obtained for a magnetic field of 0.3 T. Finally, the optical properties were measured using the Agilent-8453 (200 nm a 1100 nm) spectrophotometer.

The XRD measurements of the FTO films with different F/Sn ratios in the solution are shown in **Figure 2**.

Transporting gas	**Compressed air**
Pressure of the gas	3 kg/cm^2
Diameter of nozzle	2 mm
Distance nozzle-substrate	30 cm
Temperature of the substrate	450 ± 5°C
Tin concentration in solution	0.2 M
Atomic ratio of F/Sn in the solution	0.0–1.0
Time of a single deposition step	1 s
Time of a single cooling period	2 s

Table 1. Technological parameters used for the deposition of the FTO films.

F/Sn molar ratio	NH_4F, mg	$SnCl_4 \cdot 5H_2O$, g	$H_2O + CH_3CH_2OH$ (1:3), ml
0	0	1.519	15
0.20	22.22	1.519	15
0.35	38.89	1.519	15
0.50	55.56	1.519	15
0.65	72.23	1.519	15
0.85	94.45	1.519	15
1	111.12	1.519	15

Table 2. Content of the precursor solutions for fabrication of fluorine-doped tin oxide (FTO) films with different F/Sn ratios.

Figure 2 presents the X-ray diffraction (XRD) measurements for the FTO deposited; it can be seen, according to the American Society for Testing and Materials standard for tin oxides [15], that the films with a thickness around 500 nm have a tetragonal rutile structure since they show the more intense peak in the (200) crystallographic plane. On the other hand, those films deposited with a F/Sn ratio = 0 (pure tin oxide) shows a preferred orientation in the (211) plane and other peaks in the (101), (110) and (301) planes. These results let us know that the grain orientation is highly affected by the fluorine content that makes possible the obtaining of highly oriented films.

Actually, the preferred orientation depends not only on the precursor solutions, as reported in tin oxide films prepared using CVD, but also on the film thickness [16]. **Figure 3** shows the films topography obtained using scanning electron microscopy (SEM).

The grain size values estimated from the XRD patterns using the Scherrer's law [17], which gives the coherence length perpendicularly to the substrate, are shown in **Figure 4**. The average grain size visualized by SEM, which corresponds to the grain size parallel to the substrate, is higher than the estimation from the analysis of the XRD spectra. The root mean square

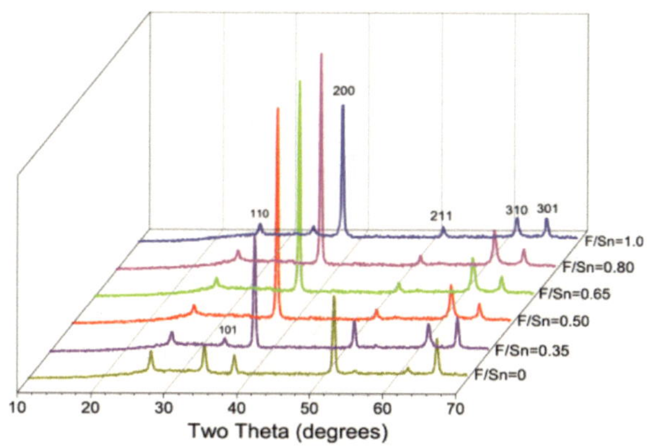

Figure 2. XRD spectra of FTO films with a thickness of around 500 nm and different F/Sn ratios in the precursor.

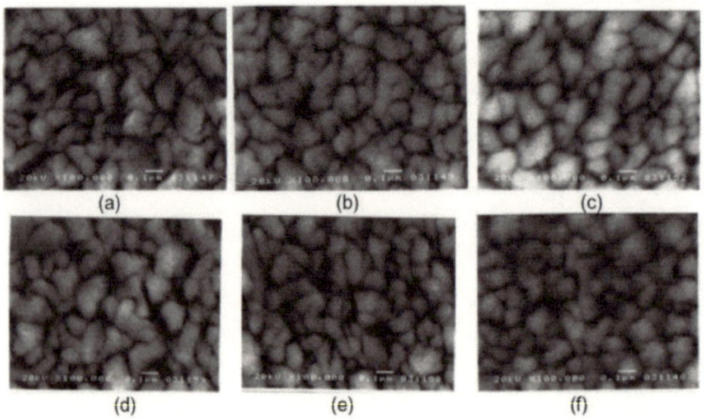

Figure 3. SEM images of the FTO films prepared using the solution with different F/Sn molar ratios. (a) F/Sn = 0; (b) F/Sn = 0.35; (c) F/Sn = 0.50; (d) F/Sn = 0.65; (e) F/Sn = 0.85; (f) F/Sn = 1.0.

(RMS) surface roughness of the FTO films for different F/Sn ratios is also shown in **Figure 4**. The highest grain size (~40 nm) and smallest RMS roughness (~8 nm) were observed for the films prepared using the precursor solution with the F/Sn ratio = 0.5. The films fabricated using the precursor solution with this fluorine concentration present a more arranged grain structure, whereas the increment of the roughness for films with a higher fluorine concentration can be connected with the etching by gaseous HF formed by the thermal decomposition of NH_4F during the growth process.

Figure 5 shows the dependence of the electrical resistivity (ρ), Hall mobility (μ), and carrier density (n) for the 500 nm thick FTO films on different F/Sn ratios. The films fabricated using the precursor with F/Sn ratio = 0.5 show the minimum resistivity, $\rho = 2.2 \times 10^{-4}$ Ω·cm (or Rs = 4.5 Ω^{-2}), as well as the higher carrier mobility and density, $\mu = 21.6$ cm^2/V s and $n = 1.7 \times 10^{21}$ cm^{-3}, respectively. Assuming an electron effective mass of 0.33 m_0 [18], the threshold carrier

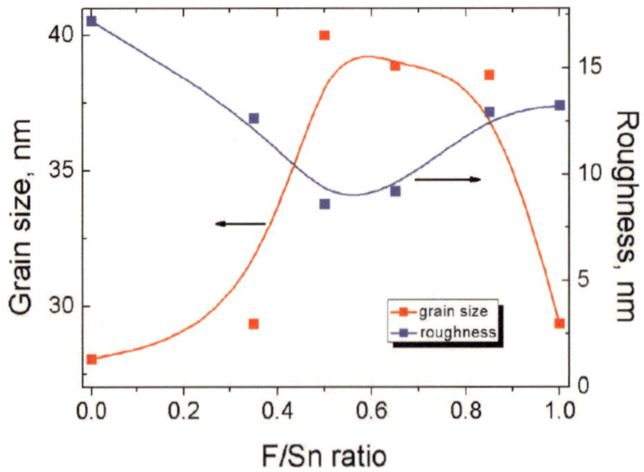

Figure 4. Mean value of the grain size for the (200) plane and the RMS surface roughness for the FTO films prepared using the solution with different F/Sn ratios.

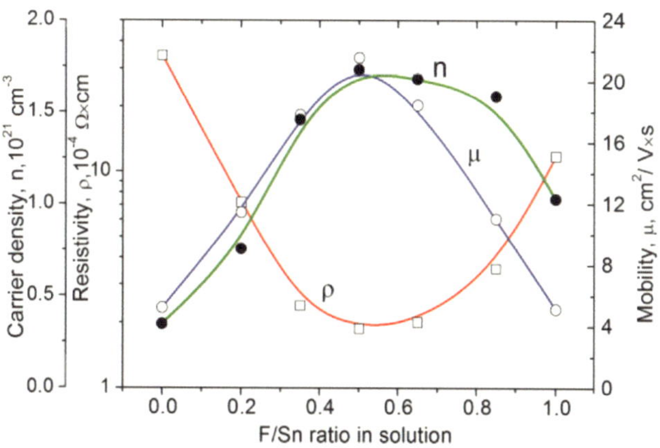

Figure 5. Resistivity (ρ), Hall mobility (μ), and carrier density (n) for the FTO films prepared using solutions with different F/Sn ratios.

density corresponding to strong degeneration in the conduction band for the FTO films is 3.6×10^{18} cm^{-3}. Electron concentration and Hall mobility increase for a higher fluorine content due to this doping effect and the improvement of the film structure. On the other hand, these parameters deteriorate when inside the film, and there is a formation of Sn-F dissipation neutral centers for carriers as well as due to a reduction in the oxygen vacancies for an increasing F/Sn ratio. In an early publication [19], we present a more detailed discussion of these results. A reduction in the mobility cannot be explained as due to the high carrier density, but as due to the electron scattering by ionized impurities and by neutral centers that are formed by the Sn-F bounds shown by XPS measurements. The mean free path of conduction electrons has the same dependence as the carrier density and shows a maximum of 5.3 nm for F/Sn = 0.5. Since the mean free path is considerably shorter than the grain size (30–40 nm), the resistivity of the films is determined by the ionized impurity scattering rather than by the grain boundary scattering.

Figure 6 shows the transmittance of the 500 nm thick FTO films deposited on Corning glass substrates using the precursor solutions with different F/Sn ratios. The transparency of the films exceeds 80% in the visible spectral range.

For highly degenerated FTO films, the value of the optical gap (**Figure 7**) is determined by a blue shift of the high-energy photon absorption edge due to the location of the Fermi level inside the conduction band. This effect is known as the Burstein-Moss shift. Hence, the lowest states in the conduction band are blocked, and the transitions of optically excited electrons take place only to energies above the Fermi level.

The films with F/Sn ratio = 0.5 that present an optical energy gap of 4.6 eV due to a high value of the Burstein-Moss shift; this is very useful for designing UV semiconductor photodetectors by increasing their conversion efficiency. Subsequently, we discussed the practical applications of this effect for designing an effective surface-barrier UV photodetectors with the FTO films as a transparent conducting active electrode.

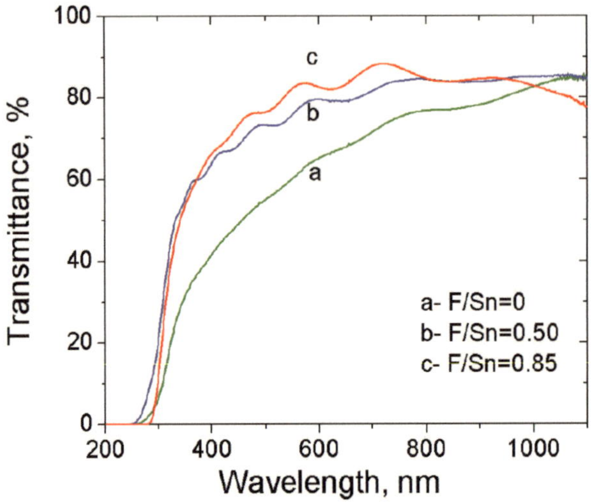

Figure 6. Transmittance of the 500 nm thick FTO films deposited using the precursor solutions with different F/Sn ratios on Corning glass substrates.

For quality estimation and comparative analysis of the TCO films fabricated by different methods and having different thicknesses, Haacke [20] proposed a revised figure of merit (FOM) defined by

$$\phi_H = \frac{T^{10}}{R_s} \tag{1}$$

where R_s and T represent the sheet resistance and transmittance, respectively.

A higher value of FOM indicates a higher performance of a film characterized simultaneously with low sheet resistance and high transparency. However, the FOM cannot be unlimitedly high. For commercial FTO films, the reported FOM is $17.4 \times 10^{-3}\ \Omega^{-1}$ [6], and the highest known Haacke FOM value for such films is $35.7 \times 10^{-3}\ \Omega^{-1}$ [7]. A different approach for the calculation

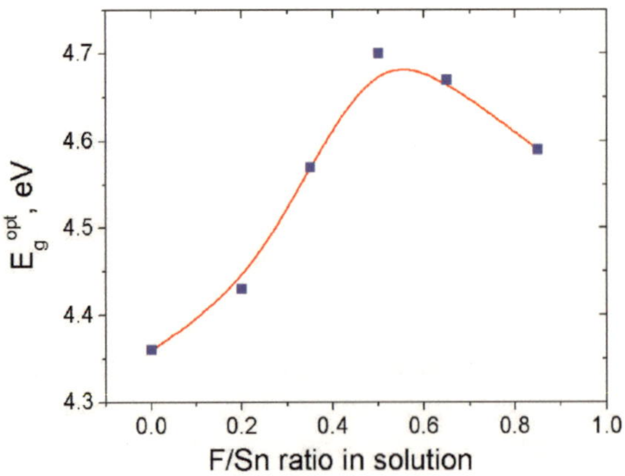

Figure 7. Value of the optical gap for FTO films fabricated from the solutions with different F/Sn ratios.

http://dx.doi.org/10.5772/67431

of the FOM was proposed by Gordon [8] to determine the FOM as the ratio of the electrical conductivity σ to the absorption coefficient α in the visible wavelength range.

$$\phi_G = \frac{\sigma}{\alpha} = \frac{1/R_s t}{\frac{-\ln(T+R)}{t}} = -[R_s \ln(T+R)]^{-1} \quad (2)$$

where t is the film thickness, R_s is the sheet resistance in ohms per square (Ω^{-2}), T is the total visible transmittance, and R is the total visible reflectance. From this point of view, a FOM = 3 Ω^{-1} is reported for CVD fluorine-doped tin oxide films with R_s = 8 Ω^{-2} and visible absorption coefficient α = 0.04 1/cm. Other authors [9] report higher values of FOM = 5.3 Ω^{-1} for FTO films fabricated using aerosol-assisted CVD. The commercial FOM standard for FTO films lies in the 1.5–1.6 Ω^{-1} range.

The best-known value of the FOM=7 Ω^{-1} is also reported by Gordon [10] for fluorine-doped zinc oxide films. The use of this method requires the measurement of the reflectance and the transmittance spectra, which is a disadvantage, however, has been adopted to establish the TCO films FOM for developing thin film solar cells modules.

In this work, we used both methods for the calculation of the FOM (ϕ), proposed by Haacke (ϕ_H) and by Gordon (ϕ_G). For using the Haacke method, the wavelength of the transmittance must be in the visible spectral range, though a specific wavelength is not determined. This leads to some uncertainty on the calculation of the FOM reported by different authors because of the strong dependence of the transmittance on the precise wavelength values. Furthermore, a strong interference effects on the films, as well as a significant difference in the maximum and minimum transmittance, strengthens this speculation.

The experimental reflectance and transmittance of the films must be integrated in order to calculate the Gordon FOM in the visible spectral range. We show in **Table 3** a comparison of the highest FOM reported from different authors with those obtained in this work using both methods.

Fabrication method	Sheet resistance, Ω^{-2}	Transmittance, %	$f_{H,}$ 10^{-3} W^{-1} (Haacke)	f_G, W^{-1}, (Gordon)	Reference
CVD, commercial	7	80–82	15.4–19.6	…	[6]
Spray pyrolysis	5.1	85 (550 nm)	35.7	…	[7]
CVD	8	~84	…	3	[8]
Aerosol-assisted	3.9	82 (400–700 nm)	35.2	5.3	[9]
Spray pyrolysis	4.5	84 (400–700 nm)	38.8	5.75	This work

Table 3. The comparison of the highest FOM values reported by different authors.

4. Spray-deposited tin-doped indium oxide films

Tin-doped indium oxide (ITO) films are more widely used for different optoelectronic applications. Spray pyrolysis is the cheapest fabrication method allowing the obtaining of ITO films with high level of electric and optical parameters. Usually, the films are fabricated by spraying solutions of $InCl_3 \cdot 4H_2O$ in ethanol. A small amount of tin chloride ($SnCl_4 \cdot 5H_2O$) is added as dopant. A detailed fabrication technique using such precursor has been described earlier [4]. **Table 4** shows the amount of the tin chloride in 25 ml 0.35 molar solution of $InCl_3 \cdot 4H_2O$ in ethanol to obtain different Sn/In ratios in the precursor solution.

The properties of the ITO films fabricated by spray pyrolysis on heated glass substrates depend strongly on the deposition parameters. Films fabricated in optimal conditions at a temperature of 480°C using the precursor solution with In/Sn ratio around 5% present the highest electric and optical parameters in comparison with films prepared by DC sputtering [20]. It was found that to obtain high-performance sputtered films, an additional annealing in an oxygen atmosphere is necessary [21]; this comparison is shown in **Table 5**.

Some works, for instance [22], were conducted for determining the dependence of the film parameters deposited by the spray pyrolysis method on the variation of the solvent in the spraying solution. It was found that organic solvents, such as ethanol and methanol, are more suitable in comparison with water for obtaining high-quality ITO films. A lower resistivity 2.5×10^{-4} Ω·cm was obtained using the methanol as solvent. The resistivity of the films fabricated using solutions in which water serves as solvent was in the range of 10^{-3} Ω·cm. However, the work function of the ITO films increases when an organic solvent mixed with water was used.

As a basic substance for preparation of the spraying solutions other indium compounds can be used, such as Indium (III) acetate, $In(OOCCH_3)_3$ dissolved in methanol. In this case, tin chloride ($SnCl_4 \cdot 5H_2O$) can be used for doping. Subsequently, the properties of the ITO films fabricated on glass substrates heated to 460°C using such precursor solutions are reported. Different Sn/In ratios in the precursors were used for the determination of the optimum tin content in the spraying solution. A ratio of Sn/In = 6%% corresponds to an optimal doping concentration for obtaining films with the minimum resistivity (**Figure 8**). For all the Sn/In ratios used in the solution, the films fabricated were nanocrystalline with a (400) columnar orientation of grains (inset in **Figure 8**).

The size of grains is around 150–180 nm. The sheet resistance of the 400 nm thick ITO films fabricated from the precursor solution with an optimal Sn/In ratio was $R_s = 5.5\ \Omega^{-2}$. The electron density and the Hall mobility were around 10^{21} cm^{-3} and 28.5 cm^2/V×s, respectively. Transmittance spectrum of such films without influence of the substrate is shown in **Figure 9**.

The figure of merit FOM = $57 \times 10^{-3}\ \Omega^{-1}$ was determined from the experimental values of R_s and integral value of the transmittance $T = 0.89$ in visible and near infra-red spectral range.

Sn/In ratio in solution	0	0.02	0.03	0.05	0.07	0.09	0.11
$SnCl_4 \cdot 5H_2O$, mg	0	61.2	92.0	153.3	214.7	276.0	337.4

Table 4. Amount of $SnCl_4 \cdot 5H_2O$ for obtaining different Sn/In ratios in the $InCl_3 \cdot 4H_2O$ ethanol solution.

Parameters	Sputtered and annealed ITO film	Spray-deposited ITO film
Annealing temperature, °C	300	–
Ratio Sn/In in solution, %	–	5
Substrate roughness (nm)	2.7	2.7
Ratio XRD peaks (222/400)	1.23	0.032
Grain size (nm)	45.0	165
ITO roughness (nm)	0.43	30
Specific resistance (10^{-4} Ω·cm)	2.8	2.3
Carrier concentration (10^{20} cm^{-3})	11.8	10
Mobility ($cm^2V^{-1}s^{-1}$)	22.0	28
Integral transparency	0.83	0.88
FOM (10^{-3} Ω^{-1})	12.5	23

Table 5. The comparison properties of sputtered and sprayed ITO films [20, 21].

Table 6 shows a comparison of this FOM value with that reported in the literature for ITO films with different thickness fabricated by different methods.

According to the high FOM obtained for ITO films deposited from a spraying solution based on indium acetate in ethanol, it is possible to conclude that this method allows for the fabrication of high-quality tin-doped indium oxide films for a number of optoelectronics applications.

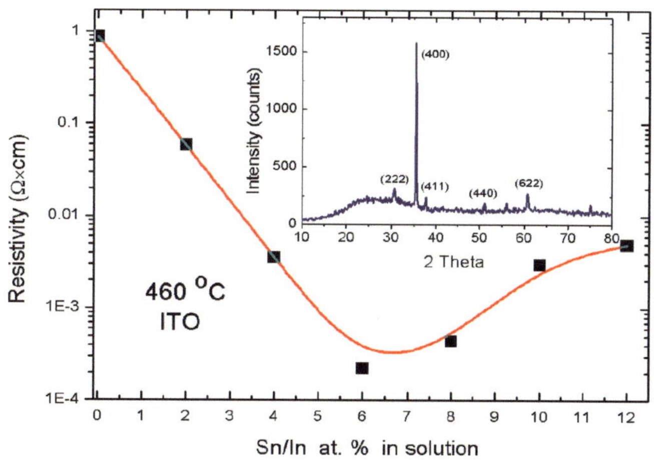

Figure 8. Resistivity of the ITO films fabricated using an indium acetate solution with different Sn/In ratios. The inset shows the XRD spectra of the ITO films with a 6 at.% of Sn/In ratio.

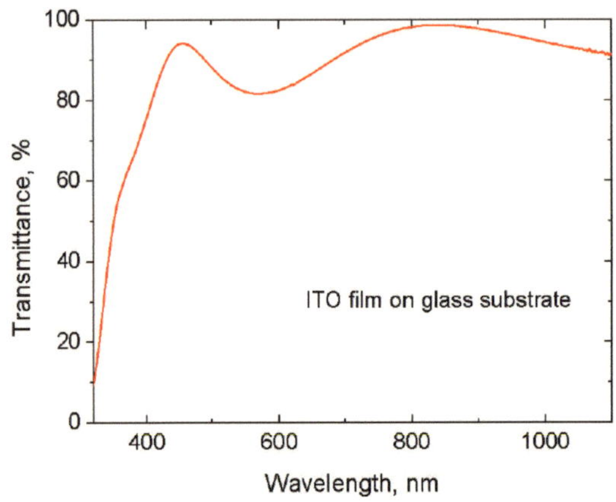

Figure 9. Transmittance of the ITO films fabricated using an indium acetate solution with a Sn/In= 6 at.% ratio.

Deposition method	R_s, Ω^{-2}	Transmittance, %	FOM, 10^{-3} Ω^{-1}	Reference
Spray pyrolysis	26	90	13.4	[23]
Spray pyrolysis	9.3	85	21	[24]
Reactive evaporation	25	98	32.7	[25]
Dip coating	7.1	78	11.7	[26]
Sputtering DC	22	92	19.7	[27]
Spin coating	30	90.2	11.9	[28]
This work	5.5	89	57	This work

Table 6. Comparison of the FOM obtained in this work using the spray pyrolysis method with that reported in literature for ITO films fabricated by other methods.

5. Spray-deposited aluminum and indium-doped zinc oxide films

The undoped zinc oxide (ZnO) thin films, with an energy band gap of 3.3 eV, present a high electrical resistivity and are used for several applications. For example, they have piezoelectric properties and their highly oriented texture may be of interest for high-frequency electroacustic transductors. When ZnO films are doped with indium [29, 30] or aluminum [31], the films present a high conductivity. Due to the high conductivity and transparency in visible region of the spectrum, the doped zinc oxide films are of great interest as transparent conductors in optoelectronic displays, photovoltaic structures, thermal reflecting layers, and also when used as a sensitive element in gas sensors. The spray pyrolysis method is very suitable for the fabrication of high transparent and low conductive ZnO films [32–36]. Precursors for the spraying alcoholic solution containing zinc ($ZnCl_2$, $Zn(NO)_2$, Zn acetate, etc.) are cheaper than those containing indium. The precursors for doping may be $InCl_3$ and $AlCl_3$ [37, 38]. However, the fabrication of In- or Al-doped ZnO films with low resistivity by spray pyrolysis without a postannealing processing does not lead to the expected results [39].

In this work, undoped and In- and Al-doped ZnO films were grown on borosilicate glass substrates by spray pyrolysis at 500°C. A 0.3 molar solution was made by dissolving zinc acetate in a 3:1 mixture of methanol and water. For doping purposes, $InCl_3$ or $AlCl_3$ was added in the spraying solution. The atomic ratio of In/Zn or Al/Zn in the solution was kept in the range 1–5%. After the deposition, some films were annealed at 300°C during 30 min in high vacuum or in argon atmosphere.

The X-ray diffraction spectra of the undoped and doped annealed ZnO films show a ZnO-single phase with an hexagonal wurtzite structure [40] as shown in **Figure 10**.

All the films are nano-crystalline with grain sizes between 40 and 60 nm, as determined by the Debye-Scherrer method [41].

Introduction of indium in ZnO matrix changes the preferred grain orientation from (101) for undoped ZnO films to (002) for In-doped ZnO films. Annealing in argon atmosphere leads to a more pronounced enhancement of the grain sizes for Al-doped ZnO films. The SEM images of the grown films are shown in **Figure 11**.

From **Figure 11**, it is clear that the In-doped films are smoother than the undoped and Al-doped ZnO films.

The variation of resistivity of the films as a function of doping and thermal treatment is shown in **Figure 12** [39]. It was found experimentally that the lowest resistivity is presented by the films with a ratio of Al/Zn or In/Zn around 3%. Annealing reduces the resistivity when this is conducted in an argon atmosphere. This fact can be explained as due to the desorption of chemisorbed oxygen at the grain boundaries [42], leading to an annihilation of the oxygen acceptor states at the grain boundaries, which act as traps for electrons.

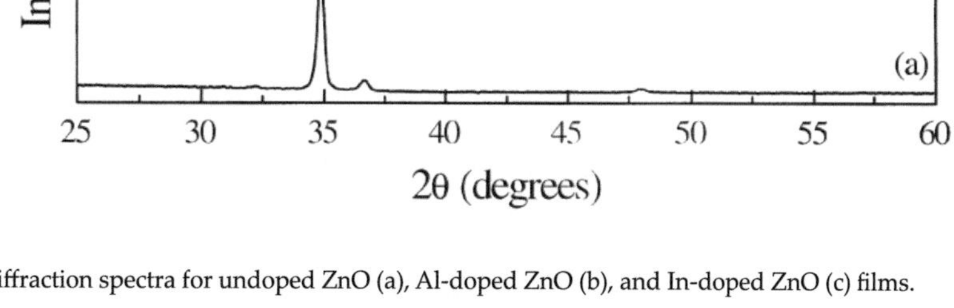

Figure 10. X-ray diffraction spectra for undoped ZnO (a), Al-doped ZnO (b), and In-doped ZnO (c) films.

Figure 11. SEM images of undoped ZnO (a), Al-doped ZnO (b), and In-doped ZnO (c) films.

The Hall effect data show that the films are n-type with a mobility ranging from 1.8 to 13 $cm^2V^{-1}s^{-1}$ when the films were annealed in argon atmosphere. The highest value of mobility is observed for the In-doped ZnO films fabricated using a solution with an In/Zn ratio of about 3%. The worse structural and electrical characteristics of Al-doped ZnO films could be explained as due to an interstitial insertion of Al atoms in the ZnO lattice leading to a distortion of the crystalline lattice and disorder in the grains orientation.

The transmittance spectra of the films in the visible and near infra-red spectral range are above 80%. A reduction in the transmittance could be explained by the light scattering (haze), and this is directly connected with the films morphology. Light scattering can be characterized by the haze factor defined as the ratio between the diffuse transmittance measured in a photometric sphere and the total transmittance (diffuse + specular). The haze factor depends on the wavelength and serves as an indicator for the light-scattering capability of diffusing

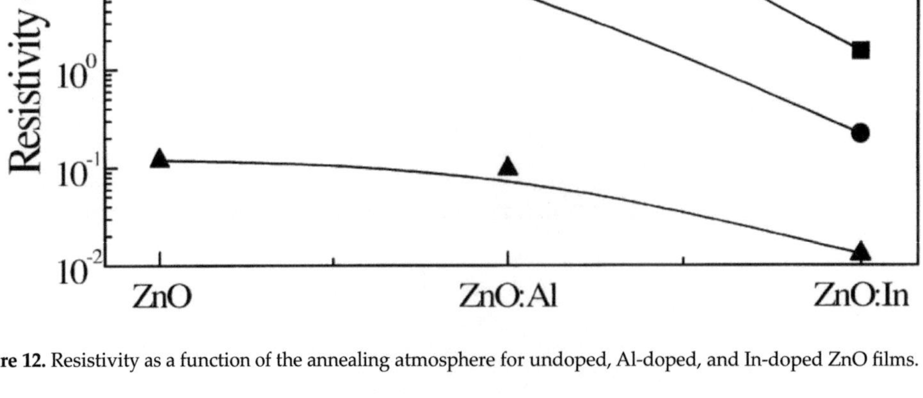

Figure 12. Resistivity as a function of the annealing atmosphere for undoped, Al-doped, and In-doped ZnO films.

transparent conducting films used in thin film solar cell applications. Textured surfaces of the films lead to an enhancement on the light collected into thin film photovoltaic structures. Al-doped ZnO films a higher haze factor, 30%, in comparison with undoped ZnO films (haze factor = 19%) and In-doped ZnO films (haze factor = 0.85 for In/Zn = 3 at.%). It is directly connected with the films morphology shown in **Figure 11**, which indicates that Al disturbs strongly the film structure.

Figure 13 shows the transmittance spectra of undoped and In-doped ZnO films. The In/Zn ratios are shown in the inset of the figure. Values of figure of merit (FOM) for the undoped and doped ZnO films differ a lot from those reported above for FTO and ITO films. As deposited, the ZnO films present low FOM values in the range of $2.5 \times 10^{-7}\ \Omega^{-1}$ for undoped films to $2.8 \times 10^{-5}\ \Omega^{-1}$ obtained for In-doped films.

The highest FOM = $3.7 \times 10^{-3}\ \Omega^{-1}$, after annealing in argon, is presented by the In-doped ZnO films. However, it is one order of magnitude below the best FOM for FTO and ITO films. Undoubtedly, spray-deposited doped ZnO films is of great interest for optoelectronic applications, but for fabrication such films with high FOM values, a further investigation is necessary.

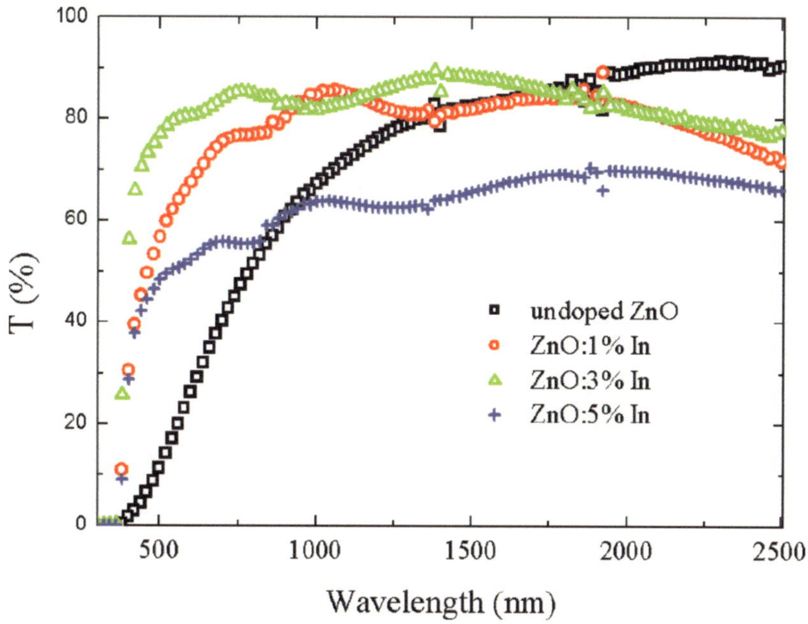

Figure 13. The transmittance spectra of undoped and In-doped ZnO films.

6. Spray-deposited fluorine-doped titanium dioxide films

In several applications such as in sensors, pigments, protective coatings, dye-sensitized solar cells, and photocatalysis [43] titanium oxide (TiO_2) has been found to be attractive. Due to its nontoxicity, high efficiency with a low cost, and biological and chemical stability, TiO_2 is probably the most studied photocatalyst. Titanium dioxide films are very attractive to optoelectronic applications, such as electrochemical solar cells, antireflection coatings, and protective layers, to prevent plasma reduction of transparent conductive oxides in thin film solar cells. Because of its simplicity, spray pyrolysis is an attractive method for the fabrication of these films [44–46].

We demonstrate the influence of fluorine on the structural and optical properties of titanium dioxide films, prepared by spray pyrolysis on glass substrates at 470°C, using $Ti(OC_3H_7)$ alcoholic solutions with an addition of fluorine in the form of NH_4F, over a large range of fluorine to titanium atomic ratio in the solution. The characterization was performed using X-ray diffraction analysis and measurements of transparency. Without NH_4F in the starting solution, the formation of titanium dioxide layers was not possible . A high amount of ammonium fluoride (F/Ti ratio in solution about of 8.5 at.%) resulted in the formation of a white hygroscopic powder, probably of polymeric nature. The formation of transparent (above 85%) and nano-crystalline TiO_2 films depends strongly on the F/Ti ratio in the spaying solution. Films prepared using a solution with F/Ti <1.5 at.% present an amorphous structure and rough surface, with an integral haze factor near 20%. Films fabricated using a solution with F/Ti ratio near 1.5–1.6 at.%, present a nanocrystalline structure, not usually observed, *brookite* structure, an orthorhombic crystalline structure with a unit cell described by the space group Pbca [47, 48], and their XRD patterns consist only in one (121) sharp peak (**Figure 14**). Such films have a smoother surface (optical haze factor <10%).

Edge absorption measurements show that the absorption coefficient above the threshold of fundamental absorption follows the $(E-E_g)^{1/2}$ energy dependence characteristic of direct allowed transitions (determined value of energy gap E_g is 3.72 eV), in contrast to the indirect

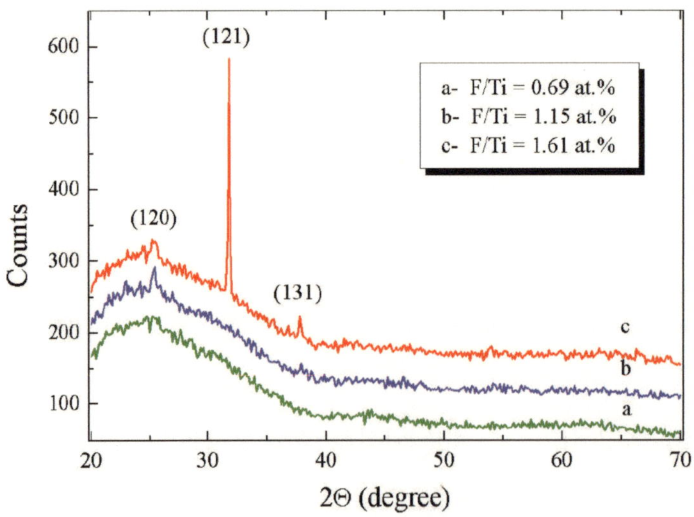

Figure 14. XRD spectra of fluorine-doped TiO2 films fabricated by spray pyrolysis.

http://dx.doi.org/10.5772/67431

transitions in TiO_2 films either with rutile or anatase crystalline structure (E_g = 3.03 eV and 3.2 eV, respectively). The synthesis of fluorine-doped TiO_2 powder was reported by using the spraying of an aqueous solution of H_2TiF_6 with F/Ti ratio from 2.76% to 9.40% at 1173 K [43]. Additional works reporting fluorine-doped titanium dioxide was not found. Undoubtedly, further investigation in this field must be of interest. Pure brookite has demonstrated to be an interesting candidate in photocatalytic applications, and the number of papers on the preparation and photocatalytic activity of the TiO_2 with brookite structure has increased exponentially by 100 times in the past 10 years [48].

7. Application of TCO films in optoelectronic devices

The spray pyrolysis method for the fabrication of high-quality TCO films has been successfully applied to the development of simple but enough effective optoelectronic devices. Readers have found a detailed description of useful effective photoelectronic devices based on surface barrier heterojunctions in our published works [49–58].

The physics of surface-barrier (SB) devices is based on the well-known Schottky barrier that is a potential energy barrier for carriers formed at a metal-semiconductor (M-S) junction [59]. In the design of SB devices, the main function of the barrier is to separate carriers photogenerated inside the semiconductor substrate. Such optoelectronic devices are very simple structures because they are unnecessary for the formation of a p-n junction by high-temperature processes. SB optical detectors operate without optical losses in the highly doped p-layer and also present a very high speed of response for a modulated optical signal. The SB structure is very useful for the fabrication of optoelectronic semiconductor devices in which a p-n junction cannot be created due to doping troubles. At the same time, in order to use M-S structures as radiation detectors, the metal electrode must be extremely thin (<15 nm) to prevent the losses of the radiation absorption and also must be chemically resistant to prevent device degradation in time (thinner metallization leads to devices susceptible to degradation). Among the several metals available, only some of them such as Au, Pt, Ni are suitable for this applications. Moreover, the high reflectivity of these metallic layers demands the use of antireflection coatings. The electrical properties of SB photodetectors based on M-S structures can be enhanced by the introduction of a very thin (<3 nm) insulating layer between the metallic film and the semiconductor (M-I-S structures). The presence of this insulator layer reduces the number of localized states at the semiconductor interface and hence serves to reduce the interface carrier recombination. Its presence can also reduce significantly the thermo ionic emission current because of an increase in the potential barrier for majority carriers [60]. In this case, the thickness of the insulating layer must not reduce the transport of minority carriers from the semiconductor to the metal.

In order to obtain the best photoelectrical properties of SB radiation detectors based on M-S and M-I-S structures, the metallic opaque layer can be changed by a thin film based on some transparent conducing oxide (TCO), such as tin-doped indium oxide (ITO) or fluorine-doped tin oxide (FTO), which are widely used for optoelectronic applications. In such structures, the TCO films operate as an active metal-like transparent conducting electrode. Thickness tuning of the TCO films allows for its use as an effective antireflection coating for reducing the radiation losses. The spray pyrolysis is the simplest method used for the deposition of such films with high electrical and optical parameters.

Detailed characteristics of optoelectronic devices will be published soon. Here, we present only some general characteristics to demonstrate the usefulness of the TCO films. **Figure 15** shows the spectral response of high-speed SB photodiodes for ultra-violet, visible, and near infra-red spectral range fabricated by the deposition of the ITO or FTO film on the surface of some epitaxial semiconductors structures, GaP, AlGaAs compounds, and Si. Density of a dark current does not exceed a value in the 10^{-9}–10^{-6} A/cm^2 range.

Figure 16 shows the current-voltage characteristic (a) and output power (b) of a solar cell under AM1 solar illumination fabricated on monocrystalline silicon by deposition of a high-quality ITO film on a chemically treated silicon surface [59, 60]. The efficiency of such low cost in the fabrication of solar cells is around 13%.

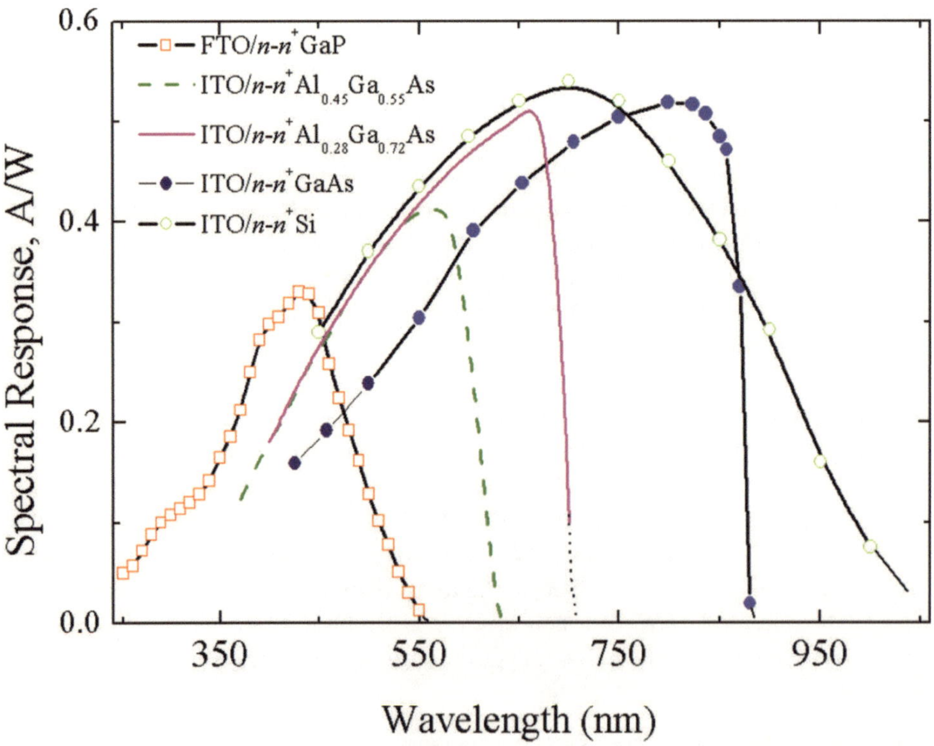

Figure 15. The spectral response of the surface-barrier photodiodes fabricated by spray deposition of the ITO and FTO films on the surface of the semiconductor epitaxial structures.

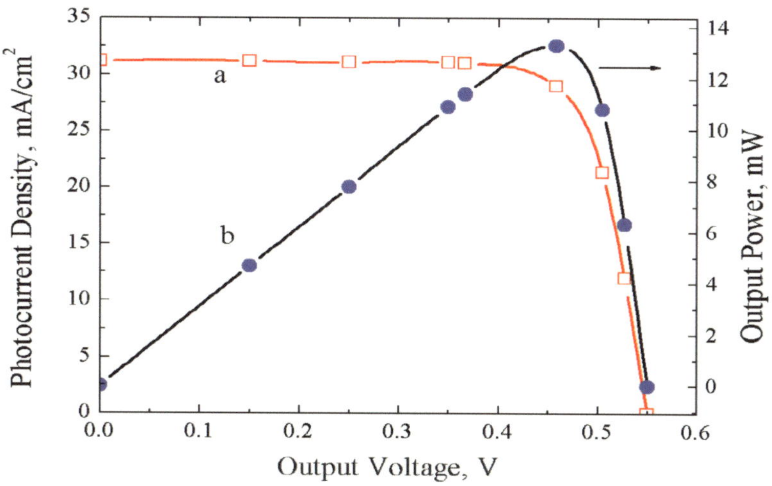

Figure 16. Current-voltage characteristic of a solar cell under AM1 solar illumination fabricated on monocrystalline silicon by deposition of high-quality ITO film on a chemically treated silicon surface.

8. Conclusion

We report some basic knowledge about the spray pyrolysis method for the fabrication of high-quality transparent conducting oxide films used for optoelectronic applications. This is a low cost though an efficient method that has been used for the deposition of fluorine-doped tin oxide, tin-doped indium oxide, indium- or aluminum-doped zinc oxide, and fluorine-doped titanium dioxide films and other semiconductor and dielectric thin films. The high-level electrical and optical parameters of these films prove that this method is very suitable for the fabrication of different optoelectronic devices.

Author details

Oleksandr Malik[1,*], Francisco Javier De La Hidalga-Wade[1] and Raquel Ramírez Amador[2]

*Address all correspondence to: amalik@inaoep.mx

1 National Institute for Astrophysics, Optics and Electronics, INAOE, Puebla, Mexico

2 Technical University of Huejotzingo, Puebla, Mexico

References

[1] Handbook of transparent conductors. D. S. Ginley, H. Hosono, and D. C. Paine, Eds. 2010. Springer, New York.

[2] R. R. Chamberlin, J. S. Skarman. Chemical spray deposition process for inorganic films-. J. Electrochem. Soc. 1966; 113(1): 86-89; doi:10.1149/1.2423871

[3] P. S. Patil. Versatility of chemical spray pyrolysis technique. Mater. Chem. Phys. 1999; 59: 185-198.

[4] O. Malik, F. J. De la Hidalga-W. Physical and technological aspects of solar cells based on metal oxide-silicon contacts with induced surface inversion layer. In: Application of Solar Energy, R. Rugescu Ed. 2013. Intech, Croatia. ISBN 978-953-51-0969-3

[5] J. B. Mooney, S. B. Radding. Spray pyrolysis processing. Annu. Rev. Mater. Sci. 1982; 12: 81-101.

[6] J. C. Viguié, J. Spitz. Chemical vapor deposition at low temperatures. J. Electrochem. Soc. 1975; 122(4): 585-588.

[7] G. J. Exarhos, X-D. Zhou. Discovery-based design of transparent conducting oxide films. Thin Solid Films. 2007; 515: 7025-7032.

[8] Y. Sawada, C. Kobayashi, S. Seki, and H. Funakubo. Highly conducting indium-tin-oxide transparent films fabricated by spray CVD using ethanol solution of indium (III) chloride and tin (IV) chloride. Thin Solid Films. 2002; 409: 46-50.

[9] T. Fukano, T. Motohiro. Low-temperature growth of highly crystallized transparent conductive fluorine-doped tin oxide films by intermittent spray pyrolysis deposition. Sol. Energ. Mater. 2004; 82: 567-575.

[10] B. Benhaoua, S. Abbas, A. Rahal, A. Benhaoua, and M. S. Aida. Effect of film thickness on the structural, optical and electrical properties of SnO_2:F thin films prepared by spray ultrasonic for solar cells applications. Superlattices Microstruct. 2015; 83: 78-88.

[11] L. Filipovic, S. Selberherr, G. C. Mutinati, E. Brunet, S. Steinhauer, A. Kock, J. Teva, J. Kraft, J. Siegert, and F. Schrank. Modeling spray pyrolysis deposition. In: Proceedings of the World Congress on Engineering. 2013; Vol II. London, UK. ISBN: 978-988-19252-8-2.

[12] A. Malik, V. Baranyuk, and V. Manasson. Solar cells based on the SnO_2-SiO_2-Si heterojunction. Appl. Sol. Energ. 1979; 2: 83-84. ISSN: 0003-701x.

[13] A. Malik, V. Baranyuk, and V. Manasson. Improved model of solar cells based on In_2O_3/SnO_2-SiO_x-nSi structures. Appl. Sol. Energ. 1980; 1: 1-2.

[14] O. Malik, F. J. De la Hidalga-Wade, and R. R. Amador. Fluorine-doped tin oxide films with a high figure of merit fabricated by spray pyrolysis. J. Mater. Res. 2015; 30(13): 2040-2045.

[15] Joint Committee on Power Diffraction Standards (JCPDS). International Centre for Diffraction Data. 1997; Card No. 41-1445.

[16] K. H. Kim and J. S. Chun: X-ray studies of SnO_2 prepared by chemical vapor deposition. Thin Solid Films. 1986; 141: 287-292.

[17] J. W. Kim, H. S. Kang and S. Y. Lee. Effect of deposition rate on the property of ZnO thin films deposited by pulsed laser deposition. J. Electr. Eng. Technol. 2006; 1; 98-105.

[18] E. Shanthi, A. Banerjee, and K. L. Chopra: Dopant effects in sprayed tin oxide films. Thin Solid Films. 1982; 88: 93 -.

[19] A. I. Martínez, L. Huerta, J. M. O-Rueda de León, D. Acosta, O. Malik, and M. Aguilar: Physicochemical characteristics of fluorine doped tin oxide films. J. Phys. D: Appl. Phys. 2006; 39: 5091-.

[20] O. Malik, F. J. De la Hidalga-W. Spray deposited thin films of tin-doped indium oxide for optoelectronic applications. Adv. Mat. Res. 2013; 677: 173-178.

[21] O. Malik, F. J. De la Hidalga-W. Comparison of tin-doped indium oxide films fabricated by spray pyrolysis and magnetron sputtering. Cryst. Res. Technol. 2015; 50: 516-520.

[22] T. Ishido, H. Kouno, H. Kobayashi, and Y. Nakato. Dependence of photovoltages of spray-deposited indium tin oxide/silicon oxide/silicon junction solar cells on spray solvents. J. Electrochem. Soc. 1994; 141(5): 1357-1361.

[23] V. Vasu, and A. Subrahmanyam. Reaction kinetics of the formation of indium tin oxide films grown by spray pyrolysis. Thin solid Films. 1990; 193/194: 696-703.

[24] J. C. Manifacier, J. P. Fillard, and J. M. Bind. Deposition of In_2O_3-SnO_2 layers on glass substrates using a spraying method. Thin Solid Films. 1981; 77: 67-80.

[25] P. Nath, R. F Bunshah. Preparation of In_2O_3 and tin-doped In_2O_3 films by a novel activated reactive evaporation technique. Thin Solid Films. 1980; 69: 63-68.

[26] S. Seki, Y. Sawada, M. Ogawa, M. Yamamoto, Y. Kagota, A. Shida, and M. Ide. Highly conducting indium tin-oxide transparent films prepared by dip-coating with an indium carboxylate salt. Surf. Coat. Technol. 2003; 169:525–527.

[27] U. Betz, M. K. Olsson, J. Marthy, M. F. Escola, F. Atamny, Thin films engineering of indium tin oxide: large area flat panel displays application. Surf. Coat. Technol. 2006; 200: 5751–5759.

[28] Z. Chen, W. Li, Ran Li, Y. Zhang, G. Xu, and H. Cheng. Fabrication of highly transparent and conductive indium tin oxide thin films with a high figure of merit via solution processing. Langmuir. 2013; 29: 13836–13842.

[29] S. N. Qiu, C. X. Qiu, and I. Shih. Air heat treatment of In-doped ZnO thin films. Sol. Energ. Mater. 1987; 15: 261-267.

[30] Sh. El Yamny, M. Abdel Rafea. Preparation and characterization of ZnO: in transparent conductor by low cost dip coating technique. J. Mod. Phys. 2012; 3: 1060-1069.

[31] C. Lennon, R. Kodama, Y. Chang, S. Sivanathan, and M. Deshpande. Al- and Al:In-doped ZnO thin films deposited by RF magnetron sputtering for spacecraft charge mitigation. J. Electron. Mater. 2008; 37(9): 1324-1328.

[32] J. Aranivich, A. Ortiz, and R. H. Bube. Optical and electrical properties of ZnO films prepared by spray pyrolysis for solar cell applications. J. Vac. Sci. Technol. 1979; 16: 994-1001.

[33] M. Krunks, E. Mellikov. Zinc oxide thin films by the spray pyrolysis method. Thin Solid Films. 1995; 270: 33-36.

[34] A. Ashour, M. A. Kaid, N. Z. El-Sayed. and A. A. Ibrahim. Physical properties of ZnO thin films deposited by spray pyrolysis technique. Appl. Surf. Sci. 2006; 252: 7844-7848.

[35] N. Lehraki, M. S. Aida, S. Abed, N. Attaf, A. Attaf, and M. Poulain. ZnO thin films deposition by spray pyrolysis: influence of precursor solution properties. Curr. Appl. Phys. 2012; 12: 1283-1287.

[36] A. Goyal, S. Kachhwaha. ZnO thin films preparation by spray pyrolysis and electrical characterization. Mater. Lett. 2012; 68: 354-356.

[37] K. Krunks, O. Bijakina, V. Mikli, T. Varema, and E. Mellikov. Zinc oxide thin films by spray pyrolysis method. Phys. Scr. 1999; T79: 209-212.

[38] M. A. Kaid, A. Ashour. Preparation of ZnO-doped Al films by spray pyrolysis technique. Appl. Surf. Sci. 2007; 253: 3029-3033.

[39] P. Nunes, A. Malik, B. Fernandes, E. Fortunato, P. Vilarinho, and R. Martins. Influence of the doping and annealing atmosphere on zinc oxide thin films deposited by spray pyrolysis. Vacuum. 1999; 52: 45-49.

[40] A. Cimino, G. Mazzone, P. Ports. Lattice parameter study of defective zinc oxide. Zinc excess and distortions in pure ZnO. Z. fuer Phys. Chem. (Muenchen, Germany). 1964; 41(3/4): 154-172. ISSN: 0044-3336.

[41] J. W. Kim, H. S. Kang, and S. Y. Lee. Effect of deposition rate on the property of ZnO thin films deposited by pulsed laser deposition. J. Electrical Eng. Technol. 2006; 1(1): 98-100.

[42] S. Major, A. Banerjee, K. L. Chopra. Annealing studies of undoped and indium-doped films of zinc oxide. Thin Solid Films. 1984; 122(1): 31-43.

[43] D. Li, H. Haneda, S. Hishita, N. Ohashi, and N. K. Labhsetwar. Fluorine-doped TiO_2 powders prepared by spray pyrolysis and their improved photocatalytic activity for decomposition of gas-phase acetaldehyde. J. Fluor. Chem. 2005; 126: 69-77.

[44] I. Oja, A. Mere, M. Krunks, C-H. Solterbeck, and M. Es-Souni. Properties of TiO_2 films prepared by the spray pyrolysis method. Sol. State Phenom. 2004; 99/100: 259-264.

[45] L. Andronic, S. Manolache, and A. Duta. TiO_2 thin films prepared by spray pyrolysis deposition (SPD) and their photocatalytic activities. J. Optoelectron. Adv. Mater. 2007; 9(5): 1403-1406.

[46] N. C. Raut, T. Mathews, S. T. Sundari, T. N. Sairam, S. Dash, and A. K. Tyagi. Structural and morphological characterization of TiO_2 thin films synthesized by spray pyrolysis technique. J. Nanosci. Nanotechnol. 2009; 9(9): 5298-5302.

[47] L. Pauling, J. H. Sturdivant. The crystal structure of brookite. Z. Kristall. 1928; 68: 239-256.

[48] I. N. Kuznetsova, V. Blaskov, I. Stambolova, L. Znaidi, and A. Kanaev, TiO2 pure phase brookite with preferred orientation, synthesized as a spin-coated film. Mater. Lett. 2005; 59: 3820-3823.

[49] A. Malik, A. Seco, E. Fortunata, and R. Martins. New UV-enhanced solar blind optical sensors based on monocrystalline zinc sulphide. Sens. Actuators A. 1998; 67: 68-71.

[50] A. I. Malik, G. G. Grushka. Optoelectronic properties of metal oxide-gallium phosphide heterojunctions. Sov. Phys. Semicond. 1991; 25(10): 1017-1020. ISSN 0038-5700.

[51] A. Malik, A. Seco, E. Fortunato, R. Martins, B. Shabashkevich, and S. Piroszenko. A new high ultraviolet sensitivity FTO-GaP Schottky photodiode fabricated by spray pyrolysis. Semicond. Sci. Technol. 1998; 13: 102-107.

[52] Yu. Vygranenko, A. Malik, M. Fernandes, R. Schwarz, and M. Vieira. UV-visible ITO/GaP photodiodes: characterization and modeling. Phys. Stat. Sol. A. 2001; 185(1): 137-144.

[53] O. Malik, F. J. De la Hidalga-Wade, C. Zuniga-Islas, and J. Abundus Patiño. UV-sensitive optical sensors based on ITO-gallium phosphide heterojunctions. Phys. Stat. Sol. C. 2010; 7(3-4): 1176-1179.

[54] A. Malik, M. Vieira, and M. Fernandes. Surface-barrier Si-based photodetectors fabricated by spray pyrolysis technique. Philos. Mag. B. 2000; 80(4): 781-790.

[55] A. I. Malik, G. G. Grushka. Self-calibrated radiometric IR photodiode based on the defect semiconductor $Hg_3In_2Te_6$ for the spectral range 0.85– 1.5 μm. Soviet. Tech. Phys. 1990; 35: 1227-1231.

[56] A. I. Malik, M. Vieira, M. Fernandes, F. Macarico, and Z. Grushka. Near-infrared photodetectors based on HgInTe-semiconductor compound. Proceedings of SPIE 3629, Photodetectors: Materials and Devices IV. 1999: 433-442.

[57] O. Malik, F. J. De la Hidalga-W, C. Zúñiga-I, and G. Ruíz-T. Efficient ITO-Si solar cells and power modules fabricated with a low temperature technology: results and perspectives. J. Non Cryst. Solids. 2008; 354: 2472-2477.

[58] O. Malik, F. J. De la Hidalga-W. Physical and technological aspects of solar cells based on metal oxide-silicon contacts with induced surface inversion layer. In: Application of Solar Energy, R. Rugescu Ed. 2013. Intech, Croatia. ISBN 978-953-51-0969-3.

[59] R. T. Tung. The physics and chemistry of the Schottky barrier height. Appl. Phys. Rev. 2014; 1: 1-55.

[60] S. J. Fonash. Solar cell device physics. Academic Press. NY 1981: 331 p. ISBN 0-12-261980-3.

Section 3

Nanomaterials and Semiconductors

Chapter 9

Fabrication of TiO_2 Nanoparticles and Thin Films by Ultrasonic Spray Pyrolysis: Design and Optimization

Raymond Taziwa and Edson Meyer

Additional information is available at the end of the chapter

http://dx.doi.org/10.5772/67866

Abstract

Ultrasonic spray pyrolysis (USP) methods offer an economical, efficient, and dependable method of depositing nanoparticles (NPs) with consistent crystalline structure and stoichiometry. The need for precise control of structural, morphological, and optical properties has stimulated researches on development spray pyrolysis (SP) methods for depositing titanium dioxide nanoparticles (NPs) and thin films (TF) to substitute traditional sophisticated and expensive wet chemistry methods and solid state techniques. SP methods, as compared to other solid state techniques, offer precise control of the stoichiometry of precursor's solutions prepared by wet chemistry methods. Moreover, SP methods offer deposition simplicity as the NPs and TFs are deposited at room temperature and pressure. The deposited NPs and TFs are produced in a single-step route without the need for laborious, expensive purification and excessive annealing procedures. The present chapter offers the experimental challenges and accomplishments experienced while working with the USP systems. Knowledge gathered was key to the development of the present USP system presented herein. This book chapter starts by presenting a review of the current methods available for fabrication of TiO_2 NPs and TFs. This chapter also provides a detailed report on the numerous experimental considerations utilized in the optimization of the novel USP system for depositing titanium dioxide NPs and TFs. Finally, the design of the USP system is presented.

Keywords: aerosol process, ultrasonic spray pyrolysis, TiO_2, nanopowders, thin films

1. Introduction

Nanostructures of titanium dioxide (TiO_2) such as nanoparticle (NPs) and thin films (TFs) have over the past century received widespread attention [1]. For both research applications and industrial applications, TiO_2 has received wide attention due to its common applications as an anode electrode for solar cells and anode electrode for photocatalytic degradation of organic waste [1, 2]. Notably, TiO_2 displays exceptional opto-electronic properties. Moreover, TiO_2 in addition to being a low-cost, wide band semiconductor material has unique chemical properties and is non-poisonous. Moreover, there are numerous stimuli for studying development of TiO_2 NPs and TFs for solar applications. TiO_2 TFs and NPs are currently employed as anode electrodes for solar cells. Moreover, the current industries involved in production of TiO_2 NPs and TFs are accustomed with the technology and will not be hesitant to adapt to synthesis techniques. Furthermore, the necessary deposition equipment is operating today in the plants. As a consequence, the development of new solar cell technology that incorporates TiO_2-processing stages could be readily implemented by the current industry without the typical long lead in time for new technology. TiO_2 NPs and TFs have been deposited by many different techniques, including the hydrothermal method [2], chemical vapor deposition [3], electrodeposition [4], pyrolysis [5], sol-gel method [6], spin methods [7], laser ablation [8], ion-assisted deposition [9], and spray pyrolysis methods [10, 11].

The structural, optical, electronic, and morphological properties of TiO_2 NPs are closely dependent on the NPs physical properties such as size and stoichiometric composition; it is necessary to have a deposition technique in which these properties are easy to deal with. Moreover, the morphology, crystalline structure, structural properties, optical properties, electronic properties, and stoichiometry of TiO_2 NPs and TFs have very sensitive deposition conditions. This is an inherent disadvantage for many physical vapor deposition methods [3, 7] such as laser ablation, where there are large variations in the optical and structural properties of the synthesized NPs and TFs [8]. These variations in NPs properties arise from small changes in the deposition conditions which are a key challenge in the development of techniques for the production of TiO_2 NPs and TFs at a large industrial scale. Therefore, the need for stoichiometric techniques in TiO_2 NPs and TFs suggests that a deposition method where the stoichiometry of the final product is controlled by a chemical reaction would enable more consistent results. Thus, much devotion has been dedicated on deposition of TiO_2 NPs and TFs in a nano-scale range with consistent crystalline structure, morphology, and chemical stoichiometric ratios of TiO_2. Spray pyrolysis methods have proven to be reliable as they are able to control the stoichiometry precursor solutions by pyrolysis by utilizing a chemical reaction to control the chemical composition of the precursor solution [5, 10, 11]. Moreover, it is an inherent requirement when developing new methods for production of TiO_2 NPs and TFs to keep the deposition system as simple as possible and keep costs at a minimum and, at the same time, maximize throughput. Spray pyrolysis methods offer such an opportunity. Moreover, in spray pyrolysis (SP) methods, NPs and TFs are produced in a one-step process without the need for further purification or excessive drying procedures, which has a negative effect on the total thermal budget and cost of production of TiO_2 [13–15]. Equally, the technique can be used to coat large substrates and presents the opportunity for industrial scaling. Spray pyrolysis method, unlike other solid state deposition methods, presents a very simple reasonable economical method for depositing TiO_2 NPs and TFs on industrial scale.

In SP methods, the precursor is produced via a precisely controlled chemical reaction to produce TiO_2 and succeeds either by hydrolysis or pyrolysis. In hydrolysis, for SP-assisted systems, separate gas line of carrier gas is bubbled through heated baths of a liquid TiO_2

precursor and water [13, 14]. The two delivery lines are then brought together close to the substrate where the reaction takes place. Hydrolysis systems have the advantage of simplicity; although, they may be relatively inflexible, as tubing diameters have to be designed around predicted flow rates. Pyrolysis systems are similar, except that a water bath is not required as the TiO_2 precursor decomposes upon reaching the heated substrate. To keep the deposition system as simple as possible, maximize throughput, and keep costs at a minimum, systems with a vacuum chamber were not considered to be a viable option. This excluded evaporation, sputtering, and the majority of chemical vapor deposition (CVD) systems. Spin-on methods were not seriously considered, as the throughput of any such system would be limited in a production environment. Screen-printing is commonly used in the photovoltaic industry PV industry for depositing metallic contacts, about 30–50 μm. Screen-printing has been used for depositing TiO_2 TFs [16, 17]. As the thickness of the metallic contacts is about 500 times thicker than the TiO_2 TFs, it is not known how the thinner films behaved with regard to reproducibility, squeegee wear, and thickness uniformity. Following the screen-printing of organometallic ink, the samples were fired in a three-step process for a 30 min duration [16, 17]. The firing is a relatively slow process as first; the thick film needs to settle for 15 min to obtain a uniform film, with subsequent drying performed at 125°C for 5 min. The final crystallization was performed in a belt furnace at temperatures between 500 and 900°C. Lengthy drying procedures are required to remove the substantial amount of organic solvents added to the TiO_2 precursor [18–21]. Therefore, to lower the thermal budget and processing costs, it would be attractive to deposit a nano-crystalline TiO_2 thin film in one step without subsequent heat treatment steps. Hence, in this book chapter, we present a novel, less-sophisticated, and economical ultrasonic spray pyrolysis (USP) system for production of high-quality TiO_2 NPs and TFs and design considerations of an ultrasonic spray pyrolysis system.

1.1. Experimental encounters with spray pyrolysis system

This section lays out practical experiences that the author's experienced while working with spray pyrolysis systems. The experience and knowledge gathered while working with SP systems played a pivotal role in designing, assembly, and optimization of the present SP system. Moreover, the knowledge gathered also helped shape the various considerations in selecting the equipment (reactor type, nebulizer size, ultrasonic vessel type, and furnace type) of the present SP system design. Furthermore, consideration is given to various material properties and deposition conditions that were used to optimize the design and construction of the present USP system. This section lays out practical huddles experienced with previous spray pyrolysis, more specifically for the production of TiO_2 NPs and TFs.

1.1.1. Use of titanium tetrachloride precursor

The previous hybrid pneumatic/ultrasonic spray pyrolysis that the author worked with consisted of (1) tube furnace, (2) pneumatic nebulizer, and (3) quartz tube reactor as the reaction zone. Spray deposition was done on glass substrates lying horizontally inside the quartz tube reactor. In the study, titanium tetrachloride was employed as the precursor. The system also employed argon gas as the carrier gas in case the ultrasonic nebulizers were used. There were a number of problems that existed with regard to deposition of TiO_2 NPs and TFs. Firstly, the use of titanium tetrachloride as a precursor for synthesis of TiO_2 created huge problems with chloride impurities. Cubic structures of sodium chloride quickly crystallized and inhibited the formation of titanium dioxide NPs on top of the glass substrates. The sodium impurities originated from the quartz glass tube that was employed as the reactor. At temperatures above 400°C, sodium impurity ions percolated from the quartz tube reactor, which caused sample contamination and inhibited growth of titanium dioxide nanostructures on top of the glass substrates. Hence, **Figure 1** shows a scanning electron microscopy (SEM) micrograph of cubic sodium chloride crystals on top of fluorine glass substrates.

Due to sodium ion impurity, it was necessary to select and design a tube reactor that could withstand deposition temperatures above 400°C and at the same time does not introduce any impurity elements to titanium dioxide nanoparticles. Secondly, titanium tetrachloride fumes at high temperatures are highly toxic, acidic, and have fast diffusion rates. The developed fumes were highly acidic and mobile at elevated temperatures. The fumes produced corroded all rubber tubing, seals, and O rings fitted on the USP system which caused serious carrier gas leakage problems. Thirdly, a dense cloud of titanium chloride fumes that formed during the deposition process further inhibited the growth of titanium dioxide nanostructures. As a result, it was difficult to deposit thin films of sufficient thickness (10 μm). This was due to low aerosol containing the titanium precursor volumes as compared to the amount of titanium tetrachloride vapors. In addition, titanium chloride in the presence of hydrous ethanol solvent was easily hydrated to form titanium hydroxide which precipitates inside the pneumatic nebulizer; this inhibited the nebulization process as the precursor quickly precipitated inside the nebulizer. This made it very difficult to atomize the precursor solution. Also, atomization of large particulates of TiO_2 hydroxides led to the development of large nano-agglomerates inside the thin film structure of amorphous TiO_2 phase inside the anatase TiO_2 thin film which is not desirable in developing thin films for photovoltaic applications. Furthermore, there was non-uniform deposition of TiO_2 thin films as some areas on the substrate were found to be thicker than others and other areas contained no TiO_2 nanoparticles at all, as shown in **Figure 2**. Hence, due to corrosive nature of the titanium chloride precursor, toxicity, and ease

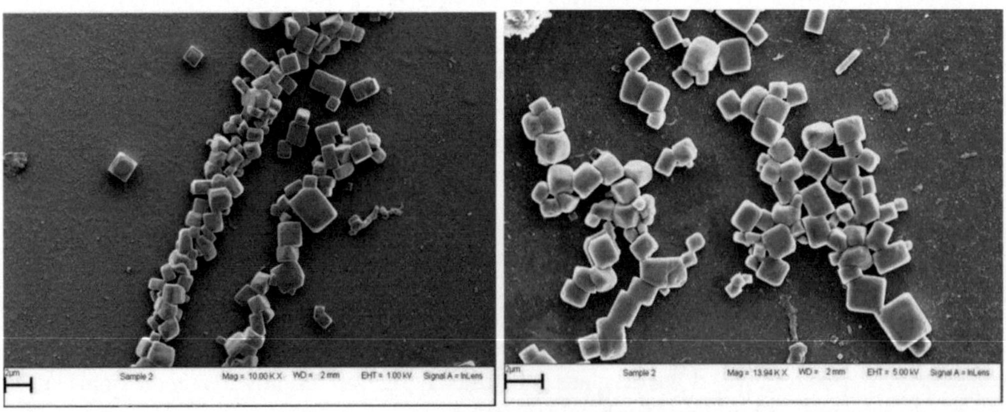

Figure 1. An SEM image showing NaCl cubic crystals that were formed on fluorine-doped tin oxide glass substrate.

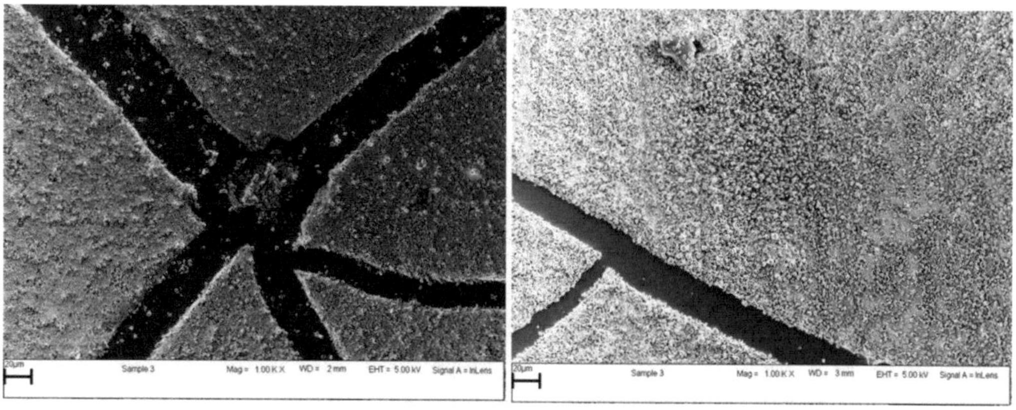

Figure 2. An SEM image of non-uniformly coated TiO_2 thin film.

at which titanium chloride precursor precipitates, when selecting a precursor for SP, it was deemed necessary to avoid chloride precursors of titanium.

1.1.2. Substrate geometry

It was also discovered that only some areas of the substrate facing the aerosol flow were fully coated. Areas on the substrate downward the aerosol flow were partially coated due to substrate orientation. The deposited thin films had cracks in them due to sudden temperature drop when the system was turned off. Fourthly, it took several hours for the system to cool down to room temperature. Fifthly, the round tube furnace employed had poor heat insulation. Hence, after some hours of operation, the temperature of the laboratory could rise to as far 38°C, which made monitoring of the spray pyrolysis deposition very difficult. Furthermore, it was impossible to mount such a spray pyrolysis system directly to a normal laboratory workbench as this presented a higher fire risk. In addition, all rubber tubings and O rings could not withstand high temperatures as they needed to be constantly replaced. This posed a further health risk due to leakage of argon and titanium precursor vapors into the working environment. Lastly, there was no way of regulating the relative humidity, and the relative humidity on rainy or overcast days (humidity up to 65%) made spraying impossible due to the formation of white TiO_2 hydroxides in the precursor solution. This made it very difficult to atomize the precursor solution.

2. USP operational conditions

2.1. Safety

The system should be enclosed so that it can safely operate on a standard laboratory bench. It also needs to have exhaust facilities into a sufficiently ventilated area. There was also a need to replace the tube furnace with a split tube furnace that had heat guides on. This made it easy and safe to operate with aluminum tubing and/or quartz tubes as reaction vessels. The furnace was built in such a way that no heat should be dissipated into the work environment. It was also desirable to discontinue the use of titanium tetrachloride as a precursor. Titanium iso-propoxide and titanium (IV) butoxide were also more attractive to use as precursors since these are safe, non-toxic liquids, and obviate the need for expensive gas handling systems.

2.2. Geometry control

The other problem that needed urgent attention was the issue of substrate orientation. It was discovered that disposition of thin films at an oblique angle of 30°or with substrate lying on the floor of the quartz tube caused inhomogeneous thin film coating. We designed an aluminum substrate holder that secured the substrate at 90° of the reaction vessel axis, facing the incoming aerosol flow. Aluminum was chosen as the metal of choice because it's stable at deposition temperatures of 400–500°C, and it is a good conductor of heat. It also allowed preheating of substrates prior to thin film deposition. In addition, significant fluctuations in the temperature at about 400°C could result in the TiO_2 film having a mixed amorphous-Anatase phase. Naturally, this outcome is undesirable.

2.3. Deposition area

Dye sensitized solar cells consist of an active area for instance titanium dioxide NPs deposited and the rest of the area on fluorine doped tin oxide (FTO) glass substrate is used for solar cell contacts. It was desirable that the substrate be masked; this allowed production of a thin film area of 0.5 cm^2, which is desirable for dye solar cell applications. To achieve accurate control of the film thickness, the system was calibrated to produce sufficient thin films in deposition times of 20–30 min.

2.4. Relative humidity

Excess humidity will result in TiO_2 particulates sticking to the deposited film. In SP spray deposition, it is always desirable to have adjustable and repeatable humidity control. It was anticipated that mixtures of dry nitrogen (N_2) and wet nitrogen ($N_2 + H_2O$) could be fed into the system in order to control the relative humidity. However, this proved to be costly, and we opted to use the carrier gas, argon, to provide dry environment with success.

3. The spray pyrolysis system

Following the operational design considerations as has been discussed in Section 2, a novel USP system was designed and constructed specifically for production of anatase TiO_2 nanostructures and thin films for solar cell applications. **Figure 3** presents a graphic diagram of the USP system that was optimized for deposition of TiO_2 NPs and TFs.

http://dx.doi.org/10.5772/67866

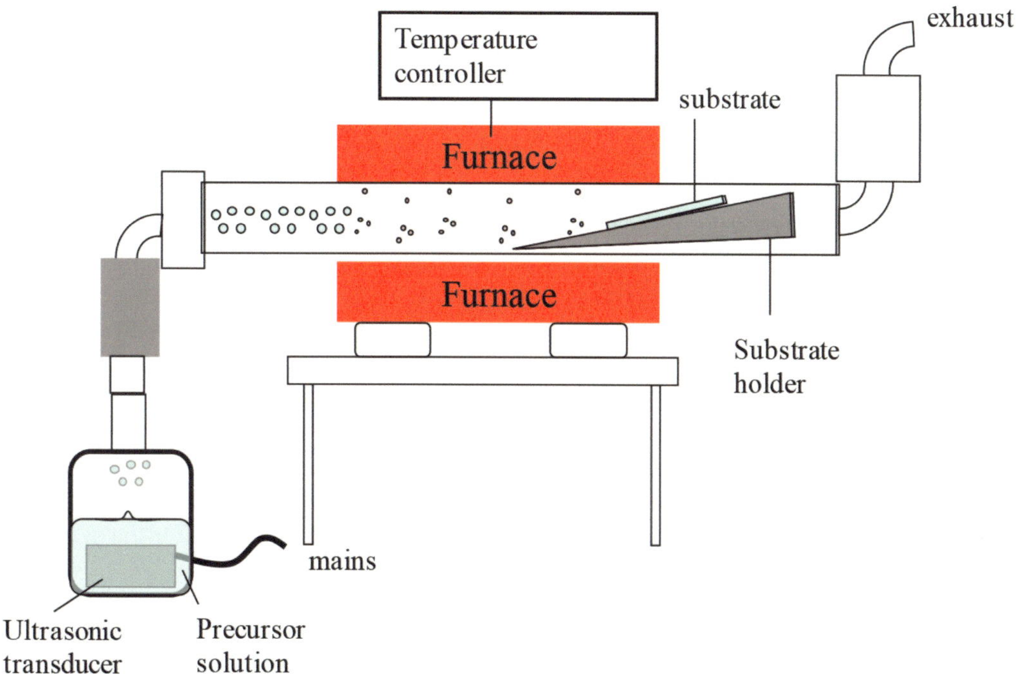

Figure 3. A graphic diagram of the USP system.

The design of a USP system comprises an (1) ultrasonic chamber, a (2) carrier gas system, (3) a heated zone or furnace with a temperature control unit, (4) substrate holder, and (5) exhaust system. The ultrasonic chamber houses a transducer, producing ultrasound waves at a frequency of 1.67 MHz that were focused to the precursor solution to produce vapors of ultrasound droplets. The argon carrier gas system was used to deliver the precursor droplets into the reaction zone that was either quartz or aluminum tube at a constant flow rate of 6 ml/min. The quartz or aluminum tube was in a temperature-controllable furnace with the substrate holder placed perpendicular to the direction of gas flow inside the tube. Both deposition temperatures and flow rate influenced the shape, morphology, and crystalline structure of the TiO_2 NPs and TFs. These TiO_2 layers were deposited on commercial viable substrates.

3.1. Selection of TiO_2 precursor

Use of titanium (IV) sulfate and titanium (IV) chloride precursor solutions has been reported [16]. Fabrication of TiO_2 NPs and TFs with titanium (IV) sulfate has been reported to have mixed crystalline phases of TiO_2. It has been reported that use of titanium (IV) sulfate always resulted in uncontrolled formation of various phases of TiO_2 due to nature of preparation method, with the amorphous phase being dominant in the dried samples. While use of titanium (IV) chloride solution resulted in formation of an undesirable rutile phase for solar cell applications, use of titanium (IV) phosphate precursors has resulted in formation of uncontrolled mixtures of anatase and rutile phases in the annealed TiO_2 NPs and TFs. It has been reported that the amorphous phase of TiO_2 preexisted prior to formation of the crystalline phases, even in situations where the produced NPs/TFs had rutile as the main dominant phase. In this regard, it is imperative to select the appropriate precursor for use in SP system. Proper choice of a precursor is a primary issue in material synthesis. Titanium iso-propoxide, also known as tetraisopropyl titanate (TPT), was chosen as the TiO_2 precursor. Apart from being the most commonly used precursor in the literature, this titanium alkoxide is also used in many solar cell production lines. Titanium tetrachloride/titanium (IV) chloride ($TiCl_4$) is another common TiO_2 precursor, which results in chlorine contamination [16]. In addition, its corrosive by-products (HCl) are produced in the reaction. It has been reported in literature that levels can be high enough to prevent nanomaterial crystallization on FTO glass substrates and can also cause thin film or poor film adhesion onto the substrate [16]. In any case, the level of contamination observed with titanium iso-propoxide is much smaller than with titanium (IV) chloride. Also, it has been reported that the use of titanium (IV) chloride results in formation in rutile and anatase phase mixtures, [17] and we also discovered that the use of titanium butoxide resulted in the formation of 95% of the anatase phase. In this work, titanium iso-propoxide and titanium butoxide were used as the precursors for the synthesis of TiO_2 NPs and TFs.

In addition, these precursors have the following added advantages:

1. They are non-corrosive and non-toxic, listed as being mild skin and eye irritants.
2. They can be highly purified and have an almost indefinite shelf life.
3. As a liquid, they are relatively easy to handle, although they should not be exposed to a naked flame. The fact that it is not dangerous makes the addition of titanium iso-propoxide system a relatively easy and safe task as no special gas handling equipment is required.
4. They are very volatile at low temperatures (50°C), which means that they will be readily decomposed.
5. They can be ultrasonically sprayed directly without dilution.
6. It has been observed that there is enough oxygen in the titanium iso-propoxide molecule that the reaction to form TiO_2 can proceed without additional oxygen in the ambient.

3.2. Pyrolysis of TiO_2 solutions

Countless processes take place concurrently before a vapor droplet forms solid nanostructures on the substrate. When the ultrasonically generated precursor vapor droplet hits the surface of the substrate, formation of solid nanostructures proceeds by evaporation of solvent molecules, decomposition of the metal salt, and spreading of the droplet. Numerous mechanisms have been presented for pyrolysis that lead to formation of solid-phase nanoparticles and thin films in spray pyrolysis. The route for each of these theories is entirely dependent on nature of the precursor solution and the experimental conditions the ultrasonically generated precursor solution is exposed to. The mechanism of the reaction of titanium iso-propoxide aerosol to form TiO_2 depends on the droplet size. When the majority of the ultrasonically generated vapor droplets hits the surface of the preheated glass substrates, depending on the experimental conditions inside the reactor such as pressure, temperature, humidity, and stoichiometry composition of the precursor droplet, smaller droplets normally burst under pressure and decompose into solid NPs. However, due to collision between the precursor droplets due to random gaseous motion, some smaller droplets tend to coalesce to form bigger vapor droplets. The bigger droplets tend to decompose into solid NPs on the surface of the preheated substrates. However, pyrolysis of precursor droplets in spray pyrolysis occurs in much the same way as chemical vapor deposition systems, which apply to all aerosol processes. **Figure 4** presents the proposed aerosol droplet transport, decomposition, and deposition on substrate at various temperatures in spray pyrolysis.

In process A, the decomposition rate at very low temperatures (< 100°C) will be slower than the deposition rate, and a liquid film will form on the surface. This layer will slowly dry; however, it will still contain many organics and probably cracks. At this stage, titanium dioxide will be present as hydrated white hydroxide and in its amorphous phase. As the temperature of the substrate increases in process B, the solvent from the droplets evaporates during its flight before reaching the surface, and a precipitate strikes the substrate where decomposition occurs. In process C, the solid precipitate melts and vaporizes (or sublimes), and the vapor diffuses to the substrate and undergoes a reaction there. This corresponds to true CVD. At higher temperatures (process D), the vapor undergoes a chemical reaction before impinging upon the substrate. The droplets in the aerosol have formed solid particles that stick to the surface of the substrate. Titanium iso-propoxide pyrolyses at temperatures greater than 350°C and films deposited in this method at 450–600°C are considerably harder than films produced by colloidal synthesis methods.

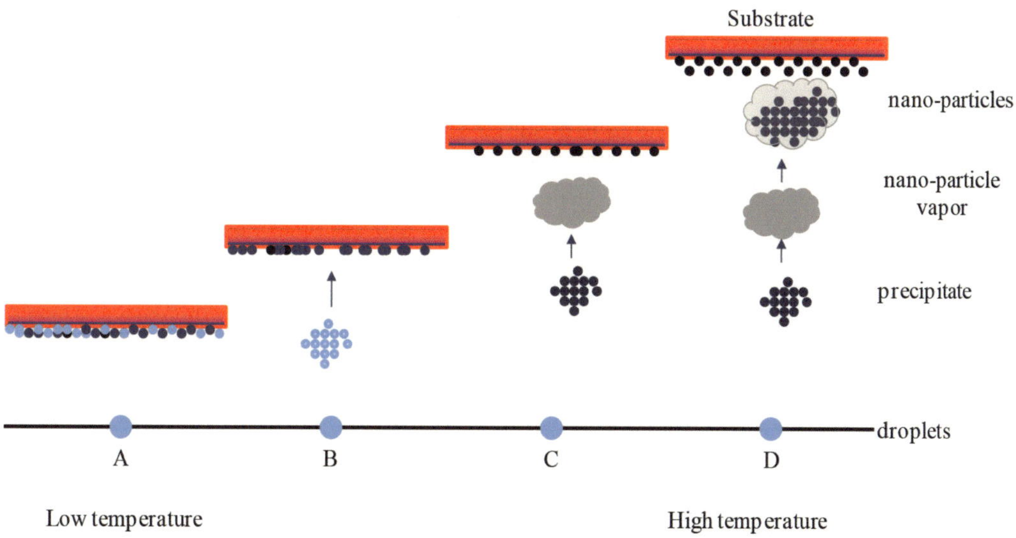

Figure 4. Illustrations of aerosol droplet transport, decomposition, and deposition on the substrate at various temperatures.

In inert and dry argon environment and assuming the absence of water vapor, the decomposition of titanium iso-propoxide occurs via pyrolysis to form TiO_2 and proceeds as follows:

$$Ti\left(O\,C_3H_7\right)_4 \xrightarrow{Pyrolysis} Ti\,O_2 \;+ Olefins. \tag{1}$$

However, such an assumption is very dangerous as there are various sources of water vapor, for instance, the solvent ethanol, the carrier gas, and the spray pyrolysis vessel itself. There is, to some extent, formation of titanium dioxide via a hydrolysis process which is governed by Eq. (1).

$$Ti\left(O\,C_3H_7\right)_4 \;+ 4\,H_2O \xrightarrow{Hydrolysis} 2Ti\,(OH)_4 \;+ 4\,C_3H_8O \tag{2}$$

$$Ti\,(OH)_4 \xrightarrow{Condestation} Ti\,O_2 + (2-x)H_2O \tag{3}$$

$$Ti\,O_2\,x\,H_2O \;\rightarrow\; Ti\,O_2 + \uparrow H_2O \tag{4}$$

One should note that the formation of titanium dioxide occurs to a large extent by Eq. (1) and to a negligible extent by Eqs. (2, 3 and 4). The size and shape of the deposited titanium dioxide NPs are determined by the molar ratio of adsorbed water in Eq. (3), that is $\left(x = \left[H_2O\right]\right)/[Ti]$. At $x < 10$, relatively spherical, monodisperse particles are formed. On the other hand, at values of $x > 10$, particles formed are unstable precipitates, and this results in aggregates which form a state dispersion of colloidal nanoparticles (sol-gel) with sizes larger than 100 nm.

n is the process responsible for the formation of stable dispersion of colloidal particles in dispersion medium. In other words, it may be defined as a process of converting a precipitate into colloidal solution by shaking it with dispersion medium in the presence of a small amount of electrolyte.

4. Design considerations and optimization of USP system

The design of the current spray pyrolysis system, the knowledge gathered in the Section 4, and Section 5 were used as guidelines to design the new system. Section 6 presents the desirable system designs and parameters taken into consideration.

4.1. Sizing of ultrasonic reaction vessel

The Bernoulli's theorem of conservation of energy of fluid flow in channels of varying cross-sectional areas was considered useful in deriving the relationship between droplet size and volume it occupied. This theorem was also applied here in determining the proper volume to house the ultrasonic nebulizer. This is mainly because the volume of the reaction vessel plays a vital role in defining the final NPs crystallite size. The Bernoulli's theorem states that:

$$p + \frac{1}{2}p\,v^2 + \rho g y \;=\; constant. \tag{5}$$

For a droplet distribution of mass (m) moving at mean speed (v), and occupying a volume (V) under pressure (P), the chain rule for $P = mV$ substitution results in

$$dp + \left[\frac{1}{2}m\,v^2 + mgy\right]dV + pvdv + g\rho dy \;=\; 0 \tag{6}$$

and assuming small variations in the droplet size change with change in velocity, which is typical in materials deposition processes, dv is approximately equal to 0. Eq. (6) then reduces to

$$dp = -EdV - g\rho dy \tag{7}$$

where **E** is the total energy of the droplet. If the droplet diameter is introduced into this equation, it is possible to rewrite Eq. 7 as

$$\frac{dD}{dp} = -\frac{dD}{EdV + g\rho dy} \tag{8}$$

Therefore, at a constant position y,

$$\frac{dD}{d\rho} = -\frac{dD}{EdV} \tag{9}$$

Similarly, at a constant volume of the system but varying position,

$$\frac{dD}{dp} = -\frac{dD}{g\rho dy}. \tag{10}$$

The theoretical data models have shown a sharp increase of droplet sizes with change in USP system volume. **Figure 5** shows the experimental evidence of droplet diameter dependence of USP volume which is in mutual agreement with theoretical models. This effect of volume change has also been observed by droplets generated by Mwakikunga *et al.*, [10, 20, 21].

The main goal for this work was to develop an ultrasonic spray pyrolysis system capable of realizing nano-sized powders, and considering previous findings by Mwakikunga *et al.*, [20, 21]. An ultrasonic reaction system with maximum volume of 0.1 m^3 was developed. **Figure 6** shows the USP reaction vessel used in this chapter. The quartz glass used for the vessel provided a non-corrosive environment for housing the acidic precursors of TiO_2.

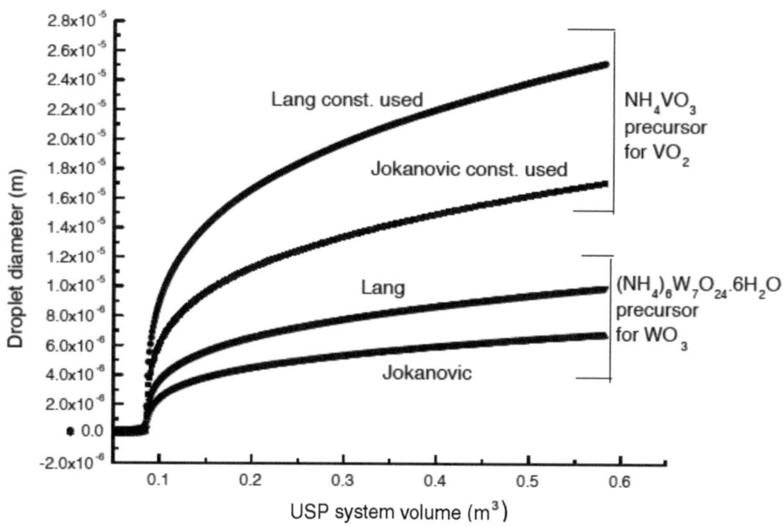

Figure 5. Experimental evidence of droplet diameter dependence on USP system volume [10, 20].

Figure 6. An actual photograph of the USP reaction vessel used.

4.2. Selection of ultrasonic nebulizers

Ultrasonic nebulizers used in this project were purchased from Time-to-Digital Converter (TDC) power products Co. Ltd, Germany. The ultrasonic nebulizer employed composed between three and five piezoelectric transducers driven at frequency of 1.7 and 4.8 MHz, each driven by a dedicated ultrasonic generator. There were several reasons for selecting an ultrasonic nebulizer, rather than pneumatic nebulizers. Firstly, pneumatic nebulizers require high pressurized gas systems that are either from a compressor or gas tank. This inherent property would increase the initial production cost of titanium dioxide. This meant that the designed system would require sophisticated equipment to control the gas flow rates, pressure regulator, high pressure valves, and tubing. We therefore preferred a nebulizer that could operate at standard atmospheric pressures. Secondly, it was believed that with proper control of nebulizer operation frequency, the reaction mechanism for droplet generation would be tuned to behave similar to a belt furnace atmospheric pressure chemical vapor deposition (APCVD) system used in the PV industry. Thirdly, there is a huge potential in scaling the project to industrial scale for fabrication of dye solar cell, considering the ease of assembly and the need for low operation skills. In addition, the ultrasonic nebulizer housing is manufactured from titanium alloy and stainless steel housing. The titanium alloy was chosen mainly because of its high mechanical durability, excellent ability to resist conversion of vibrational energy into heat (hence it does not heat up the precursor during material deposition), and high chemical resistance toward most titanium alkoxide precursors which are highly acidic. **Figure 7** shows the nebulizers used in the chapter.

After some trial runs, we discovered that the 4.8 MHz nebulizer would be unsuitable for our application shown in (b) of **Figure 7**. This was due to the extremely high volume of aerosol produced [*(300 × 7) ± 50 cc/hr*] which also caused high flow rates. The resulting titanium dioxide deposited thin films on the FTO glass substrates that were too thick. Titanium dioxide yields were just too high and uncontrollable. It was also suspected that due to the high flow aerosol rates and short residence times of the aerosol inside the heating zone of the furnace, amorphous phase of titanium dioxide was formed. We therefore opted to use the 1.7 MHz nebulizer to synthesize titanium dioxide thin films. We discovered that with controlled flow rates of 6 ml/min, we could deposit a thin film which was approximately 10 μm. To achieve longer deposition times, the lowest flow rates had to be used. Although a fine mist often emerged from the nozzle, the low volumes being pumped were not sufficient to create a cloud of droplets of several inches in diameter. Occasionally, the nebulizer would also "stall" and produce high volumes of aerosol which would cause deposition of liquid vapor onto the substrate. This resulted in high agglomerates of TiO_2 nanoparticles being incorporated into the film. These defects reduced the chemical resistance and produced high electrical resistance on carbon-doped titanium films, which were undesirable for dye solar cell applications. We therefore resorted to carry out thin deposition with

Figure 7. Ultrasonic nebulizers employed in the study: (a) shows the 1.7 MHz ultrasonic nebulizer with power supply unit and (b) shows the 4.8 MHz ultrasonic nebulizer with power supply unit.

30 min spray breaks. Spray deposition using the 1.7 MHz nebulizer over the 4.8 MHz ultrasonic nebulizer poised several advantages:

1. The 1.7 MHz nozzle enabled use of very low flow rates of almost 2×10^{-2} ml/min, which increased the residence time of the aerosol vapor inside the heating zone. This gave enough time for complete pyrolysis/hydrolysis of titanium dioxide precursors to anatase phase of titanium dioxide.

2. The use of low flow rate gives allowance for achieving, with high success, well-controlled, consistent repeatable thin films. This setup resulted in thin film deposition times at 450°C of 25–30 min.

3. In addition, the power required by 1.7 MHz ultrasonic nebulizer required from the ultrasonic generator to atomize titanium dioxide precursors was in the order of 2.5 W. **Table 1** shows the specifications of the nebulizers chosen.

4.3. Selection of the spray reactor

NPs and TFs can be deposited by wet chemistry methods [2] as well as by solid state techniques that vaporize the precursor to a gaseous phase [3, 9–12]. Use of techniques that employ an intermediate gas phase prior to formation of NPs and TFs presents numerous merits as compared to the wet chemistry methods since the purification, drying, and annealing procedures are dismissed. As techniques that employ a gaseous phase duck, the intermediate purification procedures employ high liquid volumes of expensive, corrosive, toxic, and unpleasant solvents and/or surfactants prior to formation of final high-purity NPs and TFs. Liquid phase techniques have low throughput due to losses of NPs during the purification process. Spray pyrolysis reactors that have been modernized for fabrication of nanopowders consist of flame reactors [5], furnace reactors [10, 11], and plasma reactors. Use of techniques that are assisted by a gaseous phase intermediate, it is possible produce all sorts of wide band gap semiconductors as NPs and TF, giving potential for new or low

	1.7 MHz ultrasonic nebulizer	**4.8 MHz ultrasonic nebulizer**	**Test condition 1.7 MHz nebulizer**	**Test condition 4.8 MHz nebulizer**
Resonant frequency	1.7 MHz ± 0.1 MHz	4.8 MHz ± 0.1 MHz		
Resonant impedance	5 Ω max.	3 Ω Max.		
Static capacitance	1400 pF ± 15%	2400 pF ± 15%		
Mist output	(300 × 3) ± 50 ml/hr	(300 × 7) ± 50 ml/hr	Precursor temperature of 25°C and precursor level of 38 mm	Precursor temperature of 25°C and precursor level of 38 mm
Life time	5000 hr/min	5000 hr/min		
Power in put	2.5 W	6.5 W	Standard circuit pin	Standard circuit pin

Table 1. Specifications of the 1.7 MHz ultrasonic nebulizer and 4.8 MHz ultrasonic nebulizer.

cost-efficient applications. The specific gaseous phase techniques fluctuate on how thermal energy is transferred to the precursor vapors to eliminate the solvent molecules. They also differ on how the precursor solution is transported to the reaction site, and financial aspects, final product quality, and characteristics do affect the choice of spray reactor. Currently, the PV industry employs spray pyrolysis systems with flame reactors for the production of NPs and TFs. However, use of vapor flame reactor system is restricted by choice of metal precursors. Moreover, it is not easy to deposit multi-component semiconductor NPs and TFs with consistent chemical stoichiometry. This is due to the differences in the chemical reaction rates and pressures of the reactants. Spray pyrolysis systems are equipped with tubular reactors such as a furnace, and each ultrasonically vaporized vapor molecule contains the same atomic ratios of the precursor solutions [10, 11]. The produced micro-droplets in actual fact serve as a discrete micro-reactor. This is an advantage of aerosol vapor-assisted phase as compared to flame reactors, where the stoichiometry of each of the precursor solution is not the same. This is a major advantage since in vapor flame reactors, several reactants have to be vaporized simultaneously with special care to obtain the desired stoichiometry. Furthermore, flame reactor systems are costly in operational costs and the quality of the final product in either stoichiometry or crystalline morphology leaves a lot to be desired. Spray pyrolysis systems that use furnace reactors offer superior quality at low operational cost and minimum operator skill.

Spray-assisted methods for deposition of NPs and TFs differ by how the thermal energy is delivered to the generated aerosol vapors to facilitate solidification of NPs and TFs and, at the same time, facilitating solvent evaporation. **Figure 8** presents key spray methods discussed in this work, namely spray pyrolysis in a tubular reactor (SP) [10], spray pyrolysis using a vapor flame reactor (VFSP), the emulsion combustion method (ECM), [16] and flame spray pyrolysis (FSP). In fact, in all the spray techniques presented in **Figure 8**, the precursor is generated in much the same way and vaporizes into a mist of droplets. These spray techniques reveal a significant difference in how the thermal energy is delivered to generated mist of the precursor to facilitate solidification and solvent evaporation. Additionally, the difference is also in the energy source that facilitates formation of NPs and TFs—such as in VFSP, a liquid fuel is needed to ignite the reaction. Reactors with an independent source of energy supply are not affected by the nature of the selected titanium dioxide precursors and solvents. Double distilled water is often preferred as the aqueous solvent in systems with an external energy supply due to financial advantages, ease of handling, and the outstanding solubility for metallic salts. Spray systems which have an internally supplied source of power and fuel to facilitate solvent evaporation and eventual solidification of NPs and TFs present numerous disadvantages. Mainly due to the nature of the different combustion rates of the precursor solution, there are different temperature gradients with the same reactor and large temperature differences, which in turn result in different solidification rates of the NPs and TFs. This results in differences in the crystalline morphology of the deposited TFs and NPs. In addition, flame spray pyrolysis has higher operation cost as compared to other methods. The triumph of fabricating practical metal oxides, mixed-metal oxides, and metals on metal oxides from inexpensive precursors in an inexpensive aerosol process revolutionized the design of spray pyrolysis reactors.

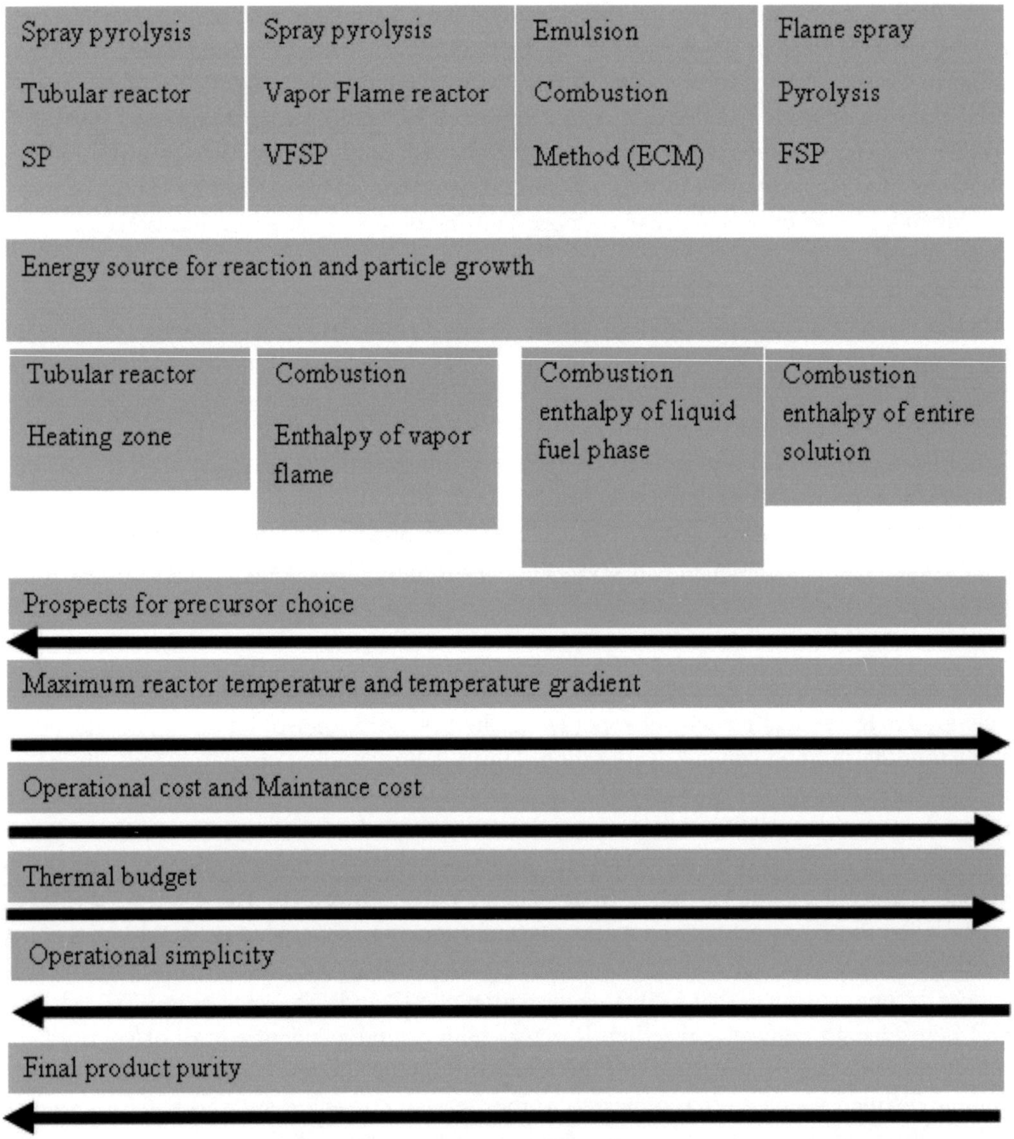

Figure 8. Main spray methods operated at ambient pressure for gas phase-assisted NPs deposition from metal precursors. Precursors are transported within a solvent into a reaction zone, where reaction and the final product formation occurs.

Spray pyrolysis systems, with reactors that employ thermal energy from the element of a coil, provide a uniform and controlled form of heat energy to evaporate solvent molecules and, at the same time, facilitate solidification of NPs and TFs of controlled consistent crystallinity, morphology, and stoichiometry as compared to other techniques. Moreover, spray pyrolysis systems that employ furnace reactors present several advantages such as (1) operational simplicity, (2) minimum throughput (in terms of consumables), (3) lower thermal budget, and (4) lower operational cost. Moreover, there is no need for subsequent heat treatment after deposition of either the NPs or TFs. Furthermore, tubular reactor systems offer the opportunity of industrial scaling. These inherent advantages were some of the drivers of the choice of the tubular reactor used at Fort Hare shown here in **Figure 9**, and **Figure 10** shows the actual aluminum reactor used in the study. **Figure 10 (a)** shows the aluminum reactor for production of thin films **(b)** production of nano-particles. Compared to liquid-phase precipitation methods, flame reactors and laser ablation systems powders produced by spray pyrolysis systems employing furnace reactors are beneficial since the deposited NPs have consistent chemical composition, consistent crystalline phase, are of high purity, and rarely require several heat treatments after deposition [8].

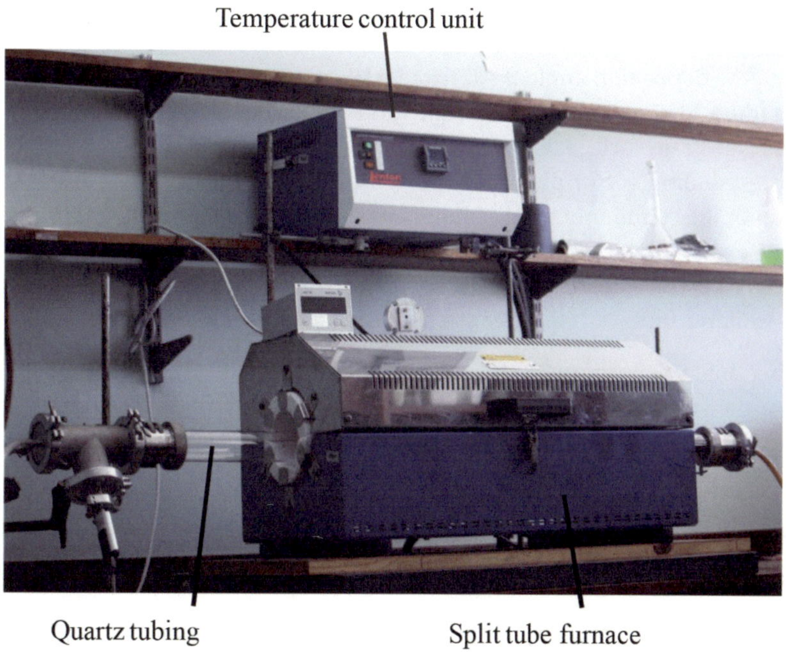

Figure 9. Shows the actual photograph of the split tube furnace employed in the synthesis of TiO_2 nano structure in the study.

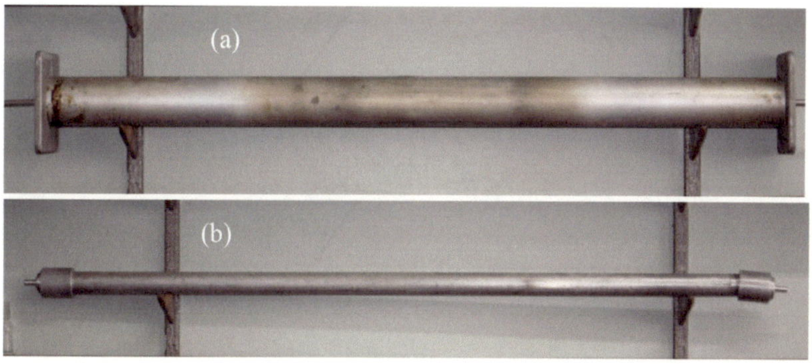

Figure 10. The aluminum reaction vessels used for ultrasonic spray pyrolysis deposition of TiO_2 NPs and TFs.

In addition, spray pyrolysis methods offer excellent spatial mixing of the precursor molecules and are capable of depositing functional multi-component semiconductor oxides.

4.4. Design of the substrate holder

The novelty of our USP system originates from the angle at which the substrate interacts with the incoming aerosol vapor as shown in **Figure 11**. Most systems that have been utilized employ a substrate holder with the substrate vertically parallel to the incoming aerosol beam. In our USP system, the substrate holder and hence the substrate are perpendicular to the aerosol stream as shown in the schematic representation of **Figure 12**. **Figure 11**, shows the aluminum substrate holder employed in the study.

The choice of the deposition angle θ = 0° can be best explained by considering schematics in **Figure 13**. **Figure 13** shows that the aerosol vapor stream is incident at an angle θ with the normal to the substrate surface. **Figure 13(a)** shows a schematic of the setup and **Figure 13(b)** shows a corresponding vector diagram. The glancing angle deposition (GLAD) technique is the extension of the commonly used oblique angle deposition (OAD) in thin film deposition industry. The experimental setup for the oblique angle deposition is shown here in **Figure 13**. In most ideal cases, the aerosol vapor stream has an incident angle, θ, with the respect to the normal substrate surface.

At a glancing angle deposition θ, the substrate interacts with the stream of the precursor equally from the vertical and lateral directions. Hence, at this angle θ of deposition, growth of NPs results in randomly formed islands on the surface of the substrate. As the spray

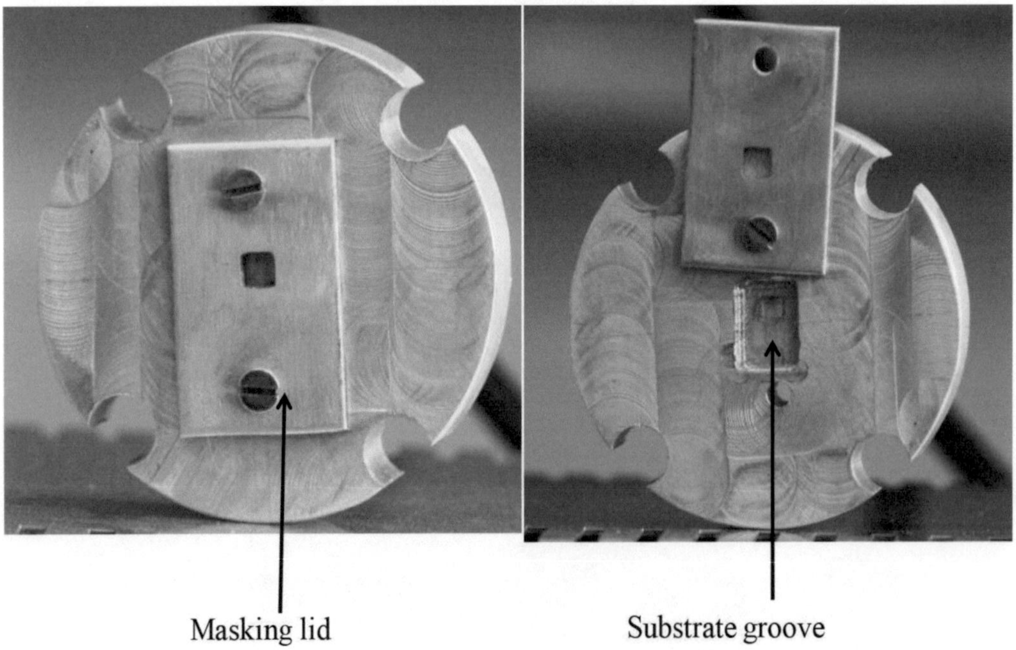

Figure 11. The aluminium substrate holder employed in open and closed configuration.

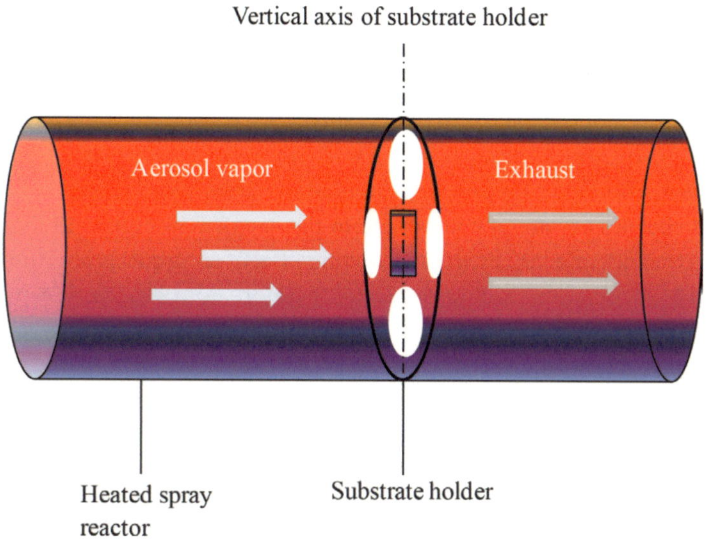

Figure 12. Schematic presentation of substrate holder orientation in the spray reactor.

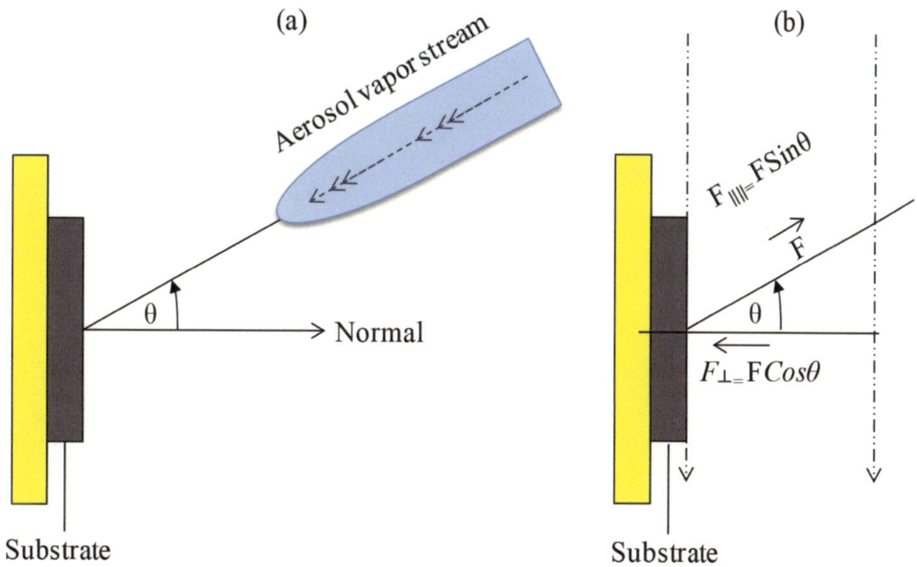

Figure 13. (a) A schematic drawing for the experimental setup for oblique angle deposition and (b) the incident aerosol vapor stream **F** can be decomposed into different components: $\mathbf{F}_{\perp}$ is perpendicular to substrate and $\mathbf{F}_{\parallel}$ is parallel to the substrate.

deposition continues, the initially formed and nucleated islands of NPs will act as shadowing centers. Hence, all of the tallest nanoparticle islands will expand into columns at the expense of other areas on the substrate that don't receive the aerosol vapor stream. Vector diagram in **Figure 13(b)** shows that the lateral component $F_{\parallel}$ is the source revealing the enhanced shadowing effect. For the oblique angle deposition, since $F_{\parallel}$ remains constant during spray deposition, a columnar film with angle β will be formed. Through scanning electron micrograph analysis, Hawkeye *et al.* [22] have shown that the shift of angle θ in deposition of silicon thin films had tremendous effects on the geometrical arrangement of the deposited NPs. They found that at θ = 30°, small columns of silicon began to grow, at θ = 60°, the columnar structures of silicon became obvious, and at θ = 80°, the columnar structures of silicon become more pronounced. They have also reported that at θ = 0°, where the incoming aerosol vapor stream was not affected by lateral component of the vector **F** in **Figure 13** (hence the growth of NPs and TFs was not affected by the shadowing effect), continuous and uniform nano-TFs were deposited as shown here in **Figure 13**. We therefore adopted this idea when designing the substrate holder for our horizontally orientated ultrasonic spray pyrolysis system as well as the aluminum reaction chamber that is capable of supporting the substrate in a vertical position. Thin film deposition was done at θ = 0° for the development of nanostructures for photovoltaic applications. **Figure 14** shows the preliminary results that were obtained with the substrate holder in this configuration.

One can observe that the developed titanium dioxide TFs by this methodology were mesoporous and have spherical NPs, which have mesoporous nanostructure morphology for dye solar cells.

Figure 14. SEM micrographs showing effects of the titanium dioxide thin film produced at θ = 0°, with an insert showing the same sample on a different spot and on a lower magnification.

http://dx.doi.org/10.5772/67866

5. Operation of the ultrasonic spray system

In its final form shown in **Figure 15**, the ultrasonic spray system incorporated all of the above component as well as those listed in Section 5. **Table 2** outlines the standard deposition parameters for the precursors chosen. The gas flow rates were optimized in order to extend thin film coverage in the forward direction. At lower flow rates of 1–2 ml/min, ultrasonic spray depositions were performed at the lowest system operating temperature of 450°C; this is mainly because of high-enough residence times spent by aerosol vapor droplets inside the heating zone. At higher flow rates of 5–6 ml/min, system operating temperature was increased to a maximum value of 500°C to accommodate the short residence times spent by the aerosol vapor droplets. Throughout this whole project, argon was used as the only carrier gas. Also, the aluminum substrate holder was used in most cases for thin film synthesis.

Figure 15. USP system in its final form.

Process parameters	Ultrasonic spray pyrolysis	
	@ 450°C	**@ 500°C**
Argon flow rate	1–2 ml/min	5–6 ml/min
Ultrasonic vessel pressure (Atm)	1	1
Precursor temperature (°C)	25	25
Precursor pH	1.95	1.95
Substrate type & temperature (°C)	FTO Glass & 450°C	FTO Glass & 550°C
Deposition time (m)	120	30
Deposition Angle (°)	0	0
Annealing temperature (°C)	450	450

Table 2. Showing operation conditions for USP system.

6. Results and discussions

6.1. SEM analysis

Figure 16 presents SEM micrographs of the undoped and N-TiO_2 NPs synthesized by ultrasonic spray pyrolysis technique with different levels of nitrogen doping. SEM images, both undoped and N-TiO_2 NPs, reveal the formation of spherical-shaped TiO_2 NPs. Moreover, the SEM micrographs reveal that the surface morphology and shape of both undoped and N-TiO_2 NPs change as the nitrogen dopant level increases. All thin films formed were mesoporous and had a multiparous network structure. Furthermore, the presence of Ti, N, and O in our PSP samples has confirmed the successful pyrolysis of our Ti precursors. Moreover, there are trace levels of Al that might have originated from the Al substrate holder or aluminum reactor that was used in the study, and they were no other contaminates detected from elemental analysis [10, 11].

Figure 17 shows the XRD spectra of the USP-fabricated TiO_2 samples. XRD analysis has that the USP-fabricated undoped and N-TiO_2 NPs have an anatase polymorph with peaks at

Figure 16. Shows SEM micrographs of (a) undoped TiO_2 (b) 5% of N-TiO_2 samples, (c) 10% of N-TiO_2 samples, and (d) 10% of N-TiO_2 samples. Additionally, insert in (a), (b), (c), and (d) reveal the elemental composition analysis by EDX of the undoped TiO_2 and N-TiO_2 samples respectively.

2θ angles and [hkl] planes of 25.67°[101], 38.19° [004], 48.56° [200], 53.91° [105], 63.14° [204], 68.98°[116], 70.72°[220], and 75.65° [215]. Moreover, close inspection of the dominant peak with 2θ angles of 25.67° has shown a peak shift to higher 2θ values, which is indicative of substitutional nitrogen doping. The crystallize size was calculated using Scherrer's equation $t = 0.9\lambda/\beta\cos\theta$ where λ is wavelength of X-ray in Å and β is full width at half maximum in

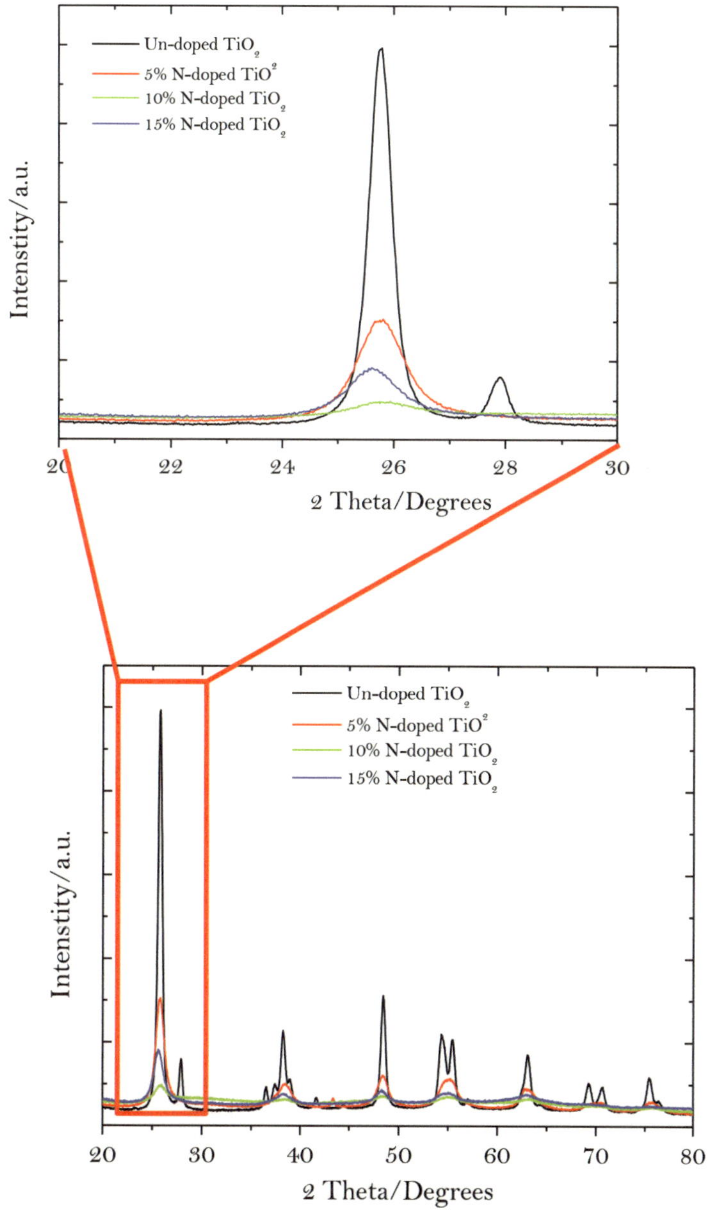

Figure 17. X-ray diffraction (XRD) spectra of the USP-fabricated TiO_2 NPs.

radian. The USP-fabricated undoped and N-doped TiO_2 NPs had nanoparticle sizes of 23.12 nm, 24.15 nm 26.2 nm, and 25.9 nm for undoped, 5% N-TiO_2, 10% N-doped TiO_2, and 15% N-doped TiO_2, respectively.

6.2. Raman spectroscopy analysis

Raman spectroscopy (RS) has revealed that the TiO_2 samples shown in **Figure 18** have Raman active bands characteristic of anatase phase at 153.43 cm^{-1}, 209.9 cm^{-1}, 401.7 cm^{-1}, 517.6 cm^{-1}, and 641.9 cm^{-1} with symmetries of Eg, Eg, B1g, A1g, B1g, and Eg, which is in mutual agreement with XRD analysis [10].

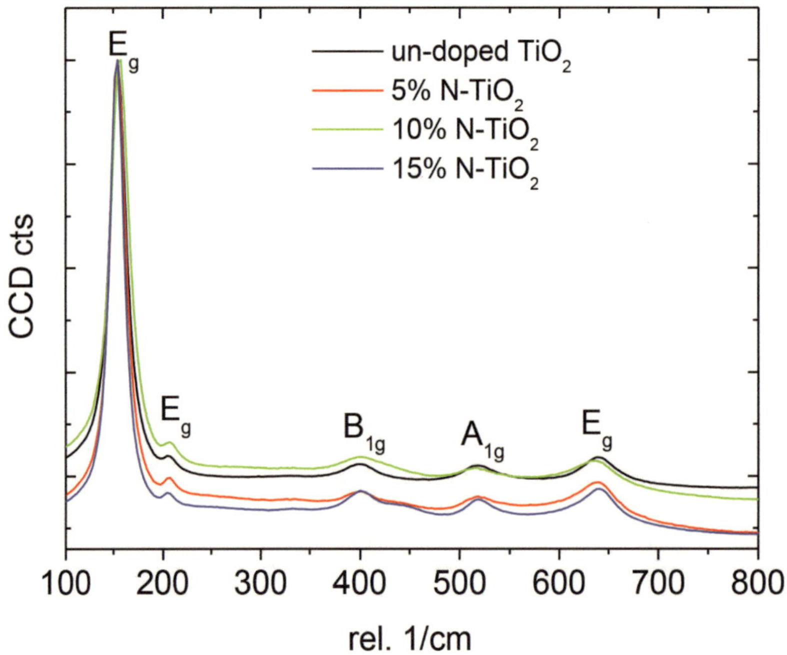

Figure 18. Raman spectra of the USP-fabricated samples.

7. Conclusions

A good understanding of basic TiO_2 material properties along with knowledge of previous experiences when working with TiO_2 spray deposition systems formed the necessary requirements for the design of the new USP spray deposition for TiO_2 thin films. Thin films were required to be dense and defect free, exhibit good thickness uniformity, possess a high refractive index and low optical absorption, and be semiconducting. Examination of deposition techniques described in the literature leads to the author designing a novel USP system. USP system developed offered unique features in material in synthesis of pure undoped TiO_2 and carbon-doped TiO_2 thin films and nanostructures. Firstly, we employed a horizontal furnace reactor as a contrast to most vertical systems in literature. Secondly, the horizontal system offered several potential advantages for deposition of thin films without any shadowing effect observed in the glancing or oblique angle depositions employed in most CVD systems. We managed to deposit samples at 0°, which allows the aerosol beam-containing precursor vapor to interact directly with the substrate uniformly. Furthermore, this allowed the system to use low deposition rates and the ability to deposit for a wide range of thin films via this method using different liquid precursors. Many of the desired TiO_2 film properties were obtained from films deposited using the USP system. XRD analysis has revealed the presence of anatase polymorph with peaks at 2θ angles and [hkl] planes of 25.67° [101], 38.19° [004], 48.56° [200], 53.91° [105], 63.14° [204], 68.98° [116], 70.72° [220], and 75.65° [215]. Additional Raman spectroscopy analysis has confirmed the presence of an anatase polymorph with Raman vibrational frequencies and symmetries 153.43 $cm^{-1}(E_g)$, 209.9 $cm^{-1}(E_g)$, 401.7 cm^{-1}(B1g), 517.6 cm^{-1}(A1g), and 641.9 cm^{-1} (Eg). SEM images, both undoped and N-TiO_2 NPs, reveal the formation of spherical-shaped TiO_2 NPs. Moreover, the SEM micrographs reveal that the surface morphology and shape of both undoped and N-TiO_2 NPs change as the nitrogen dopant level increase.

Acknowledgements

We are grateful for financial support from our sponsors, South African National Research Foundation (NRF) and Govan Mbeki Research and Development Centre (GMRDC) of the University of Fort Hare. The authors would also like to acknowledge the DST/CSIR Nanotechnology Innovation Centre, National Centre for Nanostructured Materials, and CSIR characterization of the TiO_2 NPs.

Author details

Raymond Taziwa[1,2*] and Edson Meyer[2]

*Address all correspondence to: rtaziwa@ufh.ac.za

1 Chemistry Department, University of Fort Hare, The Republic of South Africa

2 University of Fort Hare Institute of Technology, University of Fort Hare, The Republic of South Africa

References

[1] Shirke BS, Korake PV, Hankare PP, Bamane SR, Garadkar KM. Synthesis and characterization of pure anatase TiO_2 nanoparticles. J Mater Sci Mater Electron. 2011;**22**:821-824. doi:10.1007/s10854-01-0218-4.

[2] Yabing Z, Li Z, Dongping D. Single-step hydrothermal synthesis of strontium tititane nanoparticles. Ceramics International. 2015;**41**:13516-13524.

[3] Nolan MG, Pemble ME, Sheel DW, Yates HM. One step process for chemical vapour deposition of titanium dioxide thin films incorporating controlled structure nanoparticles. Thin Solid Films. 2006;**515:**1956-1962.

[4] Momeni, MM, Ghayeb Y, Gheibee S. Silver nanoparticles decorated titanium dioxide-tungsten trioxide nanotube films with enhanced visible light photo catalytic activity. Ceramics International. 2017;**43:**564-570.

[5] Yildirim Serdar, Yurddaskal Metin, Dikici Tuncay, Aritman Idil, Ertekin Kadriye, Celik Erdal. Structural and luminescence properties of undoped, $Nd3^{+}$ and $Er3^{+}$ doped TiO_2 nanoparticles synthesized by flame spray pyrolysis method. Ceramics International. 2016;**42**:10579-10586.

[6] Cotolan N, Rak M, Bele M, Cör A, Muresan LM, Milošev I. Sol-gel synthesis, characterization and properties of TiO_2 and Ag-TiO_2 coatings on titanium substrate. Surface & Coatings Technology. 2016;**307**:790-799.

[7] Hwang Heewon, Yoon Soyeon, Seok Jeesoo, Kim Kyungkon. Di(2-pyridyl)ketone stabilized titanium dioxide nanoparticles for the room temperature processed electron transporting layer in organic photovoltaics. Organic Electronics. 2016;**28**:281-286.

[8] Chaturvedia Amita, Joshi MP, Monda P, Sinhab AK, Srivastava AK. Growth of anatase and rutile phase TiO_2 nanoparticles using pulsed laser ablation in liquid: influence of surfactant addition and ablation time variation. Applied Surface Science. 2017;**396**:303-309.

[9] Liu Zhongwei, Chen Qiang, Wang Zhengduo, Yang Lizhen, Wang Chuanyue. Production of titanium dioxide powders by atmospheric pressure plasma jet. Physics Procedia. 2011;**18**:168-173.

[10] Taziwa R, Sideras-Haddad E, Erasmus RM, Manikandan E, Mwakikunga BW. Effect of carbon modification on the electrical, structural, and optical properties of TiO_2 electrodes and their performance in lab scale dye-sensitized solar cells. International Journal of Photo Energy. 2012;**2012**:904323-904354.

[11] Taziwa R, Meyer EL, Chinyama KG. Raman temperature dependence analysis of carbon doped titanium dioxide nano-particles synthesized by ultrasonic spray pyrolysis technique. Journal of Material Science. 2012, **47**:1531-1540. doi:10.1007/s10853011-5943.

[12] Ishizuka Shinnosuke, Kimura Yuki, Yamazaki Tomoya. In situ FT-IR study on the homogeneous nucleation of nanoparticles of titanium oxides from highly supersaturated vapour. Journal of Crystal Growth. 2016;**450**:168-173.

[13] Funakoshi Kunio, Nonami Toru. Influences of saturation ratios on crystallization of anatase titanium dioxide by a titanium alkoxide hydrolysis. Ceramics International. 2008;**34**:1637-1642.

[14] Kolen'ko YV, Kovnir KA, Gavrilov AI, Garshev AV, Frantti J, Lebedev OI, Churagulov BR, Tendeloo GV, Yoshimura M. Hydrothermal synthesis and characterization of nanorods of various titanates and titanium dioxide. J Phys Chem B. 2006;**110**:4030-4038.

[15] Ito S, Kitamura T, Wada Y, Yanagida S. Facile fabrication of mesoporous TiO_2 electrodes for dye solar cells chemical modification and repetitive coating. Sol Energy Mater Sol Cells. 2003;**76**:3-13.

[16] Rincón ME, Gómez-Daza O, Corripio C, Orihuel A. Sensitization of screen-printed and spray-painted TiO_2 coatings by chemically deposited CdSe thin films. Thin Solids Films. 2001;**389**:91-98.

[17] Wilska S. An X-ray diffraction study to determine the effect of the method of preparation on the crystal structure of TiO_2. Acta Chemica Scand. 1958;**8**:1796-1801.

[18] Hsiang Hsing-I, Lin Shih-Chung. Effects of aging on nanocrystalline anatase-to-rutile phase transformation kinetics. Ceramics International. 2008;**34**:557-561.

[19] Priol L, Baudel P, Louste C, Romat H. Theoretical and experimental study (linear stability and Malvern granulometry) on electrified jets of diesel oil in atomization regime. J Electrostatics. 2005; **63**: 899-904.

[20] Mwakikunga BW. Nano-size effects optical, structural and phononic properties of VO2 and WO3 by ultrasonic nebulizer spray pyrolysis technique dissertation. Republic of South Africa University of 109, Johannesburg.

[21] Harra J, Nikkanen JP, Aromaa M, Suhonen H, Honkanen M, Salminen T, Heinonen S, Levänen E, Mäkelä JM. Gas phase synthesis of encapsulated iron oxide-titanium dioxide composite nanoparticles by spray pyrolysis, julkaisussa. Powder Technology, Vuosikerta. 2013;**243**:46-52.

[22] Hawkeye MM, Brett MJ. Glancing angle deposition: fabrication, properties, and applications of micro- and nanostructured thin films. Journal of Vaccum Science & Technology A. Vaccum, Surface and Films. 2007;**25**:1317-1335.

Chapter 10

Semiconductor Materials by Ultrasonic Spray Pyrolysis and Their Application in Electronic Devices

Miguel Dominguez, Jose A. Luna-Lopez and
Francisco J. Flores

Additional information is available at the end of the chapter

http://dx.doi.org/10.5772/67548

Abstract

Ultrasonic spray pyrolysis is a deposition technique that enables a fine mist of the precursor solution in order to deposit higher-density thin films. This characteristic makes of great potential the use of ultrasonically spray-deposited semiconductors films for low-cost, transparent, flexible and large-area applications. In this chapter, low-temperature deposition and characterization of ultrasonically spray-deposited zinc oxide (ZnO) films are presented. The ZnO films deposited by ultrasonic spray pyrolysis at 200°C were characterized by optical transmittance, photoluminescence spectroscopy, X-ray diffraction and Fourier transform infrared spectroscopy. The study of low-temperature annealing of ZnO films is also presented. Moreover, the characterization of aluminum-doped ZnO films deposited by ultrasonic spray pyrolysis at 200°C is presented. Finally, applications of these ultrasonic spray-deposited films in electronic devices are presented.

Keywords: spray pyrolysis, low temperature, electronic devices

1. Introduction

Currently, semiconductor thin films are attractive to enable novel electronic applications. Deposition techniques such as pulsed laser deposition, chemical vapor deposition and sputtering have been used [1–4]. However, these deposition techniques present technical limitations such as low compatibility with large-area substrates, high cost and need of high or ultra-high vacuum. On the other hand, solution process techniques offer a solution to these problems at low cost and the possibility to deposit films under air ambient [5, 6]. The conventional solution-processed thin films by spin-coating have a low density due to pores during the annealing of the films [7]. However, ultrasonic spray pyrolysis is a deposition technique that enables a fine mist of the precursor solution in order to deposit higher-density thin films. This characteristic makes of great potential the use of ultrasonic spray-deposited semiconductors films for low-cost, transparent, flexible and large-area applications.

Metal-oxide semiconductors are promising materials to be used in novel electronic applications. Applications, such as thin-film transistors (TFTs), sensors, electronic circuits, among others, have been reported [8–10]. The results reported by other authors show that the main limitation is the need to use high deposition temperatures to achieve high-quality semiconductor films, since low-temperature deposition may lead to an incomplete pyrolysis of the precursor solutions [11].

In this chapter, low-temperature deposition and characterization of ultrasonic spray-deposited zinc oxide (ZnO) films are presented. The ZnO films deposited by ultrasonic spray pyrolysis at 200°C were characterized by optical transmittance, photoluminescence (PL) spectroscopy, X-ray diffraction (XRD) and Fourier transform infrared (FTIR) spectroscopy. The study of low-temperature annealing of ZnO films is also presented. Moreover, the characterization of aluminum-doped ZnO films deposited by ultrasonic spray pyrolysis at 200°C is presented. Finally, applications of these ultrasonic spray-deposited films in electronic devices are presented.

2. Zinc oxide films deposited by ultrasonic spray pyrolysis at 200°C

Previous reports [11] show a dependency of the zinc oxide structure with deposition temperature. The crystallinity of the film is increased as the deposition temperature is higher. **Figure 1** shows the XRD pattern of the as-deposited ZnO film at 200°C. The pattern shows three-weak peaks at 2θ = 31.72°, 34.42° and 56.64°, which are associated with the (100), (002) and (110) planes, respectively. From the overall XRD diffractogram, the ZnO film tends to present some polycrystallinity and agree with the JCPDS Card No. 36-1451.

Figure 2 shows the optical transmittance of as-deposited ZnO films. It can be seen the highly transparent in the visible range of the film. The gap energy is extracted by the extrapolation of the linear region in the $(\alpha h\nu)^2 \sim A(h\nu\text{-}Eg)$ characteristics, where hν is the photon energy, α is the absorption coefficient, Eg is the gap energy and A is the constant. The extracted gap energy was 3.26 eV, which is similar than the commonly reported [5].

Figure 3 shows the FTIR spectra of the ZnO film. It can be observed the peak at 415 cm^{-1} related to Zn—O stretching modes [11, 12]. The peak at 1627 cm^{-1} is attributed to O—H bending modes. Also, the peaks at 2500–3500 cm^{-1} are due to C—H and O—H stretching modes [11, 12]. The peaks at 1413 cm^{-1} and 1530 cm^{-1} are attributed to C—O stretching modes, while at 1750 cm^{-1} to C=O bonds [11, 12]. The presence of C—O bonds suggests an incomplete precursor pyrolysis due to the low deposition temperature [11]. However, the characteristic Zn—O band approximately at 415 cm^{-1} indicates the ZnO formation. On the other hand, it is important to mention that bonds related to O—H stretching have been previously reported in ZnO films, indicating the presence of O—H complexes that are associated with different defects and increased free-carrier concentration [11–14].

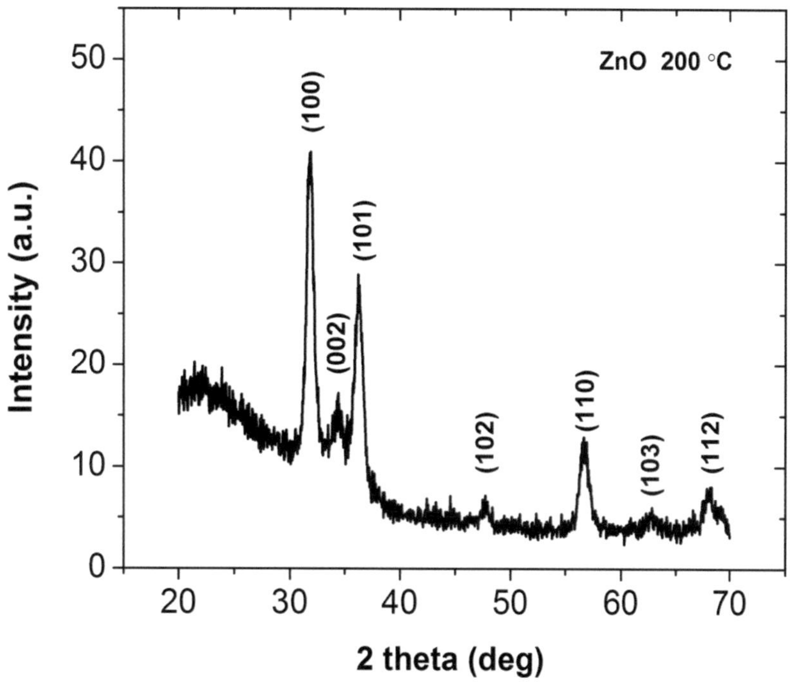

Figure 1. XRD pattern of the as-deposited ZnO films by ultrasonic spray pyrolysis.

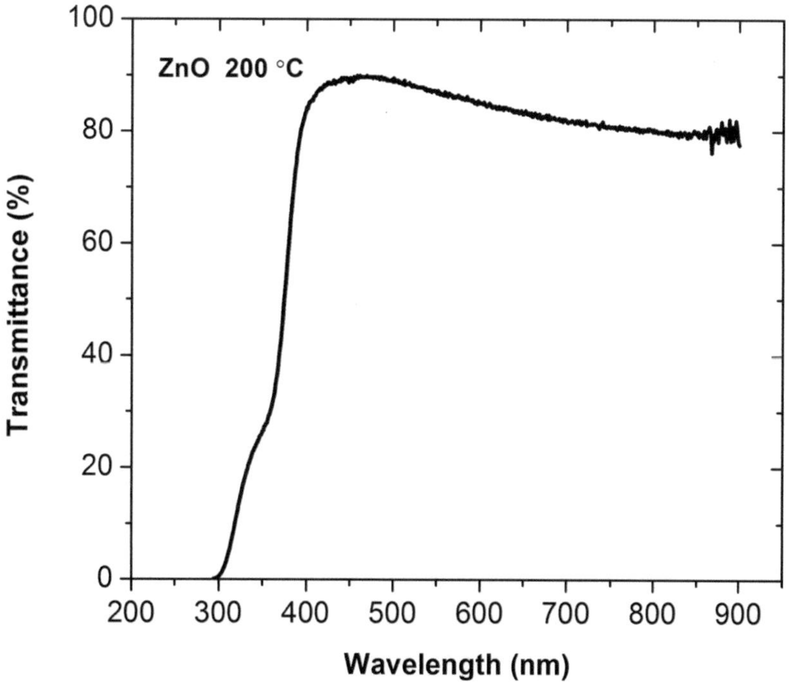

Figure 2. Transmittance of the as-deposited ZnO films by ultrasonic spray pyrolysis.

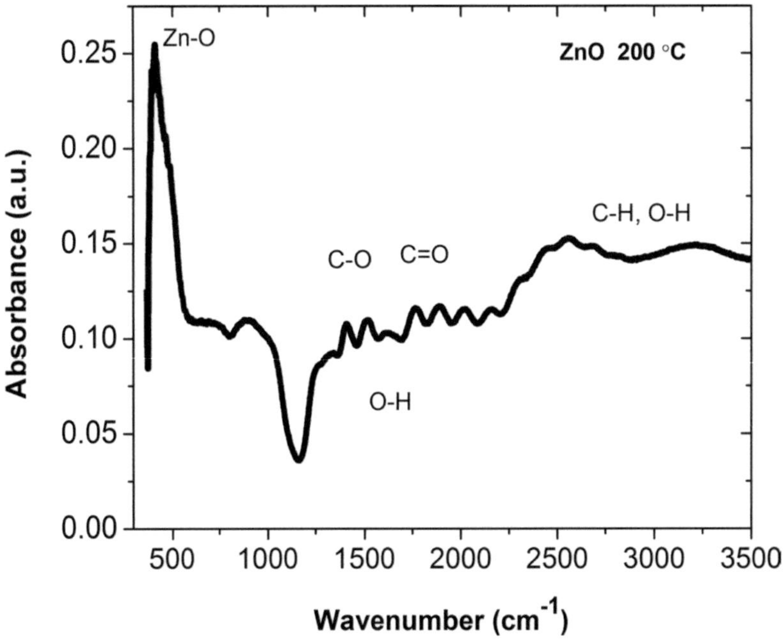

Figure 3. FTIR spectra of the as-deposited ZnO films by ultrasonic spray pyrolysis.

Figure 4 shows the photoluminescence (PL) spectra of the as-deposited ZnO film. It exhibits a PL spectrum with a peak centered at 390 nm of high intensity and a broad band from 450 to 700 nm. The peak at 390 nm (UV emission) is typically associated with the near band-edge (NBE) emission of the gap, which is attributed to the recombination of the free excitons [15, 16]. The broad band from 450 to 700 nm (visible emission) is typically associated with the impurities and defects, which are considerable in our case. The origin of this visible emission band has been related to zinc and oxygen vacancies, zinc and oxygen antisites, and to zinc and oxygen interstitials [16, 17]. Several authors have reported the use of photoluminescence spectroscopy to study the role of the defects and impurities in ZnO. However, even with the same experimental conditions they have reported contradictory results [17]. Therefore, since the impurities and defects are highly dependent on the deposition technique and its conditions, their role in the electronic properties of ZnO is still controversial.

2.1. Thermal annealing effects

The use of metal-oxide semiconductors in semiconductor devices is constantly increased, since thin-film transistors, solar cells, optical sensors, among others. In these devices, a high-quality metal-semiconductor interface is desired in order to avoid voltage drops which result in loss of performance. Ideally, a metal-semiconductor contact no exhibit barriers for the carrier flow in whatever polarization (positive or negative). This is true when there are no interface states, and the metal and semiconductor work functions are similar. However, obtaining

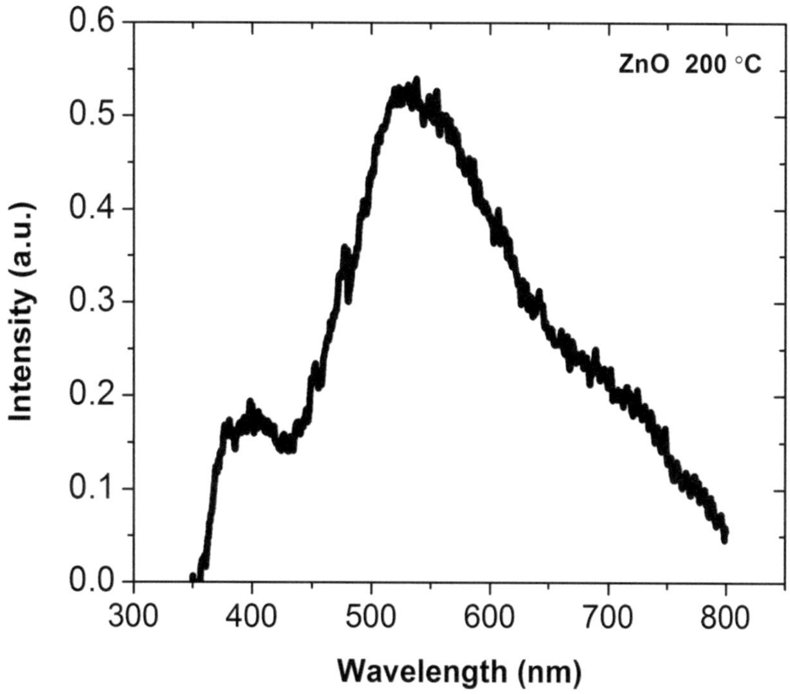

Figure 4. Photoluminescence of the as-deposited ZnO films by ultrasonic spray pyrolysis.

a metal-semiconductor contact without interface states is difficult. Moreover, matching the metal and semiconductor work functions is nearly impossible. It is well known that metal-oxide films are highly dependent on the deposition and post-treatments conditions, resulting in different surface conditions and defect density distribution. For this reason, different results have been reported.

Figure 5 shows the conductivity and contact resistance of Al-ZnO contacts annealed at 180°C as a function of annealing time. It exhibits a reduction of more than two orders of magnitude in contact resistance in samples annealed for 30 min. This improvement can be associated with a higher carrier injection through the Al-ZnO (metal-semiconductor) interface. Nunes et al. [6, 18, 19] reported that this is due to desorption of oxygen present at the grain boundaries. As result of this oxygen loss, there is an increase in the effective carrier concentration near to the Al-ZnO interface.

Also, an increase in the contact resistance with longer annealing time than 30 min is appreciated in **Figure 5**. This increase can be associated with a change in the ZnO film quality, as the reduction in conductivity shows. One can conclude that there is an optimal annealing time and after this time, the metal-semiconductor interface deteriorates. Using FTIR spectroscopy and XRD, the effects of the low temperature annealing on the ZnO film can be studied.

Figure 6 shows the FTIR spectra of ZnO films at different annealing times. The peak at 415 cm^{-1} was previously related to Zn—O stretching modes. The characteristic Zn—O peak

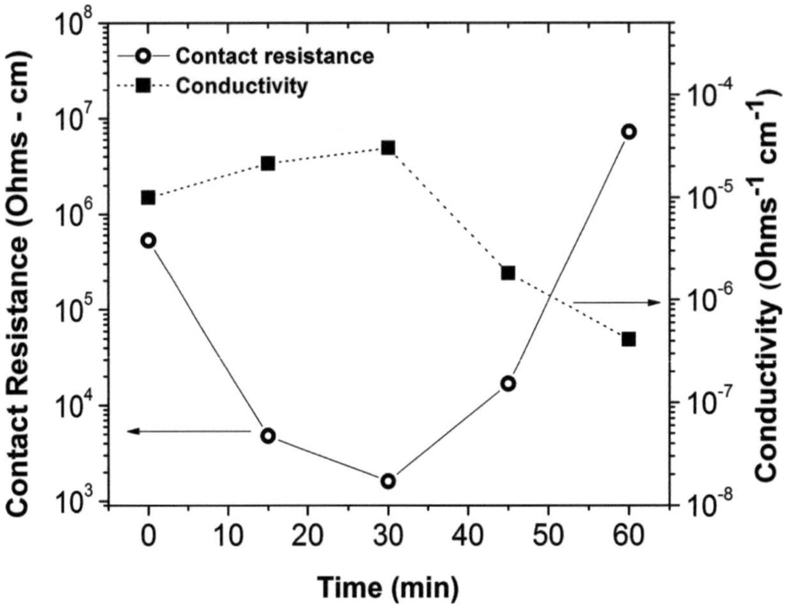

Figure 5. Al-ZnO contact resistance and conductivity extracted by TLM as a function of annealing time.

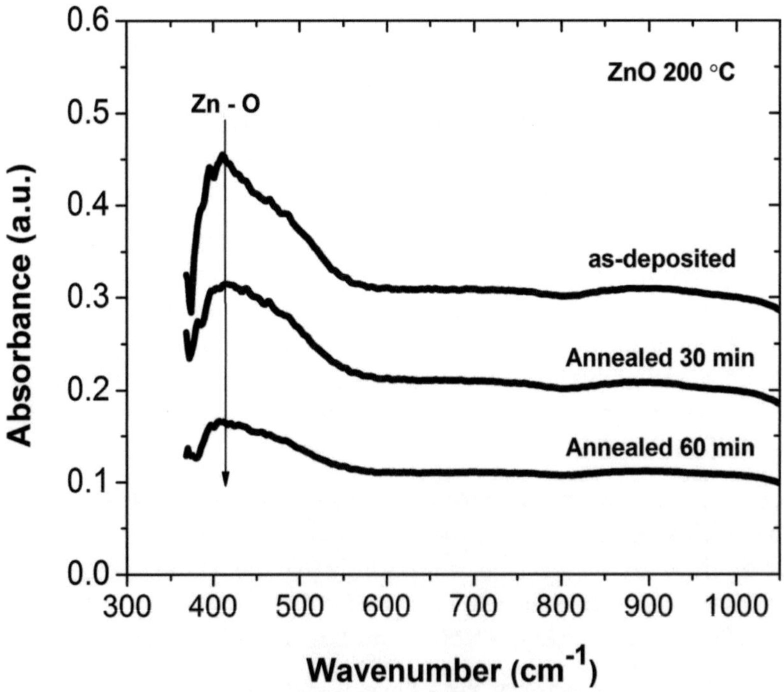

Figure 6. FTIR spectra of as-deposited and annealed ZnO films.

indicates the ZnO formation. A reduction in Zn—O bonds is clearly exhibited as the annealing time is increased. This agrees with the results of contact resistance and conductivity, where the carrier concentration is increased by oxygen vacancies. However, the progressive reduction in Zn—O bonds at 60 min of annealing may not explain the increase in contact resistance. In this case, **Figure 7** shows the XRD patterns of as-deposited and annealed ZnO films. In the as-deposited ZnO films, a strong peak associated with the (100) plane can be appreciated, and also peaks related to the planes (002), (101) and (110) can be identified. At 30 min of annealing, the strongest peak is now associated with the (002) plane. Also, peaks related to the planes (100), (101), (102), (103), (112) and (110) can be identified (in agreement with the JCPDS Card No. 36-1451). A better polycrystalline nature of ZnO films for 30 min of annealing can be confirmed. The preferential orientation in (002) plane has been reported in high-quality ZnO films [11, 20]. Finally, at 60 min of annealing, the preferential orientation in (002) plane disappeared and also the peaks related to the planes (102) and (103). A reduction in the peaks related to the planes (100) and (101) is appreciated. The FTIR spectroscopy and XRD results confirm the degradation of the ZnO film after 60 min of annealing and agree with the contact resistance and conductivity results.

2.2. Aluminum-doped zinc oxide films

Typically, the zinc oxide film is doped with different impurities such as aluminum (Al), cadmium or gallium in order to increase its conductivity. In this case, aluminum was used as a doping source. **Figure 8** shows the optical transmittance of the AZO films at different

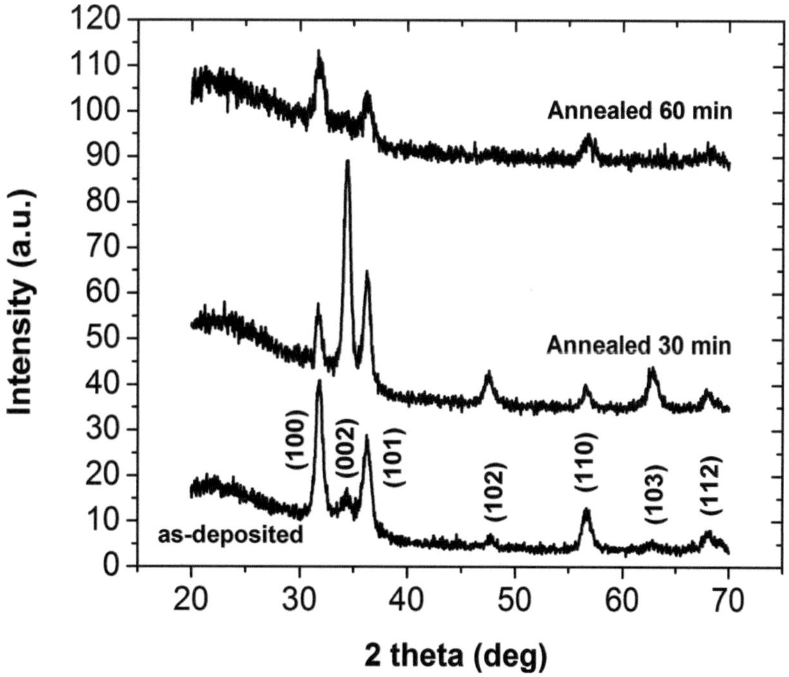

Figure 7. XRD patterns of as-deposited and annealed ZnO films.

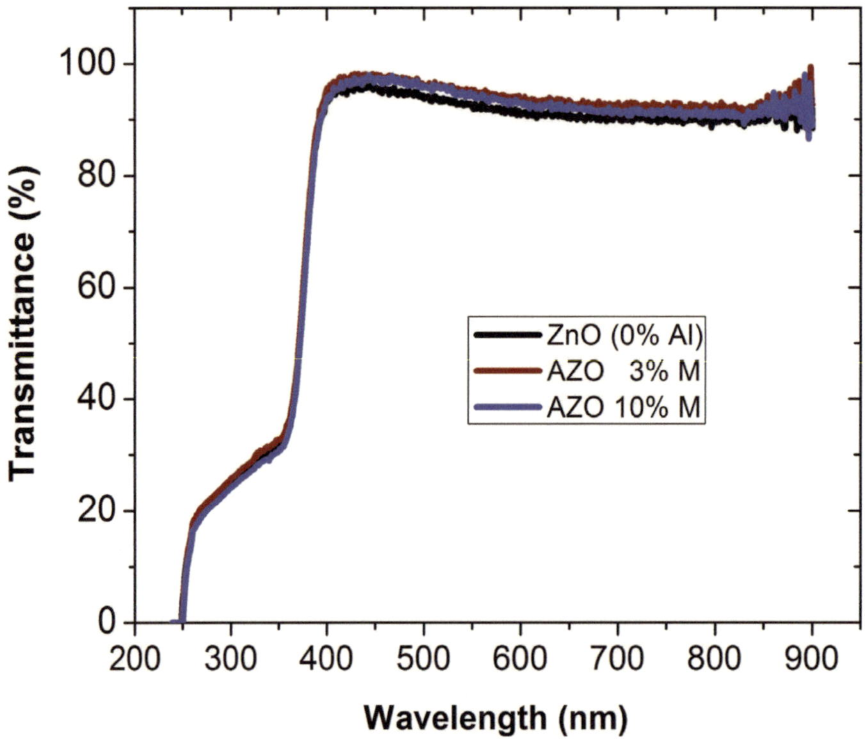

Figure 8. Optical transmittance of the AZO films at different aluminum-doping concentrations.

aluminum-doping concentrations. The films are highly transparent in the visible range. It is important to note that the increase in the doping concentration has no effect on the optical transmittance. This agrees with the reported information in Ref. [20].

Figure 9 shows the XRD pattern of the AZO films at different aluminum-doping concentrations. The AZO films show an amorphous structure regardless of the Al-doping concentration, since no presence of peaks is exhibited. This is expected due to the low temperature of deposition [11].

Figure 10 shows the resistivity of the AZO films at different aluminum-doping concentrations. It can be observed that the resistivity decreases as the doping concentration increases at 3 M%. This decrease in resistivity is considered as result of the increase in carrier concentration. The increase in carrier concentration of AZO films is due to the substitutional incorporation of Al^{3+} ions at Zn^{2+} cation sites or the incorporation of Al ions in interstitial positions [21]. However, as the doping concentration increases above 3 M%, the AZO resistivity also increases. This increase in resistivity is attributed to a decrease in the mobility of carriers resulted by ionized impurity scattering [22]. This agrees with that reported by other authors [21, 23], where the excessive Al-doping deteriorates the properties of the AZO films due to the formation of stress by the smaller radius of Al^{3+} ions compared with Zn^{2+} ions.

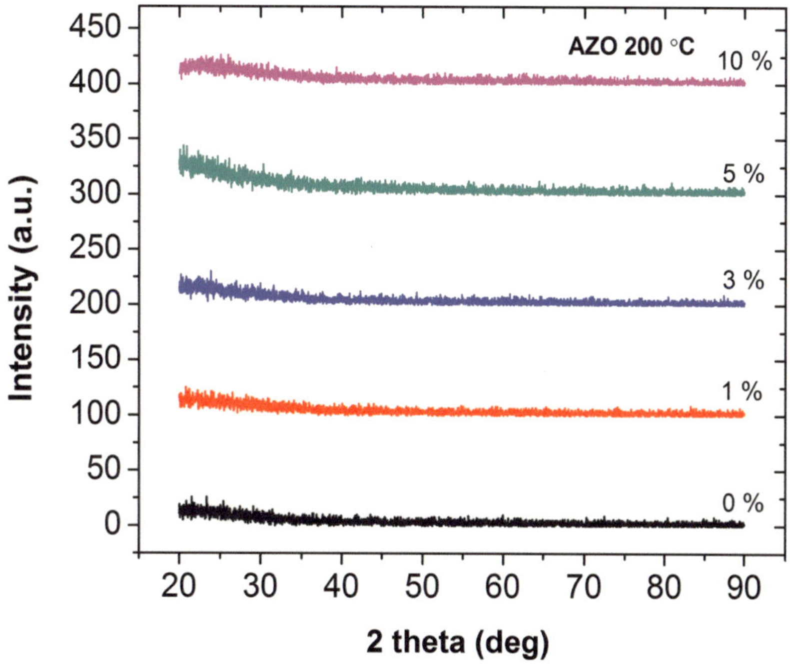

Figure 9. XRD patterns of the AZO films at different aluminum-doping concentrations.

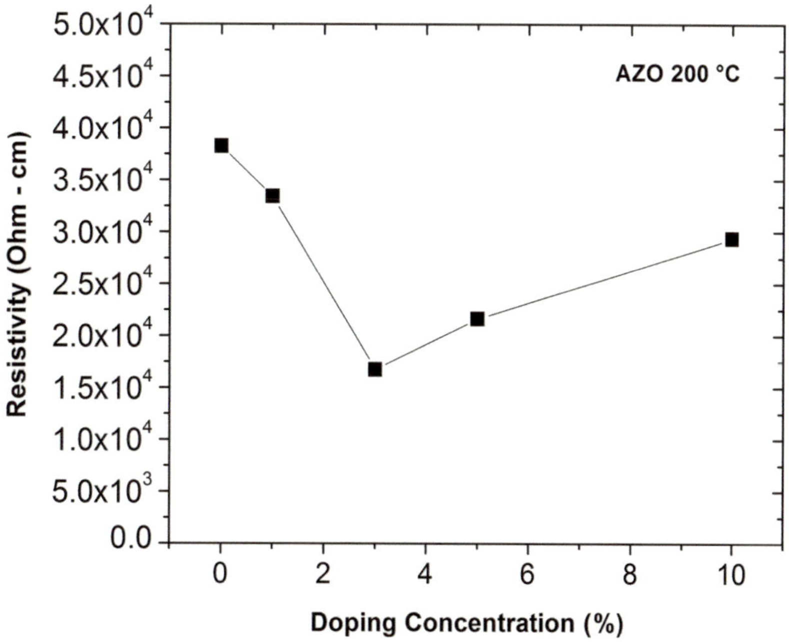

Figure 10. Resistivity of the AZO films at different aluminum-doping concentrations.

3. Applications in electronic devices

An important area of the materials science is the application of the materials obtained. The advantages of spray pyrolysis make of great potential the use of semiconductors as active layers in electronic devices. In recent years, the development of low-cost electronics has achieved a considerable progress, since wearable electronics, transparent circuitry, e-paper, solar cells and more.

3.1. Fully solution-processed transparent MIS capacitors

This basic device can work as energy storage from solar cells, sensor, memory device and charge-discharge capacitor in active-matrix displays and also can be very useful to analyze the quality of the dielectric-semiconductor interface in field-effect devices. In this section, we present fully solution-processed capacitors employing the ultrasonic spray pyrolysis and spin-coating techniques. Zinc oxide films deposited by ultrasonic spray pyrolysis were used as an active layer and spin-on glass (SOG) is used as dielectric. The maximum fabrication temperature used was set at 200°C. **Figure 11** shows a top-view photograph of the MIS capacitors. The MIS capacitors are highly transparent, which is used in transparent electronics.

Figure 12 shows the capacitance-voltage characteristics at 10 KHz for the MIS capacitors. When a negative voltage is applied at the top contact, there is an accumulation layer of electrons in the ZnO film at the SOG/DI-ZnO (dielectric-semiconductor) interface; as a result, the capacitance-voltage characteristics exhibit the dielectric capacitance (C_{ox}). On the other hand, when a positive voltage is applied at the top contact, there is a depletion region in the ZnO film at the SOG/DI-ZnO interface; then, the capacitance-voltage characteristics show a

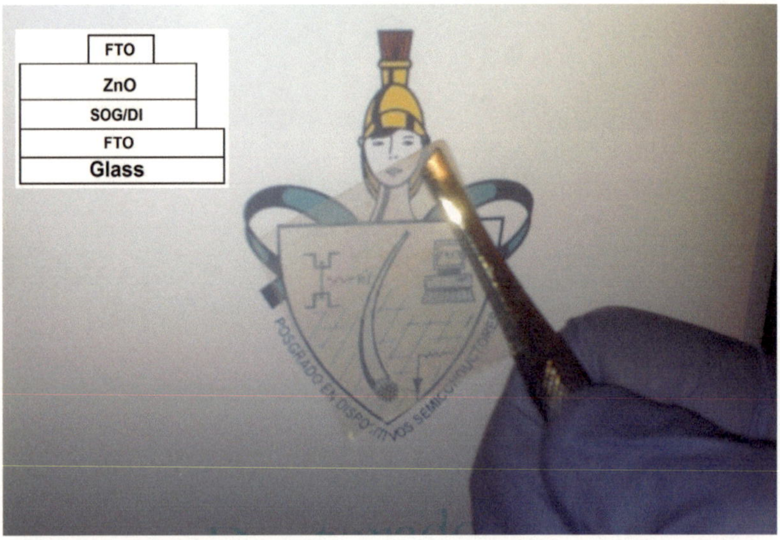

Figure 11. Top-view photograph of the MIS capacitors.

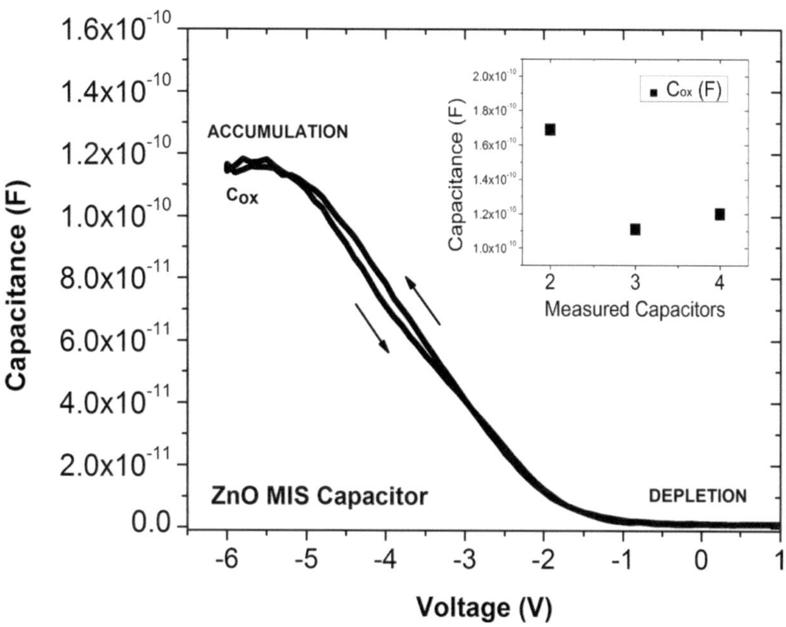

Figure 12. Forward and reverse capacitance-voltage curves of the ZnO MIS capacitors.

minimum capacitance. The capacitance-voltage measurements exhibit a very low hysteresis when the characteristics are forward and reverse measured. However, at negative voltage values, the accumulation region presents effects of interface states [24]. This is attributed to the defects in the ZnO film near at the dielectric-semiconductor interface, as the photoluminescence and FTIR spectroscopies show.

On the other hand, **Figure 13** shows the current density of the ZnO MIS capacitors. The breakdown voltage of the MIS capacitor can be appreciated at 36 V, which is due to the dielectric breakdown. The values of current density and capacitor breakdown voltage obtained are reliable for electronic device applications [25, 26].

The ZnO MIS capacitors were fabricated employing simple and low-cost solution process techniques with no-vacuum need, using feasible and easily prepared precursor solutions.

3.2. Flexible metal-insulator-semiconductor diodes

The semiconductor diodes are known as devices that allow the flow of current in only one voltage bias (positive or negative). These devices are extensively used in electronics such as circuit protection, rectifiers, mixing, isolating and detection signals. Recently, Son et al. [27] reported for the first time Schottky diodes using solution-processed zinc tin oxide on corning glass. Alternatively to PN and Schottky diodes, metal-insulator-semiconductor (MIS) diodes are devices, which used a thin insulator film, that allow the tunneling of carriers in only one voltage bias. For this reason, these devices are also known as MIS tunnel diodes [28–31]. Güllü et al. [32] reported the application of MIS diodes as temperature sensors.

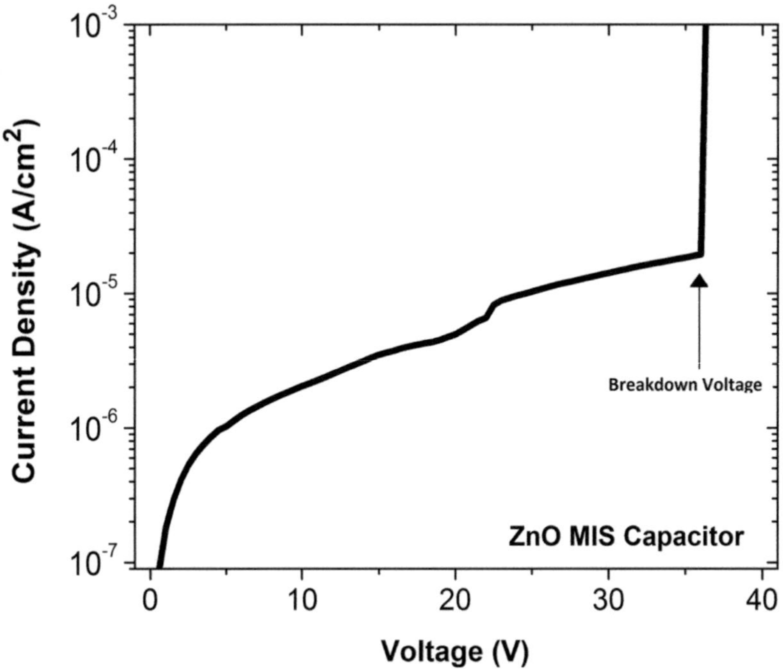

Figure 13. Current density of the ZnO MIS capacitors.

In this section, the fabrication and characterization of fully solution-processed flexible metal-insulator-semiconductor diodes is presented. As an active layer, aluminum-doped zinc oxide (AZO) thin film by ultrasonic spray pyrolysis was used. As an insulator, a silicon oxide thin film by spin-on glass (SOG/DI) was used. The maximum temperature used was 200°C. As far as the authors know, this is the first flexible solution-processed MIS diode using amorphous oxide semiconductors.

To fabricate the flexible MIS diodes, the AZO film with Al-doping concentration at 3 M% was used, which exhibits the best doping efficiency.

Figure 14 shows the electrical characteristics of the flexible MIS diodes. A good rectifying behavior is observed when negative voltage is applied to the ITO contact (reverse bias), and the current is independent of the voltage applied. When a positive voltage is applied to the ITO contact (forward bias), the injected carriers are collected by tunneling through the insulator; then, the current increases exponentially with increasing bias. The inset in **Figure 14** shows the inverse saturation current, which is close to 150 pA. An on/off current ratio of 10^2 is reached at |4|V. This value is similar to other diodes reported by other authors at higher fabrication temperature [27, 31]. The forward bias current in a MIS diode is assumed to be due to thermionic emission and can be expressed as [32]:

$$I = I_0 \exp\left(qV/nkT\right) \quad (1)$$

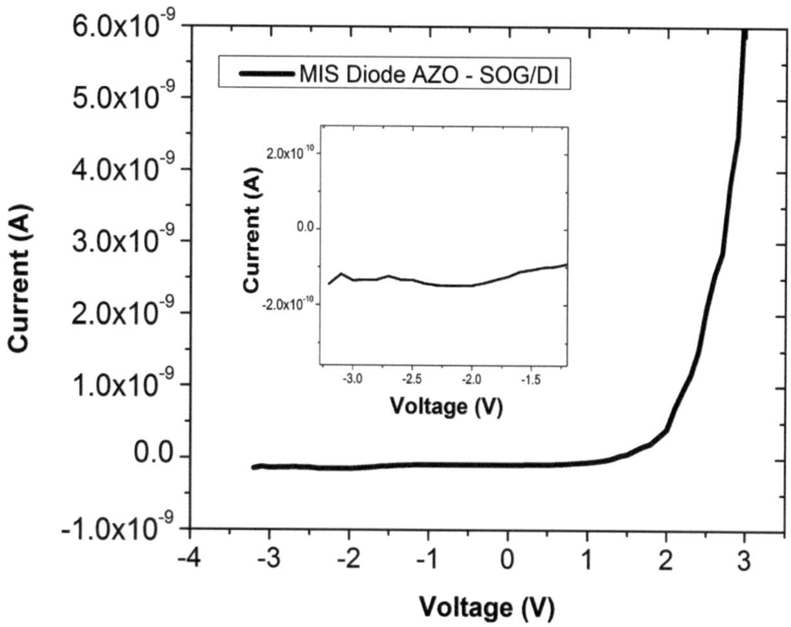

Figure 14. Electrical characteristics of the flexible MIS diodes. Inset: inverse saturation current.

where I_0 is the saturation current, q is the electron charge, *n* is the ideality factor, *k* is the Boltzmann constant and T is the temperature in K. The saturation current I_0 can be expressed as:

$$I_0 = AA^* T^2 \exp\left(-q\Phi_B / kT\right) \quad (2)$$

where A is the diode area, A^* is the effective Richardson constant of 32 A/cm^{2}K^2 for ZnO [33], and Φ_B is the barrier height.

Typically, the ideality factor and barrier height can be extracted from the extrapolation to 0 V and its slope of the linear region of the forward bias of ln(I)-V characteristics [30, 32]. The extracted values of ideality factor and barrier height were 2.9 and 0.88 eV, respectively.

According with the thermionic emission theory, the ideality factor should be close to 1.01, a higher value of ideality factor indicates a secondary transport mechanism. Possible mechanisms may include interface dipoles or fabrication-induced defects at the interface [32]. Also, this higher value can be attributed to an insulating layer in the metal-semiconductor interface [30]. These extracted values are similar to those reported by other authors [30, 32].

In order to compare the effect of bending on the flexible devices, a flexible substrate was attached around a rigid plastic rod of 5 mm radius. The bent to a tensile radius of 5 mm is equivalent to a mechanical strain of ~0.5% [34]. **Figure 15** shows the bending effect of the flexible MIS devices. The electrical characteristics are very similar with and without bending, where the minor variations in current may be induced by an increase in tunneling under tensile strain. It is important to mention that these electrical characteristics are reversible after removal of the tensile strain.

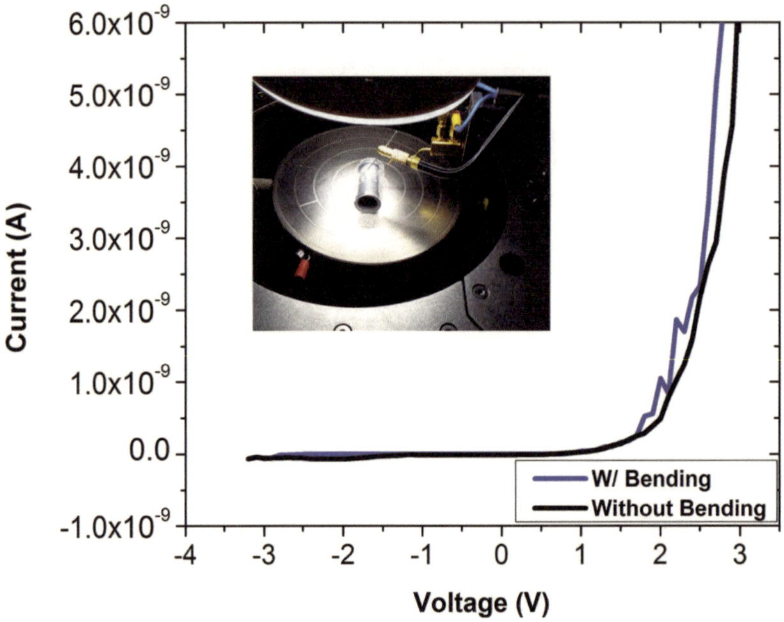

Figure 15. Electrical characteristics of the flexible devices with and without bending. Inset: Picture of the bent flexible MIS diodes.

These results show the potential of solution-processed AZO films to fabricate flexible semiconductor devices. The precursor solutions may be combined or replaced by new solutions in order to optimize the properties of the deposited thin films. The flexible MIS diodes were fabricated employing simple and low-cost solution process techniques under air ambient, using easily prepared precursor solutions.

3.3. Thin-film transistors

Although metal-oxide thin-film transistors (TFTs) fabricated by spray pyrolysis have already been demonstrated, the temperature of deposition to obtain high performance devices is still high to be compatible with most of the low-cost plastic substrates used in flexible and low-cost electronics. Then, it is necessary to reduce the temperature of deposition at values about 200°C or less in order to be a real alternative for low-cost applications. Moreover, yet, the role of the impurities and defects distribution in electronic properties of ZnO is still controversial, since they are highly dependent on the deposition technique and its conditions [10, 15, 35]. Therefore, the extraction of the density of states (DOS) within the gap of the ZnO is a great challenge and, typically, reflects contributions from the measurement techniques and interfaces from the TFT device [36–39].

Using this film as an active layer, inverted coplanar ZnO TFTs were fabricated. The transfer characteristics of the ZnO TFTs are presented in **Figure 16**. The electron field-effect mobility and threshold voltage were extracted from the square root of Ids versus Vgs, using Eq. (3) of the saturation regime [40].

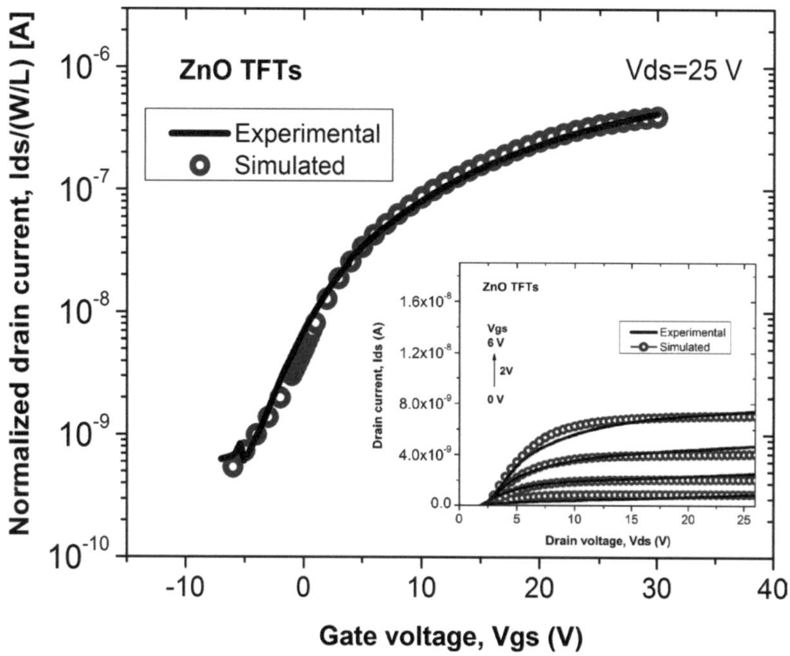

Figure 16. Experimental and simulated transfer characteristics of the ZnO thin-film transistors. Inset: Experimental and simulated output characteristics.

$$Ids = \mu FE \cdot C_{ox} (W/2L) (Vgs - V_T)^2 \quad (3)$$

where μ_{FE} is the electron field-effect mobility, C_{ox} is the capacitance per unit area of the gate insulator, W and L are the channel width and length, respectively, and V_T is the threshold voltage. The average value extracted was approximately 0.011 cm^2/Vs and 2.6 V for field-effect mobility and threshold voltage, respectively. The obtained results for ZnO TFTs with 50-nm-thick gate dielectric are better than those reported by Adamopoulos et al. [11]. They reported carrier mobilities from 0.003 to 0.001 cm^2/Vs and on/off-current ratios from 10^2 to 10^1 at deposition temperature of 200°C. Also, **Figure 16** shows the simulated transfer characteristic. The simulated data reproduces very well the experimental electrical characteristics of the device. The inset in **Figure 16** shows the experimental and simulated output characteristics. In order to reproduce the experimental electrical characteristics by physically based simulations, it is necessary to estimate approximately the density of states (DOS) within the gap of the ZnO film, commonly correlated to defects in the ZnO film. The typical DOS is composed of acceptor-like states (near the conduction band) given by the sum of tail states and deep states, and donor-like states (near the valence band) given by the sum of tail states and deep states [41].

The mathematical model of Silvaco simulator involves the Poisson's equation, the continuity equations and the transport equations to simulate any semiconductor device. In this mathematical model, using the TFT module, one can incorporate the DOS distribution g(E) proposed. The total charge caused by the presence of traps or defects is added into the right-hand side of Poisson's equation. Also, the recombination/generation rate in the carrier continuity equations is modified by g(E). For accurate description of the model used in Silvaco simulator, please see Ref. [41]. The parameters used for the ZnO film and DOS were extracted from previous measurements (Section 2) and other Refs. [42, 43]. The DOS parameters were adjusted meanwhile the simulation fitted the experimental data.

The parameters used in the simulation for the DOS and ZnO film are listed in **Table 1**. As can be seen, the DOS obtained is higher than those previously reported by different authors [44, 45]. This can be due to more defects in the ZnO films originated by the incomplete precursor pyrolysis, corroborated by the presence of O—H complexes in the FTIR spectra and defects as the photoluminescence spectroscopy shows.

Parameter	Value	Description
N_C	5×10^{18} (cm^{-3})	Effective conduction band states
N_V	5×10^{18} (cm^{-3})	Effective valence band states
Eg	3.26 (eV)	Energy gap
Affinity	4.29 (eV)	Electron affinity
Permittivity	8.12	Dielectric constant
N_{TA}	1.95×10^{20} ($cm^{-3}eV^{-1}$)	Density of tail-acceptor states
N_{TD}	1.85×10^{20} ($cm^{-3}eV^{-1}$)	Density of tail-donor states
W_{TD}	0.385 (eV)	Decay energy of tail-donor
W_{TA}	0.105 (eV)	Decay energy of tail-acceptor
N_{GA}	1.2×10^{18} ($cm^{-3}eV^{-1}$)	Density of deep-acceptor states
E_{GA}	1 (eV)	Peak energy of deep-acceptor
W_{GA}	0.9 (eV)	Decay energy of deep-acceptor
N_d	9×10^{17} (cm^{-3})	Donor density
Rc	875 (Ωcm)	Contact resistance
μe	6 (cm^2/Vs)	Electron band mobility
μh	0.1 (cm^2/Vs)	Hole band mobility

Table 1. Main parameters used in the ZnO TFTs simulation.

4. Experimental section

For deposition of ZnO films, a home-made ultrasonic spray pyrolysis deposition system using air as carrier gas at flow rate of 467 sccm was used. The deposition system was adapted from an ultrasonic humidifier (Heaven Fresh). The precursor solution consists of zinc acetate (0.2 M) in methanol. The AZO films were deposited using a home-made ultrasonic spray pyrolysis deposition system adapted from an ultrasonic nebulizer (CITIZEN CUN-60) using air as the carrier gas, from 0.2 M precursor solution of zinc nitrate (Sigma-Aldrich) in distilled water, using aluminum nitrate (Sigma-Aldrich) as doping source at different molar concentration percentages with respect to zinc nitrate (M%). The samples were on a hotplate at 200°C during deposition. The SOG/DI was obtained by spin-on glass (SOG700B Filmtronics) diluted 2:1 with deionized water (DI). The transparent electrodes were obtained using fluorine tin oxide (FTO) from 0.2 M precursor solution of tin tetrachloride pentahydrate in ethanol with ammonium fluoride diluted in deionized water prepared with Fluor/tin ratio of 0.52.

The optical transmittance of the thin films above corning glass was measured from 200 to 900 nm. The resistivity of the films was measured by four-point probe. The orientation of the films was obtained using an X-ray diffractometer (XRD) (Discover D8-Bruker axs) at 2θ range between 20° and 80° and 0.002° step. For the Photoluminescence spectroscopy, it was used as an exciting source, a laser of He-Cd with 325 nm line. The photoluminescence measurement was performed using a silicon PIN Thorlabs (DET-210) detector with a spectral response of 200–1100 nm. The IR absorption spectra of the films were measured with a "BRUCKER" FTIR spectrometer, Model Vertex-70. The IR spectrum was observed for wave numbers between 3500 and 400 cm^{-1}.

The flexible MIS diodes were fabricated above ITO-coated PET substrates (Sigma-Aldrich). First, the SOG/DI film was spin-coated at 5000 RPM for 30 sec and cured at 200°C for 1 h. Then, the AZO film was ultrasonic spray deposited at 200°C. As top electrodes, silver ink (AgIC Inks) was patterned. The contact area was 0.012 cm^2. The fabrication procedure of the inverted coplanar ZnO TFTs (bottom-contact bottom-gate) and transparent MIS capacitors can be found elsewhere [46, 47].

The electrical characteristics of the devices were measured using the Keithley-4200 Semiconductor Characterization System at room temperature and under dark conditions.

Acknowledgements

M. Dominguez wants to thank the financial support from NPTC-PRODEP by SEP-Mexico.

Author details

Miguel Dominguez*, Jose A. Luna-Lopez and Francisco J. Flores

*Address all correspondence to: madominguezj@gmail.com

Centro de Investigaciones en Dispositivos Semiconductores, Instituto de Ciencias, Benemerita Universidad Autonoma de Puebla (BUAP), Puebla, Mexico

References

[1] S. Fay, U. Kroll, C. Bucher, E. Vallat-Sauvain and A. Shah. Low pressure chemical vapour deposition of ZnO layers for thin-film solar cells: Temperature-induced morphological changes. Sol. Energy Mater. Sol. Cells. 2005;**86**:385.

[2] S.H. Mohamed and R. Drese. Structural and optical properties of direct current sputtered zinc aluminium oxides with a high Al concentration. Thin Solid Films. 2006;**513**:64.

[3] N. Bouhssira, S. Abed, E. Tomasella, J. Cellier, A. Mosbah, M. Aida and M. Jacquet. Influence of annealing temperature on the properties of ZnO thin films deposited by thermal evaporation. Appl. Surf. Sci. 2006;**252**:5594.

[4] J. Nishii, F.M. Hossain, A. Takagi, T. Aita, K. Saikusa, Y. Ohmaki, I. Ohkubo, S. Kishimoto, A. Ohtomo, T. Fukumura, F. Matsukura, Y. Ohno, H. Koinuma, H. Ohno and M. Kawasaki. High mobility thin film transistors with transparent ZnO channels. Jpn. J. Appl. Phys. 2003;**42**:L347.

[5] M. Olvera, H. Gomez and A. Maldonado. Doping, vacuum annealing, and thickness effect on the physical properties of zinc oxide films deposited by spray pyrolysis. Sol. Energy Mater. Sol. Cells. 2007;**91**:1449.

[6] P. Nunes, B. Fernandes, E. Fortunato, P. Vilarinho and R. Martins. Performances presented by zinc oxide thin films deposited by spray pyrolysis. Thin Solid Films. 1999;**337**:176.

[7] M. Dominguez, O. Obregon and J. Luna-Lopez. Study of stability of solution-processed dielectric film under electrical stress. J. Alloys Comp. 2016;**688**:893-896.

[8] A. Bashir, P. Wobkenberg, J. Smith, J. Ball, G. Adamopoulos, D. Bradley and T. Anthopoulos. High-performance zinc oxide transistors and circuits fabricated by spray pyrolysis in ambient atmosphere. Adv. Mater. 2009;**21**:2226-2231.

[9] G. Zhang, H. Wu, C. Chen, T. Wang, J. Yue and C. Liu. Transparent and flexible capacitors based on nanolaminate $Al_2O_3/TiO_2/Al_2O_3$. Nanoscale Res. Lett. 2015;**10**:76.

[10] D. Padilla, J. Vadillo and J. Laserna. Room temperature pulsed laser deposited ZnO thin films as photoluminescence gas sensors. Appl. Surf. Sci. 2012;**259**:806.

[11] G. Adamopoulos, A. Bashir, W. Gillin, S. Georgakopoulos, M. Shkunov, M. Baklar, N. Stingelin, D. Bradley and T. Anthopoulos. Structural and electrical characterization of ZnO films grown by spray pyrolysis and their application in thin- film transistors. Adv. Funct. Mater. 2011;**21**:525-531.

[12] R. Gayen, K. Sarkar, S. Hussain, R. Bhar and A. Pal. Zno films prepared by modified sol-gel technique. Indian J. Pure Appl. Phys. 2011;**49**:470-477.

[13] M.D. McCluskey, S.J. Jokela, K.K. Zhuravlev, P.J. Simpson and K.G. Lynn. Infrared spectroscopy of hydrogen in ZnO. Appl. Phys. Lett. 2002;**81**:3807.

[14] G. Shi, M. Stavola, S. Pearton, M. Thieme, E. Lavrov and J. Weber. Hydrogen local modes and shallow donors in ZnO. Phys. Rev. B. 2005;**72**:195211.

[15] S. Xiao, L. Zhao, Y. Liu and J. Lian. Nanocrystalline ZnO films prepared by pulsed laser deposition and their abnormal optical properties. Appl. Surf. Sci. 2013;**283**:781.

[16] S. Santra, P. Guha, S. Ali, P. Hiralal, H. Unalan, J. Covington, G. Amaratunga, W. Milne, J. Gardner and F. Udrea. ZnO nanowires grown on SOI CMOS substrate for ethanol sensing. Sensor Actuat. B Chem. 2010;**146**:559.

[17] D. Wang, H. Seo, C. Tin, M. Bozack, J. Williams, M. Park, N. Sathitsuksanoh, A. Cheng and Y. Tzeng. Effects of postgrowth annealing treatment on the photoluminescence of zinc oxide nanorods. J. Appl. Phys. 2006;**99**:113509.

[18] P. Nunes, A. Malik, B. Fernandes, E. Fortunato, P. Vilarinho and R. Martins. Influence of the doping and annealing atmosphere on zinc oxide thin films deposited by spray pyrolysis. Vacuum. 1999;**52**:45-49.

[19] P. Nunes, E. Fortunato and R. Martins. Influence of the annealing conditions on the properties of ZnO thin films. Inter. J. Inorg. Mater. 2001;**3**:1125-1128.

[20] J. Lee and B. Park. Characteristics of Al-doped ZnO thin films obtained by ultrasonic spray pyrolysis: Effects of Al doping and an annealing treatment. Mater. Sci. Eng. B. 2004;**106**:242-245.

[21] M. Jun, S. Park and J. Koh. Comparative studies of Al-doped ZnO and Ga-doped ZnO transparent conducting oxide thin films. Nanoscale Res. Lett. 2012;**7**:639.

[22] G. Sanon, R. Rup and A. Mansingh. Growth and characterization of tin oxide films prepared by chemical vapour deposition. Thin Solid Films. 1990;**190**:287.

[23] S. Kuo, W. Chen, F. Lai, C. Cheng, H. Kuo, S. Wang and W. Hsieh. Effects of doping concentration and annealing temperature on properties of highly-oriented Al-doped ZnO films. J Cryst Growth. 2006;**287**:78.

[24] D. Neamen. Semiconductor physics and devices. 3th ed. McGraw Hill, New York; 2003.

[25] Y. Yoo, J. Park, K. Lee, H. Lee, K. Song, S. Lee and H. Baik. Solution-processed high-k HfO_2 gate dielectric processed under softening temperature of polymer substrates. J. Mater. Chem. C. 2013;**1**:1651-1658.

[26] J. Park, Y. Yoo, K. Lee, W. Jang, J. Oh, S. Chae and H. Baik. Low-temperature, high-performance solution-processed thin-film transistors with peroxo-zirconium oxide dielectric. ACS Appl. Mater. Interf. 2013;**5**:410-417.

[27] Y. Son, J. Li and R. Peterson. In situ chemical modification of Schottky barrier in solution-processed zinc tin oxide diode. ACS Appl. Mater. Interf. 2016;**8**:23801.

[28] M. Green and J. Shewchun. Current multiplication in metal-insulator-semiconductor (mis) tunnel diodes. Solid State Electron. 1974;**17**:349.

[29] M. Hudait and S. Krupanidhi. Effects of thin oxide in metal-semiconductor and metal-insulator-semiconductor epi-GaAs Schottky diodes. Solid State Electron. 2000;**44**:1089.

[30] A. Tataroglu and S. Altındal. The analysis of the series resistance and interface states of MIS Schottky diodes at high temperatures using I–V characteristics. J. Alloys Comp. 2009;**484**:405.

[31] H. Jeong, H. Oh, S. Bang, H. Jeong, S. An, G. Han, H. Kim, S. Yun, K. Kim, J. Park, Y. Lee, G. Lerondel and M. Jeong. Metal–insulator–semiconductor diode consisting of two-dimensional nanomaterials. Nano Lett. 2016;**16**:1858.

[32] Ö. Güllü and A. Türüt. Electronic properties of Al/DNA/p-Si MIS diode: Application as temperature sensor. J. Alloys Comp. 2011;**509**:571.

[33] Zhe Chuan Feng. Handbook of Zinc Oxide and related materials Vol. 2. CRC Press, Florida; 2012.

[34] J. Jang, K. Cho, S. Lee and S. Kim. Transparent and flexible thin-film transistors with channel layers composed of sintered HgTe nanocrystals. Nanotechnology. 2008;**19**:015204.

[35] A. Janotti and C. Van de Walle. Native point defects in ZnO. Phys. Rev. B. 2007;**76**:165202.

[36] S. Baranovski. Charge transport in disordered solids with applications in electronics. 1st ed. Chichester: John Wiley & Sons Ltd; 2006.

[37] M. Dominguez, P. Rosales, A. Torres, F. Flores, J. Molina, M. Moreno, J. Luna and A. Orduña. Planarized Ambipolar a-SiGe:H thin-film transistors: Influence of the sequence of fabrication process. Solid State Electron. 2014;**99**:45-50.

[38] A. Nathan, S. Lee, S. Jeon and J. Robertson. Amorphous oxide semiconductor TFTs for displays and imaging. J. Display Technol. 2014;**10**:917-927.

[39] M. Dominguez, S. Alcantara and S. Soto. Physically-based simulation of zinc oxide thin-film transistors: Contact resistance contribution on density of states. Solid State Electron. 2016;**120**:41-46.

[40] M. Dominguez, P. Rosales and A. Torres. Performance improvement of low-temperature a-SiGe:H thin-film transistors. Solid State Electron. 2012;**76**:44-47.

[41] ATLAS User's manual. Silvaco International, California; 2004.

[42] T. Fung, C. Chuang, C. Chen, K. Abe, R. Cottle, M. Townsend, H. Kumomi and J. Kanicki. Two-dimensional numerical simulation of radio frequency sputter amorphous In–Ga–Zn–O thin-film transistors. J. Appl. Phys. 2009;**106**:106:084511-1.

[43] F. Hossain, J. Nishii, S. Takagi, A. Ohtomo, T. Fukumura, H. Fujioka, H. Ohno, H. Koinuma and M. Kawasaki. Modeling and simulation of polycrystalline ZnO thin-film transistors. J. Appl. Phys. 2003;**94**:7768-7777.

[44] S. Bubel and M. Chabinyc. Model for determination of mid-gap states in amorphous metal oxides from thin film transistors. J. Appl. Phys. 2013;**113**:234507-1.

[45] S. Lee, D. Kim, E. Chong, Y. Jeon and D. Kim. Effect of channel thickness on density of states in amorphous InGaZnO thin film transistor. Appl. Phys. Lett. 2011;**98**:122105-1.

[46] M. Dominguez, F. Flores, A. Luna, J. Martinez, J. Luna-Lopez, S. Alcantara, P. Rosales, C. Reyes and A. Orduña. Impact of active layer thickness in thin-film transistors based on zinc oxide by ultrasonic spray pyrolysis. Solid State Electron. 2015;**109**:33-36.

[47] M. Dominguez and A. Orduna. Fully solution-processed zinc oxide MIS capacitors by ultrasonic spray pyrolysis in air ambient. J. Appl. Res. Technol. Forthcoming.

Section 4

Pyrolysis of Fossil Fuels and Petrochemicals

Chapter 11

Interaction between Polyethylene and Petroleum Coke Substrate during Pyrolysis

Sharifah Shahnaz

Additional information is available at the end of the chapter

http://dx.doi.org/10.5772/67568

Abstract

There has been a constant growth for plastic demand globally in the past decades, and the continuing expanding trend with rapid emerging economies has increased the concerns of many parties. Various approaches of recycling waste plastic including chemical recycling, thermal recycling, and mechanical recycling has been practiced. As chemical recycling is known to be a promising method in recovering hydrocarbon compounds, which can be used in high-end product, new avenues for waste recycling need to be established. Consumable carbon anodes are a major requirement for process used for producing primary aluminum. Since carbon is a main constituent of waste plastics, which have very low impurity levels, these clearly have the potential as a cheap readily available auxiliary source of carbon in carbon anodes. Coal tar pitch, a major by-product produced in petroleum refining, is the binder of choice for carbon anodes. Pitch penetrates the pores of petroleum coke-binding particulates and gets carbonized during the baking process. In-depth wettability and interfacial phenomena investigation was carried out to study interactions between polyethylene (PE) and petroleum coke (PC). The effect pyrolysis parameters on degradation process of PE have been characterized. The wettability study of polyethylene polymer on PC substrates has been carried out.

Keywords: chemical recycling, waste plastic , heat treatment, pyrolysis, wettability

1. Introduction

Plastic industry has grown extensively in the past 30 years and is expected to grow steadily in future due to their product versatility. However, the relatively short life span of plastic goods has caused abundancy of waste plastic globally where only 60 % of waste plastics are incinerated or buried into landfill. These methods have caused serious environmental problems and had led to the necessity of more efficient and novel recycling approaches that will not hurt the environment at the effective cost. According to the statistic, waste plastic component only summed up to about 10 % of municipal solid waste (MSW), but due to its high volume to plastic ratio properties and also high resistance to chemical, weather, and harsh conditions, the amount of waste plastics piled up are worrying [1, 2]. Only 12 % of these waste plastic is incinerated while the remainder is end up to be landfilled. The waste landfilling method requires an active extraction industry, proper location that is close to waste generation, low cost transportation, and also must adhere to the policy requirements [3, 4]. The incineration of waste plastic will diminish the solid piled up, while recovering energy from the burning of waste but simultaneously emitting various environment pollutants [5].

Recycling plastic materials has become utmost serious business globally. There are variety of programs that were implemented to enhance the effectiveness of recycling and increase the awareness among consumers. Since plastics are mainly composed of hydrocarbon components, waste plastic are readily rich in carbon and have low impurity levels, which has the potential to be used as a cheap, abundantly available, auxiliary source of carbon. Owing to the high value of carbon in waste plastic, it can be used to produce consumable carbon anodes for industrial applications. Carbon anodes are manufactured by baking blends of some varieties of coke with hydrocarbon binder, which is generally a coal-tar pitch [6]. Prolonged treatment of waste plastic at high temperature (up to 1000 °C) has the potential of breaking down the hydrocarbon chain where the yield at the particular temperature and time can be used as a precursor for carbonization process for the anodes.

1.1. Recycling of waste plastic

Plastic is generally prepared from petroleum by-products and natural gas. These are composed of high molecular polymeric compounds containing primarily carbon, hydrogen, and a few other elements such as nitrogen and oxygen. Crude oil from petroleum is processed and refined to produce raw materials for plastics. Statistics have reported that total global plastic production has increased by an average of almost 10 % annually since 1950. The total plastic production has grown from around 1.5 million tons (MT) in 1950 to 322 MT in 2015 globally. China is the largest plastic producer, accounted up to quarter of world plastics, followed by Europe and former Soviet Union and North America [1]. Plastics have been one of the materials with the fastest growth because of their wide range of applications, but the duration of plastic life cycle is relatively small, which causes serious environmental problem every year.

There are three methods of recycling waste plastic that include material or mechanical recycling, thermal recycling, and chemical recycling. Chemical recycling or feedstock recycling allows the conversion of waste plastic into low molecular weight materials, which consist of their liquid or gaseous hydrocarbons for chemical industries. Chemical recycling can be classified into thermal decomposition, depolymerization (monomerization), and gasification (partial oxidation) [7]. Chemical recycling has been proposed as one of the most attractive methods for sustainable developments in the field. Waste plastics can be converted back into their original monomers or other valuable chemicals. Pyrolysis, one of the key chemical recycling techniques, has been used to produce a series of refined petrochemical products and particularly liquid fractions similar to commercial gasoline [8]. Appropriate technique, including applied temperature and time of pyrolysis, is crucial in determining the desired polymeric product.

1.2. Pyrolysis of waste plastic

Pyrolysis as a process consists of chemical and thermal reactions, generally leading to the smaller molecules. Pyrolysis may be performed using a range of parameters such as temperature, reaction time, pressure, the presence or absence of reactive gases or liquids, and use of catalysts. Plastic pyrolysis can be conducted in three ranges of temperatures, which are low (<400 °C), medium (400–600 °C), and high (>600 °C). Pyrolysis temperature and heating time are known to enhance bond breaking and favor the production of smaller molecules while longer residence time increase the formation of secondary primary products, yielding more coke, tar, as well as thermally stable products, thus gradually obscuring the effect of original polymer structure. Pressure applied on the other hand determines the condensation rate of reactive fragments forming coke and heavy chain products.

Pyrolysis of two major plastics, PE and PP, has shown major conversion of plastic into oil with a low concentration of gas and no solid residue. The emitted hydrocarbon gases observed from the pyrolysis consist of alkane gases, methane, ethane, propane, and butane. Studies on the pyrolysis done at 430 °C of waste plastic mixtures have also been reported by Bhaskar et al. [9], describing the yield of liquid, gas, and residue from municipal waste plastic as being 59, 25, and 16 % weight, respectively. There is a significant level of liquid yield from pyrolysis; similar results have also been reported by Lee and Shin [2] in their pyrolysis experiments conducted at 350 °C and 400 °C. The liquid products obtained from the pyrolysis of waste plastics mainly consisted of liquid paraffin, liquid olefin, liquid naphthene, and liquid aromatics, with their relative proportions varying with polymer types, temperature, and lapse time. McIlveen-Wright et al. [10] have reported that the pyrolysis of waste plastic is important as it can provide oil and wax feedstock for the production of new plastics or refined fuels. The treatment can also generate a range of gases such as hydrogen, methane, ethane, and propane at higher temperatures. The main products are gas, oil/wax, and char products in some cases as its production depends on the types of plastic, reactor type, and process conditions [11].

2. Experiment

The research was focused on the polyethylene (PE) polymer as it is one of the mainstream waste plastic. PE was obtained from ExxonMobile Chemical. LL 6201 contains heat stabilizer, is high flow PE grades of 50 g/10 min melt-flow index (190 °C/2.16 kg) (based on ASTM D1238) with 0.926 g/cm^3 density, and 123 °C melting temperature. PE has high calorific value of up to 80.30 % carbon, 19 % hydrogen, 0.03 % sulfur, and 0.52 % ash. **Figure 1** shows raw polyethylene samples used in the experiment.

Petroleum coke (PC) clumps was supplied by Rio Tinto Australia. Samples were then sieved to segregates for particular particle size for further analysis and then ground by ring mill into fine powder (**Figure 2**).

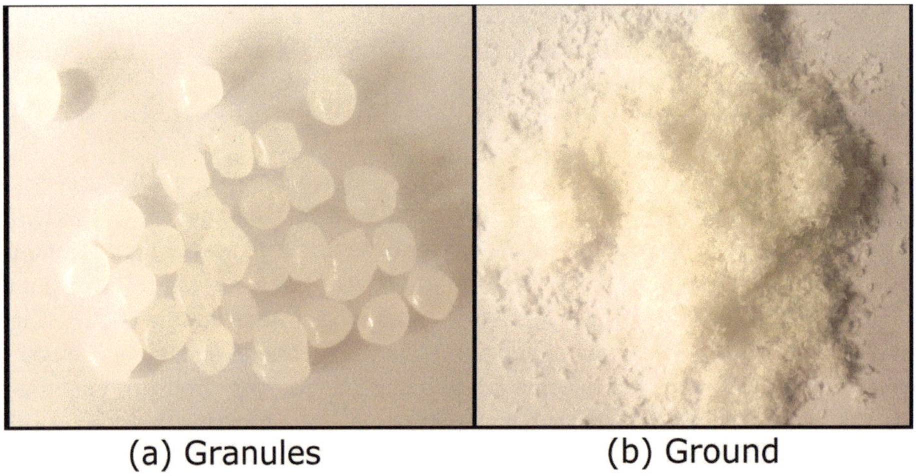

Figure 1. Raw polyethylene samples in (a) granules and (b) powder form.

2.1. Wettability studies

Petroleum coke powder were mixed with 5 wt% phenol formaldehyde binder and put onto roller milling machine for 24 h to ensure homogeneous mixing for preparing cylindrical substrates. The substrates sized 20 mm diameter and 5 mm thickness were prepared and baked at 180 °C for 24 h to carbonize the binder and harden the substrates. Small amount of ground polymer was put on the petroleum coke substrate and the assembly was charged into horizontal tube furnace under 1 L/min argon flow. Samples were then heated to 150, 200, 250, 300, and 350 °C and let soaked for 15, 30, and 60 min.

Figure 2. Raw petroleum coke samples in (a) granules and (b) powder form.

http://dx.doi.org/10.5772/67568

2.2. Heat treatment on the PE and PC mixtures

Ground PE and PC were weighted to obtain the desired weight percentages range from 10 to 50 %. The mixed samples were placed on roller milling machine for 24 h to obtain homogeneous mixtures. The mixture samples were placed in the tube furnace under purging of argon gas and undergoes three cycles of heat treatment as stated in **Table 1**.

Heating cycle	Baking method (1 L/min argon flow)
Cycle 1	Ambient temperature—heated to 150 °C and dwell for 30 min—cooled to room temperature
Cycle 2	Ambient temperature—heated to 150 °C and dwell for 30 min—heated up to 600 °C and dwell for 30 min—cooled to room temperature
Cycle 3	Ambient temperature—heated to 150 °C and dwell for 30 min—heated up to 600 °C and dwell for 30 min—heated up to 1000 °C—cooled to room temperature

Table 1. Heat treatment cycles method of polyethylene and petroleum coke mixtures.

3. Results

At 150 °C, PE starts to fuse together only after 30 min and liquefied after 60 min but no penetration occurs at this temperature. Similar observation was seen for up to 250 °C of heating where raw ground PE can still be seen, but as soaking time was increased to 60 min, the PE sample started to fuse together and adhere to the surface of petroleum coke substrate. PE has completely melted after 30 min of soaking at 300 °C where contact angle was calculated at 52.10 ° at this stage. However, the melt color still exhibited opaque raw PE. Prolonged soaking time to 60 min has decreased the contact angle between the melt PE and PC substrate to 36.30 °. The color of the PE melt started to change to translucent yellow at 350 °C, showing some bubble inside the melt. Increasing time has decreased the contact angle between PE and PC substrate (**Figure 3**).

This increment in the height becomes more obvious in samples treated at 300 °C. This increase may be due to the bubbles foaming from inside the PE melt. With the increasing temperature up to 300 °C, PE polymer starts to volatilize and starts releasing gases. The formation of bubbles was observed inside the PE melts from the SEM images. However, these bubbles formation are low and it may be due to the high viscosity nature of PE. The length of contact area increased significantly as the time was increased. These increasing lengths were due to the PE melt becoming less viscous which resulted in better spreading.

The mechanism of polyethylene degradation has been explained in a number of studies [12–14]. The degradation step is initiated by random scission reaction. Upon this reaction the polyethylene backbone may be further depolymerized by two competing reactions, e.g., (1) the propagation (unzipping) to yield monomers, and also (2) free radicals transfer which involves hydrogen transfer yielding the formation of unsaturated end and new free radicals [15].

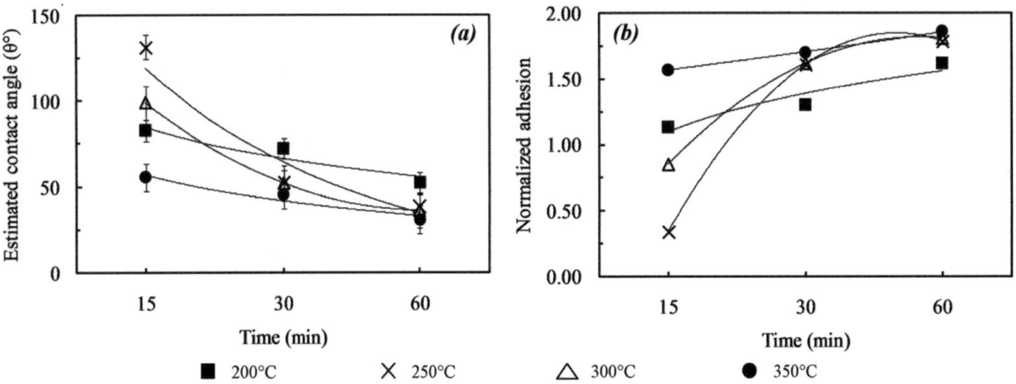

Figure 3. Effect of time and temperature on (a) estimated contact angle and (b) normalized adhesion of PE melt on PC substrate.

Temperature and residence time of the volatiles in the hot region of the furnace are the important parameters in determining the end-products of pyrolysis. This can be explained as the reaction products are produced from raw material decomposition, which is the primary reaction and also these primary volatiles product may be further depolymerized by secondary reactions that result in smaller monomers that mainly are in gases forms [15]. These primary reaction products are highly vulnerable to temperature and time of pyrolysis as they may undergo secondary reaction. Polyethylene is stable up to 290 °C but starts to reduce its molecular weight with increasing temperature. Polyethylene products varies according to the temperature where at mild degradation from 290 °C to 400 °C yield plastic similar to original polyethylene or hard waxes; and at extensive degradation results in semisolids pastes or liquids [16]. Pyrolysis of PE at 400–450 °C yields high liquid fraction of 69–84 and 9–13 wt% of gases [17, 18].

The weight loss of PE and PC sample mixtures undergoing heat treatment is plotted in **Figure 4**. The increase of PE ratio in the mixtures has decreased the percentage of residues obtained. The increasing of heat treatment temperature (Cycles 1–3) also lessens the weight of residues measured. This trend is expected as PE is composed of high volatiles that have been released during the heating cycles. Highest temperature of heating cycles (Cycle 3) results in the lowest residue left for both mixtures ratios, whereas the highest residues are obtained from Cycle 1, since at this temperature (150 °C) PE had not melted.

The effect of mixing and heat treatment cycles were also investigated by FTIR analysis. Each sample spectrum was stacked and compared in **Figures 5** and **6**, and the corresponding peak vibration has been characterized in **Table 2**.

The CH_2 stretching around 2913–2918 cm^{-1} shared by raw PC and raw PE is visible in all residues after the heat treatments. In contrast, another vibration of symmetric CH_2 at 2850 cm^{-1} in PE is lost after mixing with PC. The double bond stretching absorbed by raw PC (1685 cm^{-1}) also has been reduced after the mixing. Another alkene C=C absorption, which is visible in PC and PE around 1645–1655 cm^{-1}, was intact throughout the heating cycles of up to 1000 °C (Cycle 3). Similar outcomes on the absorption of CH_3 bending (1457–1459 cm^{-1}) and $C{=}CH_3$ bending (1059–1072 cm^{-1}), which originated from their raw samples, also

http://dx.doi.org/10.5772/67568

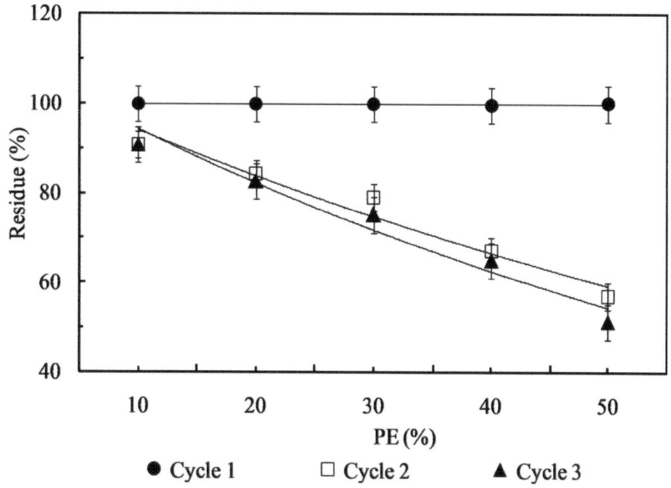

Figure 4. Residue left (%) of PE and PC mixtures after heat treatment cycles.

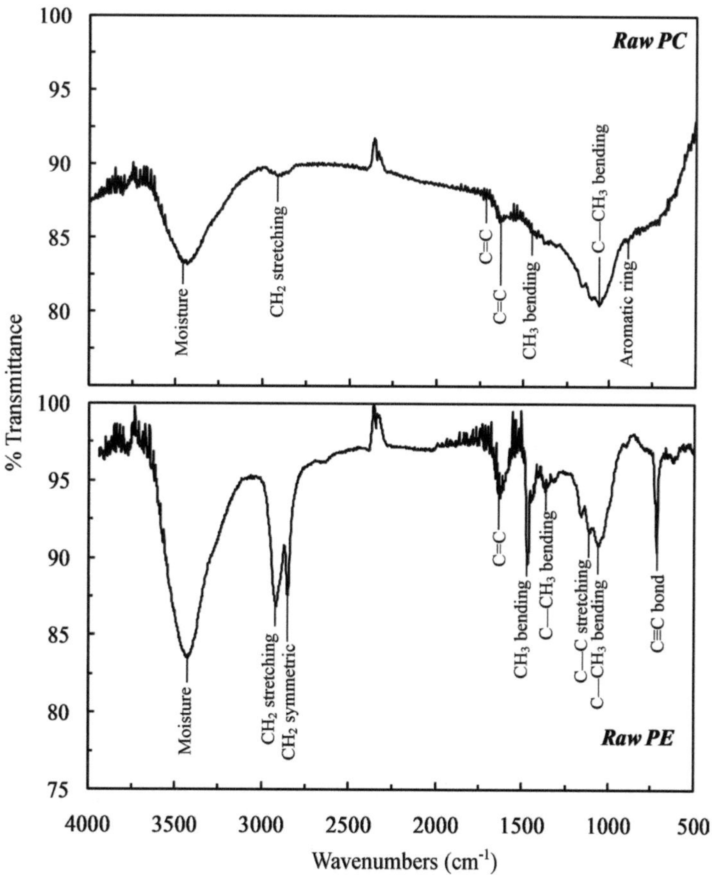

Figure 5. FTIR peaks of raw PE and PC.

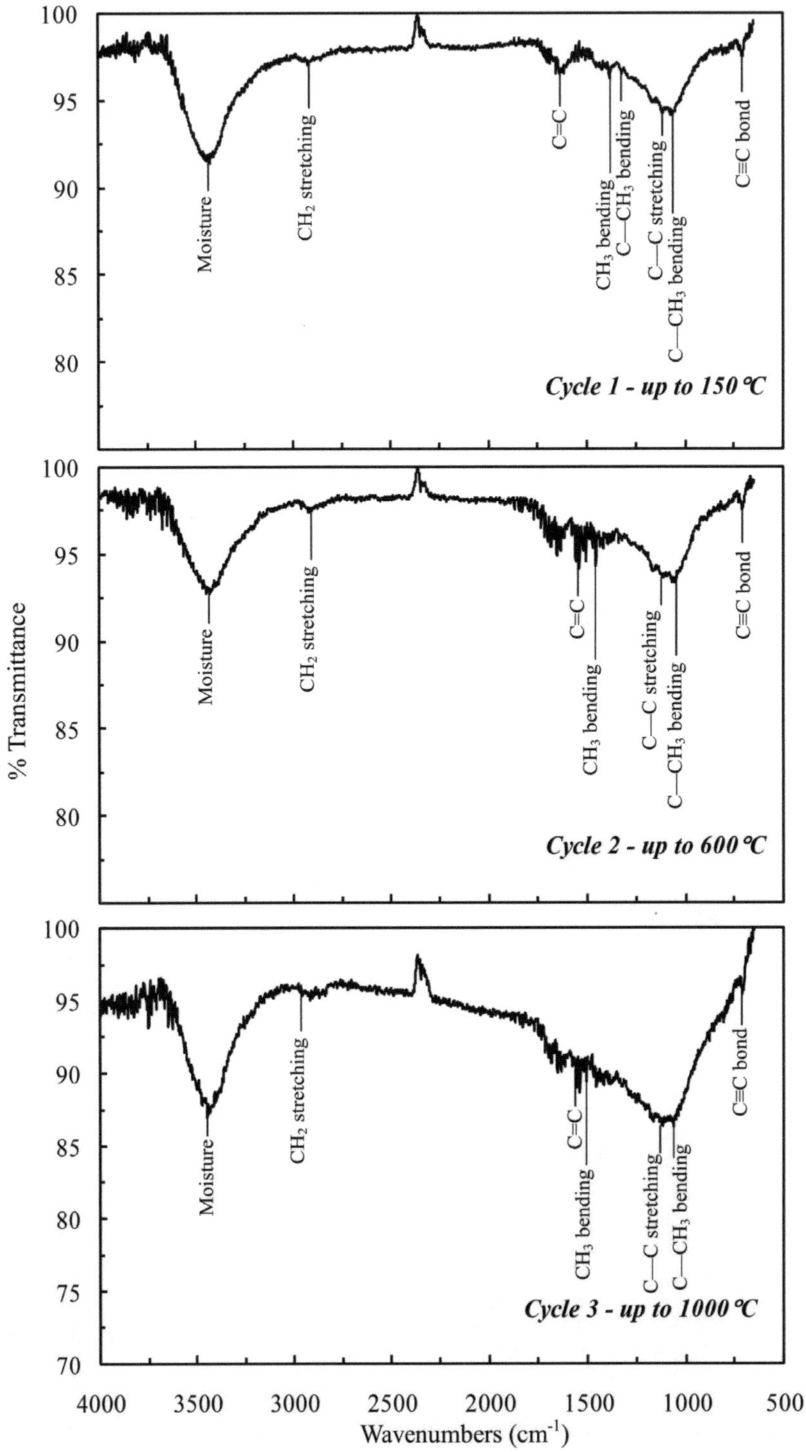

Figure 6. FTIR peaks of PE and PC mixtures at 50:50.

http://dx.doi.org/10.5772/67568

Raw PC	Raw PE	Cycle 1	Cycle 2	Cycle 3	Possible functional groups
3484.12	3477.22	3478.10	3489.31	3475.89	Moisture
2913.98	2917.98	2915.61	2918.56	2915.61	CH_2 stretching
-	2850.54	-	-	-	CH_2 symmetric
1685.49	-	-	-	-	C=C stretching 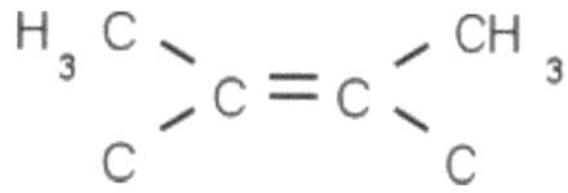
1653.83	1645.71	1651.96	1653.58	1653.37	C=C (alkene absorption)
1457.52	1459.20	1457.58	1457.67	1457.61	CH_3 bending
-	1379.27	1383.55	-	-	$C{=}CH_3$ bending
-	1162.30	1165.86	1168.48	1167.06	C=C stretching
1060.47	1059.33	1072.79	1069.93	1065.25	$C{=}CH_3$ bending
893.93	-	-	-	-	Aromatic ring
-	718.30	711.80	708.86	708.86	Alkynes (triple C bond)

Table 2. Characterization of FTIR peak profiles of PE, PC, and their mixtures at 50:50.

survived in the heat treatment cycles. However, aromatic ring vibration at 893 cm^{-1} from the raw PC has been significantly reduced after the mixture and heat treatment. Most of the peaks (raw PE) became significantly broader with reduced intensities after mixing and heating.

4. Summary

PE has a higher viscosity that limits its flow and penetration, where it has been shown that PE melted at high temperature, 250 °C after 60 min residence time. As mentioned earlier, PE has high viscosity that is influenced by temperature and time of the heat treatment. At the highest treatment parameters, of 350 °C temperature and 60 min time, PE melt has flattened on the substrate with contact angle of 30.90 °. The higher contact angle showed by PE was measured by its melt, without much disruption, while the bubbling and blistering of PE melt have caused it to increase the contact angle on the substrate surface.

The interaction behavior between PE and PC was further investigated by mixing PE in 10–50 % with PC and subjected the mixtures into heating cycle's heat treatment. The treatment of Cycle 1 (150 °C) has showed no changes in mass loss, whereas heating of Cycle 2 (600 °C) has showed some traces of PE, where the percentage of loss is about 33 and 43 % in 40 and 50 % PE mixtures, respectively. However, high temperature heating of Cycle 3 (up to 1000 °C) has totally decomposed PE in the mixtures. This result is in agreement with SEM images where the EDS analysis detected the presence of PE until Cycle 2 (600 °C) and was absent after Cycle 3. The carbon content of the mixtures residues was in the range of 92–97 %, with the decreasing value with increasing time and temperature.

The residues of 50 % PE blend were also analyzed by FTIR to investigate the effect of blend, temperature, and time on the chemicals' bonding of its residues. PE is composed of many functional groups, some of which were reduced in intensity and became broad with increasing time and temperature. The CH_2 symmetric vibration at 2850 cm^{-1} was gone after the mixture, whereas C—CH_3 vibration around 1380 cm^{-1} was lost after Cycle 2. The reduced intensities and broader peak occurred after the heat treatment indicates that the temperature has decomposed those functional groups, and they gradually disappeared after progressive heating.

Author details

Sharifah Shahnaz

Address all correspondence to: shahnaz@unimap.edu.my

School of Materials Engineering, Universiti Malaysia Perlis (UniMAP), Arau, Perlis, Malaysia

References

[1] Gourmelon G. (2015). Global Plastic Production Rises, Recycling Lags. New Worldwatch Institute analysis explores trends in global plastic consumption and recycling. Available from: http://www. worldwatch.org.[Accessed: 2017-03-04]

[2] Lee K.-H. and Shin D.-H. Characteristics of Liquid Product from the Pyrolysis of Waste Plastic Mixture at Low and High Temperatures: Influence of Lapse Time of Reaction. Waste Management. 2007;**27**(2):168-176.

[3] Miskolczi N., Bartha L. and Angyal A. High Energy Containing Fractions from Plastic Wastes by Their Chemical Recycling. Macromelecular Symposia. 2006;**245-246**(1):599-606.

[4] Delattre C., Forissier M., Forissier M., Pitault I., Schweich D. and Bernard J.R. Improvement of the Microactivity Test for Kinetic and Deactivation Studies Involved in Catalytic Cracking. Chemical Engineering Science. 2001;**56**(4):1337-1345.

[5] Curlee R.T. Plastic Recycling: Economic and Institutional Issues. Conservation & Recycling. 1986;**9**(4):335-350.

[6] Couderc P., Hyvernat P. and Lemarchand J.L. Correlations between Ability of Pitch to Penetrate Coke and the Physical Characteristics of Prebaked Anodes for the Aluminum Industry. Fuel. 1986;**65**(2):281-287.

[7] Iwaya T., Sasaki M. and Goto M. Kinetic Analysis for Hydrothermal Depolymerization of Nylon 6. Polymer Degradation and Stability. 2006;**91**:1989-1995.

[8] Achilias D.S., Roupakias C., Megalokonomos P., Lappas A.A. and Antonakou V. Chemical Recycling of Plastic Wastes Made from Polyethylene (LDPE and HDPE). Journal of Hazardous Materials. 2007;**149**(3):536-542.

[9] Bhaskar T., Uddin M.A., Murai K., Kaneko J., Hamano K., Kusaba T., Muto A. and Sakata Y. Comparison of Thermal Degradation Products from Real Municipal Waste Plastic and Model Mixed Plastics. Journal of Analytical and Applied Pyrolysis. 2003;**70**(2):579-587.

[10] McIlveen-Wright D.R., Pinto F., Armesto L., Caballero M.A., Aznar M.P., Cabanillas A., Huang Y., Franco C., Gulyurtlu I. and McMullan J.T. A Comparison of Circulating Fluidised Bed Combustion and Gasification Power Plant Technologies for Processing Mixtures of Coal, Biomass and Plastic Waste. Fuel Processing Technology. 2006;**87**(9):793-801.

[11] Williams P.T. and Slaney E. Analysis of Products from the Pyrolysis and Liquefaction of Single Plastics and Waste Plastic Mixtures. Resources, Conservation and Recycling. 2007;**51**(4):754-769.

[12] Peterson J.D., Sergey V. and Charles A.W. Kinetics of the Thermal and Thermo-Oxidative Degradation of Polystyrene, Polyethylene and Poly(propylene). Macromolecular Chemistry and Physics. 2001;**202**(6):775-784.

[13] Wall L.A., Madorsky S.L., Brown D.W., Straus S. and Simha R. The Depolymerization of Polymethylene and Polyethylene. Journal of the American Chemical Society. 1954;**76**(13):3430-3437.

[14] Wampler T.P. Thermometric Behavior of Polyolefins. Journal of Analytical and Applied Pyrolysis. 1989;**15**:187-195.

[15] Conesa J.A., Font R., Marcilla A. and Garcia A.N. Pyrolysis of Polyethylene in a Fluidized Bed Reactor. Energy and Fuels. 1994;**8**(6):1238-1246.

[16] Oakes W.G. and Richards R.B. The Thermal Degradation of Ethylene Polymers. Journal of Chemical Society. 1949;0:2929-2935.

[17] Scheirs J and Kaminsky W. Feedstock Recycling and Pyrolysis of Waste Plastics: Converting Waste Plastics into Diesel and Other Fuels. John Wiley & Sons Ltd, Chichester, UK; 2006.

[18] Feng Z., Zhao J., Rockwell J., Bailey D. and Huffman G. Direct Liquefaction of Waste Plastics and Coliquefaction of Coal-Plastic Mixtures. Fuel Processing Technology. 1996;**49**(1-3):17-30.

Chapter 12

Pyrolysis of Polyolefins in a Conical Spouted Bed Reactor: A Way to Obtain Valuable Products

Miriam Arabiourrutia, Gorka Elordi,
Martin Olazar and Javier Bilbao

Additional information is available at the end of the chapter

http://dx.doi.org/10.5772/67706

Abstract

The amount of waste plastic increases every single year, which causes a critical environmental issue. Polyolefins (mainly high- and low-density polyethylene and polypropylene) are the most common types of plastics, accounting for 60 wt% of the plastic waste. Pyrolysis, the thermal degradation in an inert atmosphere, is considered to be one of the most appealing technologies for the recycling of plastic materials. The conical spouted bed reactor is suitable for the pyrolysis of plastic waste due to its ability to avoid agglomeration problems that may be caused by the melted plastic. The pyrolysis process may be carried out at different temperatures and with or without the presence of catalysts in the reaction environment in order to streamline product distribution. The resulting products are hydrocarbons: non-condensable gases (C_1–C_4), gasoline fraction, diesel fraction, and waxes. These fractions might be used directly as feed streams for refinery units or as pools to be mixed with other streams from refineries.

Keywords: spouted bed, pyrolysis, catalytic pyrolysis, zeolites, pyrolysis products

1. Introduction

The polyolefins are polymers produced from the polymerization of olefins. The polyolefins are the most common and cheapest plastics. Polyolefins (particularly polyethylene and polypropylene (PP)) are the most important group of plastics.

The world's plastic production reached 311 million tonnes in 2014 (thermoplastics and polyurethanes and other plastics such as thermosets, adhesives, coatings, and sealants) [1].

Polyolefins (particularly polyethylene and polypropylene) are the most important group of plastics. The advantages of polyolefins include the following: they are odourless, they are non-toxic, and they have relatively good chemical resistance. The worldwide market for polypropylene will reach 62.4 million metric tonnes by 2020, driven by growth in the Asia Pacific region, the Middle East, and Africa [2]. This increase in the production of plastics is accompanied by a growth in the generation of plastic wastes, of which about 60 wt % in weight corresponds to polyolefinic plastics.

The interest in the valorization of plastic wastes arises from the need to avoid their stockpiling, where they cause serious environmental problems and deteriorate the landscape. In the EU member states, directive 2008/98/EC of the European Parliament states that no less than 30% of plastic solid waste (PSW) should be utilized for re-use or recycled use. The directive clearly states that by 2020, all solid waste streams (including plastics) should be diverted towards thermal and/or mechanical treatment and energy recovery, reducing the percentage of solid waste being landfilled to a minimum. Consequenty, the recycling of plastics is the objective of various studies and brings the proposition of various technological alternatives. In this sense, the thermal degradation of plastics allows obtaining monomers, combustible gases, and/or energy, with the reduction of landfills as an added advantage [3].

1.1. Properties

High-density polyethylene (HDPE) and low-density polypropylene (LDPE) are thermoplastic polymers formed by repetitive units of ethylene. Their chemical formula is $(-CH_2-CH_2-)n$.

HDPE is a polymer of a linear chain without ramification, so its density is high and also the intermolecular forces.

LDPE is a polymer of a branched chain, so its density is lower.

PP is a thermoplastic polymer formed by repetitive units of propylene. Its chemical formula is

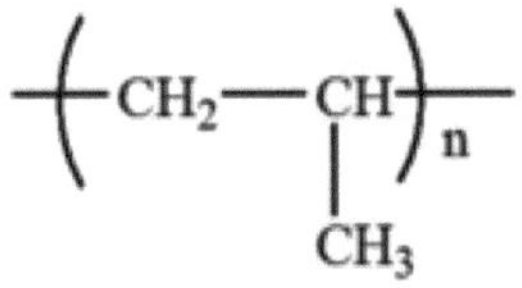

The properties of polypropylene are similar to the ones of polyethylenes. The major differences are that it has a lower density, a higher fusion temperature (160–170°C as opposed to 135°C), and it has a higher grade of crystallinity (it is almost crystalline, compared with around 90% crystallinity for HDPE and 40—50% crystallinity for LDPE). The crystallinity is related to the branched structure and the molecular weight, as the more branched the structure and the higher the molecular weight are, the lower the crystallinity is.

In **Table 1**, the following properties of the polyolefins are shown: molecular weight, polydispersity, density and higher heating value.

Property	HDPE	LDPE	PP
Mw (g/mol)	46,200	92,200	70,000
Polydispersity	2.89	5.13	2.00
ϱ (kg/m^3)	940	923	890
Higher heating value (kJ/kg)	46.2	46.5	46.05

Table 1. Properties of polyolefins.

The polydispersity value (ratio between the weight average molecular weight and the number average molecular weight) indicates the heterogeneity of the molecular weight of macromolecules that form the polymer. Any industrial polymer consists of fractions of different macromolecules of different molecular weights. The polydispersity value depends on the conditions in which polymerization has been carried out. Its minimum value is 1 for monodispersed species.

The Mw, polydispersity, and density are provided by the manufacturer of the materials (Dow Chemical, Spain), and the higher heating value is measured in a calorimetric pump (Parr 1356).

The LDPE has the highest average molecular weight, whereas HDPE has the lowest one among the three. The polydispersity value is the highest for LDPE due to its more branched structure. The PP has the lowest density, and the three polyolefins have similar higher heating Values.

1.2. The recycling of polyolefins

The options to recycle plastic wastes are: primary recycling (re-extrusion), secondary recycling (mechanical recycling), tertiary recycling (chemical recycling), and quaternary recycling (energy recovery) processes.

Primary recycling: involves the re-introduction of clean scrap of a single polymers to the extrusion cycle in order to produce products of similar material. It is commonly applied to the processing line itself.

Mechanical recycling: waste polymers are sorted, ground, sometimes washed and dried, and then granulated. The applications of these granulates are limited to the production of the lower quality products (pipelines, rubbish bag, garden furniture, protective slope containers on roads or motorways, synthetic grass, pavements, etc.) [4].

Chemical recycling: It is the thermo-chemical treatment aimed at degrading the polymer under controlled temperatures in inert atmospheres. After carrying out this process (pyrolysis), the monomers that form the polymer or other products are obtained. The catalytic cracking of polyolefins leads to combustibles for automobiles, and by means of gasification (at high temperatures and with a gasification agent, oxygen, water steam, etc.) synthesis gas is obtained. Synthesis gas is the raw material to obtain methanol, DME, and fuels for automobiles (Fischer-Tropsch way). These processes can be incorporated into a refinery, contributing to the development of the waste-Refinery concept, which allows for the use of the refinery unites in the valorization processes of intermediate products to commercial products.

Energy recovery: This implies burning waste to produce energy in the form of heat, steam, and electricity. The disadvantage of this method is the irretrievable loss of hydrocarbon materials [5].

Other method of polyolefin waste reutilization is by using them in construction and building materials, such as polymer-bitumen mixtures [6] or sand bricks and cement concentrates [7, 8] and different polymer-based composites [9, 10]. Among these options, pyrolysis is the method of valorization of greater interest, especially for addition polymers (which are the most consumed plastics, and within them are the polyolefins). Pyrolysis presents some advantages: operational, environmental, and economical.

The operational advantages include the joint treatment of waste and energetic integration. The main environmental advantage is the valorization of dangerous or disagreeable waste. Among the economic advantages, on the one hand, the the pyrolysis process could be integrated into a refinery with amortized units and on the other hand, the process is energetically efficient.

Proposed pyrolysis processes are flexible and can treat mixtures of plastics such as those obtained from municipal solid wastes [11, 12] or mixtures of these with other residual materials such as biomass [13] and plastic materials derived from car recycling [14], electronics and computers [15], construction (such as polymethyl metacrylate [16], and healthcare [17].

1.2.1. Pyrolysis of polyolefins

Pyrolysis of polyolefins consists of treating them in the presence of heat under controlled temperatures in an inert atmosphere without catalysts. As a result, three fractions of products can be obtained: gas fraction (composed mainly of the monomers that form the polyolefins), liquid fraction (composed of hydrocarbons larger than C_5), and solid fraction (char) formed at temperatures higher than 700°C.

Various technologies have been proposed with different types of reactors. In the processes carried out at high temperatures (650–850°C), the rotary kilns were a primitive solution to solve the problems related to the physical use of materials of different physical properties and of mixtures of wastes (biomass, plastics, tires, etc.). The fluidized bed reactor (with sand to help the fluidization) provides a good heat and mass transfer rate, and as a consequence, a uniform temperature is attained in the reactor. The principal international reference is the "Hamburg process" [18, 19].

The processes at low temperatures (350–550°C) are carried out in fluidized beds (G-S), with the previously mentioned advantages. The products are predominantly liquids and/or waxes with yields in the interval of 60–85 wt%. The temperature is one of the most important variables, and an increase in temperature accelerates the degradation process, resulting in a higher yield of gases.

The conical spouted bed reactor is an appropriate reactor for working at low and high temperatures to carry out the pyrolysis of plastic materials. It is suitable because the cyclic movements described by the sand particles in the bed prevent the bed agglomeration that may be produced by the plastic materials when they melt. Furthermore, it makes it feasible to treat materials of different densities and granulometries.

1.2.1.1. Conical spouted bed reactor

The advantages that the conical spouted bed reactor offers are characteristic of conventional spouted beds (cylindrical with a conical base), and they are improved by the exclusively conical geometry of the reactor: construction and design simplicity, low charge loss, vigorous contact between particles, high heat and mass transfer rate between phases, versatility in the gas flux, low residence time of gas, and ease of the operation in a continuous regime with solids.

Working at low temperatures, the problems of particle agglomeration with melted plastic are aggravated. The conical spouted bed reactor provides an appropriate contact to avoid this problem, [20] and pyrolysis occurs according to the following stages, as shown in **Figure 1** [21]: (i) melting of plastic, which coats the sand particles; (ii) cyclic movement of the sand particles coated by plastic during their devolatilization.

The conical spouted bed reactor has also been used for the catalytic cracking of polyolefins (or catalytic pyrolysis) using catalysts in situ in the bed. Contrary to what happens in the thermal process, in this case, the reaction does not progress through intermediate radicals but through carbocationic ones due to the Bronsted-Lewis acidity of the catalysts. The advantages of catalytic cracking with respect to the thermal one are the following:

(i) Reduces the required temperature (decreases the activation energy of the C—C link breakage) with subsequent energy saving.

(ii) Decreases the residence time of plastics (increasing the production per reactor unit volume and the energy consumed per unit).

(iii) Improves the selectivity of the products of interest as well as controls molecular weight intervals.

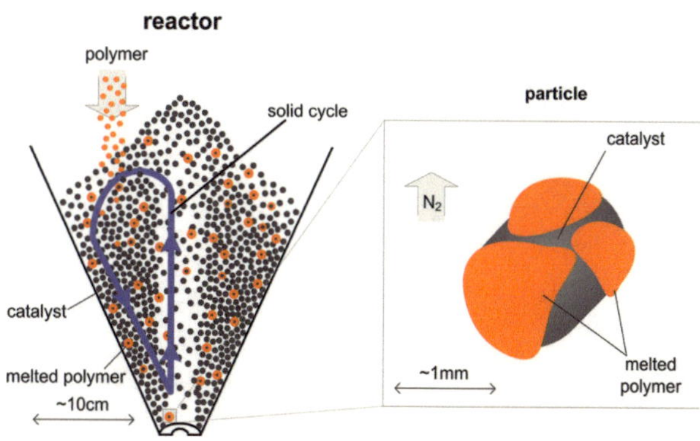

Figure 1. Scheme of the stages of the polyolefin pyrolysis in the conical spouted bed reactor.

The viability of pyrolysis and catalytic pyrolysis has the drawback of the high fixed asset necessary to install a new industrial unit. This viability is increased if the valorization process is integrated into the operations and processes of a refinery (waste Refinery), which allows for the use of amortizated facilities and the treatment and posterior commercialization with as common products.

2. Experimental

2.1. Pyrolysis pilot plant

The thermal and catalytic pyrolysis of high-density polyethylene has been carried out in a conical spouted bed reactor.

The general scheme of the plant used to carry out pyrolysis is shown in **Figure 2**. The principal components of the plant are the following: (1) the solid feeding system, (2) gas mixture and feeding system, (3) preheater of gases, (4) reactor, and (5) condensers and filter system.

The most important component is the conical spouted bed reactor. It has a conical form at the bottom and a cylindrical section at the top for the development of the fountain. The angle of the conical zone is 28°. The diameter of the cylindrical part, Dc, is 2.3 cm; the diameter of the base, Di, is 2 cm; and the diameter of the gas entrance, Do, is 1 cm (**Figure 3**).

The feeding system consists of a hopper, a hollow ball valve where the plastic to be fed is located, an inlet tube cooled by tap water, and a support. Below the reactor is a cartridge

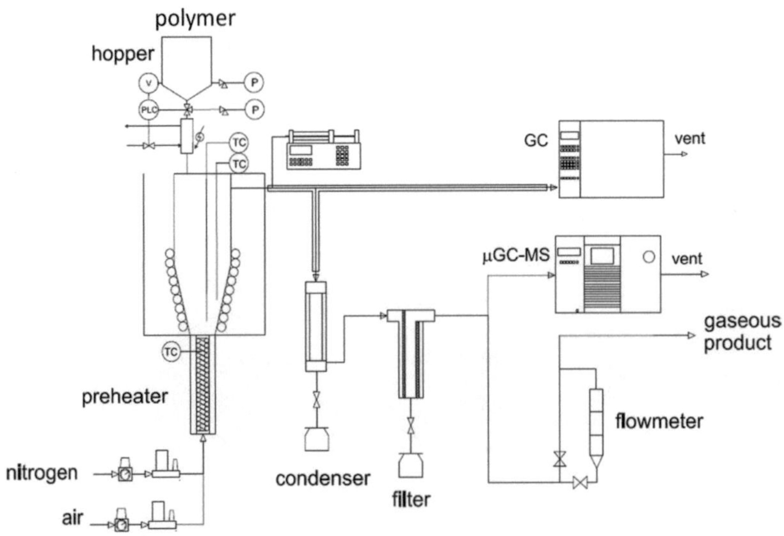

Figure 2. General scheme of the pilot plant used.

http://dx.doi.org/10.5772/67706

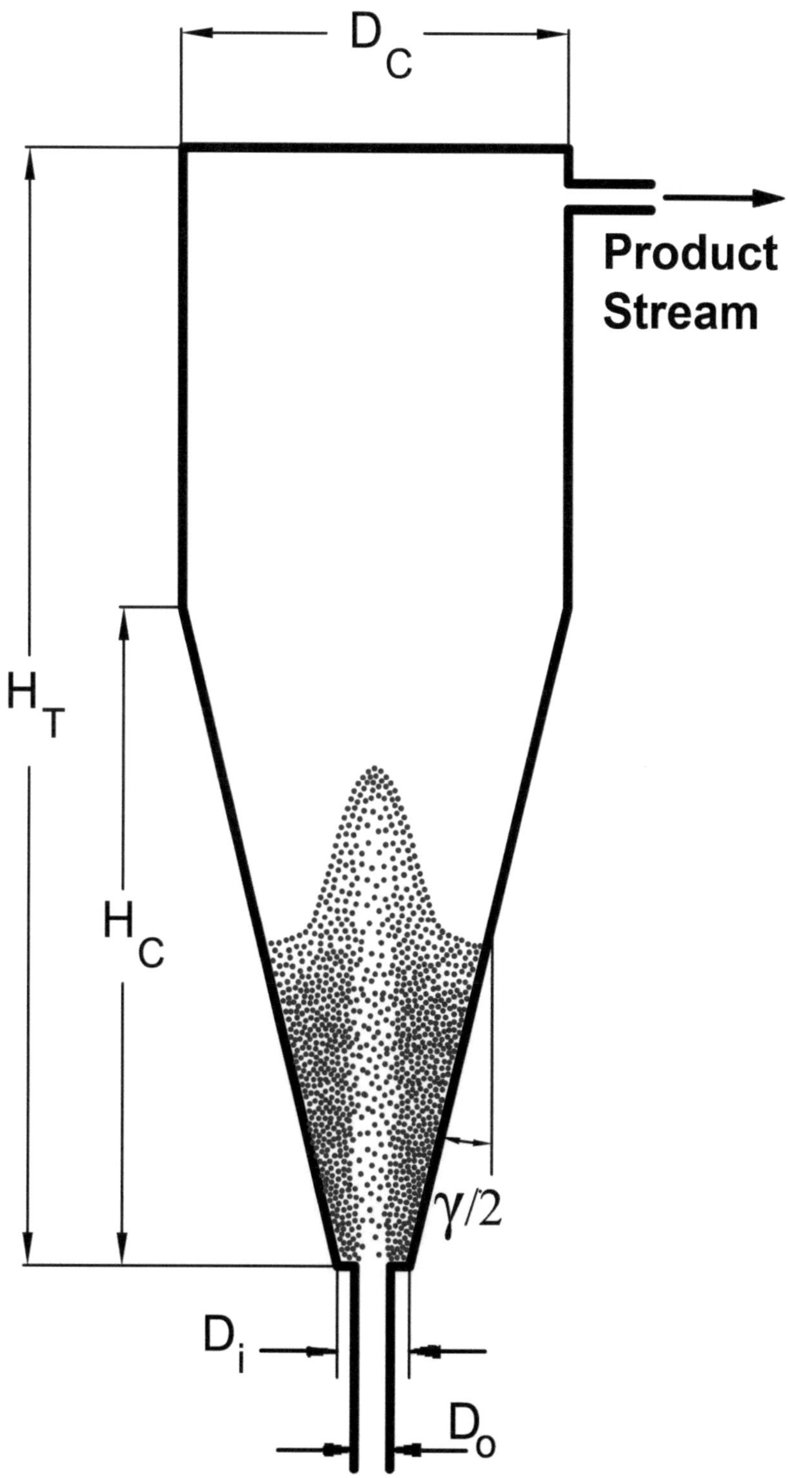

Figure 3. Design parameters of the conical spouted bed reactor.

containing a ceramic resistance, which is inside a metallic casing and thermally insulated. This resistance heats the N_2 stream to the reaction temperature.

The condensation system is formed by a typical condenser refrigerated by water, and by a coalescence filter to coalesce the fog formed in the volatiles of their collision against the surface.

2.2. Characterization techniques

2.2.1. *Characterization techniques for volatiles*

The volatiles formed during pyrolysis are analysed by means of an Agilent 6890 gas chromatograph provided with the flame ionization detector (FID). This chromatograph is connected online to the reactor by means of a calorifugated line at 290°C. Moreover, non-condensable gases have been analysed and identified by means of an mGC–MS spectrometer (Agilent MSD5975B) connected online at the exit of the coalescence filter. The products condensed in the condenser have been identified by means of a mass spectrometer (Shimadzu GC-MS-QP2010S).

2.2.2. *Characterization techniques for waxes*

2.2.2.1. *Simulated distillation*

This analysis was carried out in an Agilent 6890 gas chromatograph provided with an FID and a SimDis 2887 fast column (10 m × 0.53 mm × 0.88 μm), the maximum operation temperature is being 375°C. This column has been previously calibrated using sample patterns of the known boiling points in order to obtain a relationship between the time at which peaks that appear and the boiling temperature.

2.2.2.2. *Fourier Transform Infrared (FTIR) spectrophotometry*

The nature of the bonds was analyzed using a Nicolet 6700 spectrometer. A 200 mg KBr pellet was pressed, and a drop of wax dissolved in tetrahydrofurane added over it. The pellet impregnated with wax was located over the sample holder and analyzed under wavelengths between 500–4000 cm^{-1}.

2.2.2.3. *Heating value*

The calorific power of waxes has been measured in an isoperibolic calorimetric bomb PARR 1356. It is composed of an oxygen bomb where the sample is introduced for its combustion, a recipient with 2 kg of distilled water, and a jacket situated into the calorimeter.

3. Results

3.1. Pyrolysis

In the pyrolysis of high-density polyethylene carried out in a conical spouted bed reactor, the products obtained are grouped in the following fractions: gas fraction (C_1–C_4), gasoline fraction (C_5–C_{11}), diesel fraction (C_{12}–C_{20}), and wax fraction (>C_{21}). The yields obtained are shown in **Figure 4** [22].

The thermal pyrolysis was carried out in the temperature range of 500–700°C, using 30 g of sand (particle diameter between 0.6 and 1.2 mm) in the bed and using N_2 as a fluidization agent. The HDPE pellets with a diameter of 4mm were fed continuously at a rate of 1 g/min.

The most abundant gas fractions are C_2, C_3, and C_4 lumps. More specifically ethylene, propylene, and 1-butene, in a minor proportion are the main components, especially at 700°C.

In the gasoline fraction (C_5–C_{11}), the distribution of different lumps is affected by the temperature. At lower temperatures, the yields of all lumps are quite similar; at medium temperatures (600–650°C), the C_6–C_7 and C_{10}–C_{11} lumps are the most abundant, increasing their yield from around 3 to 5 wt%; and at higher temperatures the lighter lumps (C_5, C_6, and C_7) are the most abundant (7–9 wt%) as a consequence of the cracking reactions that take place. With respect to the nature of the compounds, the olefins are by far the most abundant compounds, followed by the paraffins.

In the diesel fraction, (C_{12}–C_{20}), at 500°C, the C_{18}–C_{20} lumps are the most abundant; at 600°C, the yields of these lumps decrease while that of C_{12}–C_{13} increase; and at 700°C, the yields of C_{18}–C_{20} lumps decrease again but their yield is the most abundant, around 2 wt%.

As seen in **Figure 4**, among all the fractions, waxes are those of the highest yield, is obtained except at 700°C. These waxes are collected mostly in the coalescence filter (**Figure 2**). This

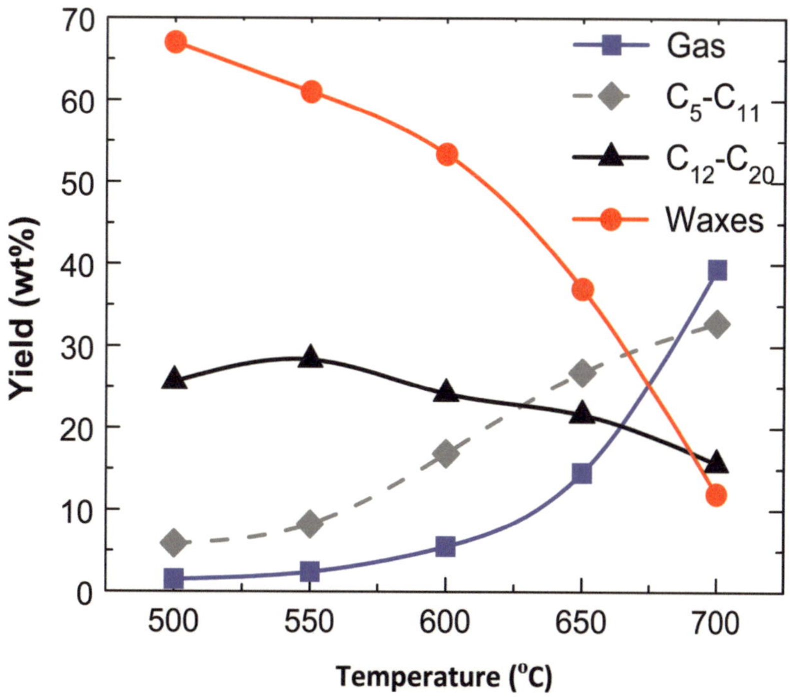

Figure 4. Yields of products obtained (wt %) at 500–700°C temperature range.

fraction is formed by hydrocarbons of long chains which are paraffinic and olefinic in nature, the latter being more abundant at higher temperatures. The characteristic gas-solid contact of this reactor is especially suitable for selectively obtaining waxes due to the high linear velocity of the gas, the vigorous gas-solid contact, and the low residence time of the gas in the reactor [23]. These characteristics allow the production of waxes at low temperatures and minimize the secondary reactions of wax transformation.

The yields of waxes obtained from the HDPE pyrolysis are similar to those obtained from the LDPE pyrolysis, while the yields obtained from the PP pyrolysis are higher, especially at low temperatures [24]. This result can be explained as a consequence of the more branched structure of polypropylene, compared to that of polyethylenes. At low temperatures such as 450 and 500°C, cracking takes place firstly in the branched chain of PP, and the cracking of the principal chain is minimal due to the very short residence times resulting from the contact method used for wax production.

In thermal degradation process that was carried out in a batch reactor at 450°C, Hájeková and Bajus [25] also obtained a higher mass yield of the oil/wax from PP than from LDPE, 87.8% as opposed to 75%. The yield of waxes obtained for polyethylene pyrolysis is lower than that reported by Predel and Kaminsky [26] in a fluidized bed reactor at 510°C, which is between 85 and 88 wt% when polyolefins are continuously fed into the reactor. Nevertheless, it is higher than the value of 71 wt% obtained by Chaala et al. [27], by vacuum pyrolysis carried out at 450°C and 20 kPa in a batch reactor.

3.1.1. *Characterization of waxes*

The waxes obtained in the pyrolysis of polyolefins are classified as "synthetic waxes," and they are an alternative to those obtained from petroleum products.

A simulated distillation analysis, an FTIR analysis, and a heating value measurement were carried out.

3.1.1.1. *Simulated distillation*

In order to analyse the possible inclusion of waxes as part of the feed of the Fluid Catalytic Cracking (FCC) unit of a refinery, a simulated distillation analysis was carried out in order to know the temperature interval at which the waxes boil. The simulated distillation curves obtained are shown in **Figure 5** [22].

As it is observed, gasoline and diesel fractions are dissolved in the waxes. Based on the paraffin standard used, the two fractions have been quantified (delimited by the discontinuous line in the figures): (i) light waxes, C_{21}–C_{40}, corresponding to compounds with a boiling points between 343 and 525°C and (ii) heavy waxes or the remaining fraction, with higher boiling temperatures. It should be noted that 343°C is considered as the minimum temperature to define the range of heavy cycle oil (HCO) fraction.

Predel and Kaminsky [26] delimited the light wax fraction in the 300–500°C range (corresponding to C_{17}–C_{36} compounds). It was observed that, as the reaction temperature was

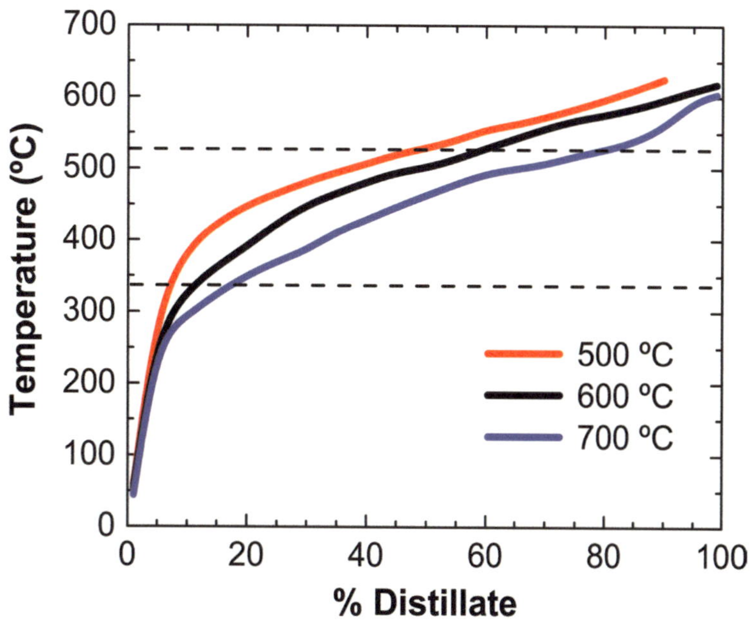

Figure 5. Simulated distillation curves for waxes.

increased, the content of the light fraction compounds (C_{20}–) that were dissolved in the waxes increased as well: 7, 12, and 19 wt % for 500, 600, and 700°C, respectively. Waxes with a boiling point lower than 500°C are a valuable feedstock for a steam cracking unit combined with conventional naphtha to produce reusable olefins [25]. The remaining waxes, with a boiling point higher than 500°C, can be combined with the usual feeding of catalytic cracking (FCC) units to produce gasoline, or they can be upgraded in a hydrocracker [27]. This latter strategy has good industry perspectives with the goal of meeting the demand for new feeds in refineries.

3.1.1.2. Fourier Transform Infrared spectrophotometry (FTIR) analysis

An FTIR analysis was carried out to determine the type of bonds that appears in the waxes and to demonstrate its paraffinic and olefinic nature. The FTIR spectra is shown in **Figure 6**.

The peaks that correspond to paraffinic bonds are the following: the characteristic symmetric and asymmetric stretching bands of C—H bonds corresponding to $-CH_2-$ groups at 2855 and 2920 cm^{-1} [28] are similar to the bonds of the same group in commercial waxes [29]. Likewise, the double peak at 725 cm^{-1} corresponds to the skeletal vibration of these groups. The shoulders observed at 2960 and 2900 cm^{-1} correspond to the $-CH_3$ terminal group bonds, although in the case of commercial waxes, these shoulders are more pronounced, which is evidence that the waxes obtained in this chapter are lesser branched chains than those found in commercial waxes. Other bands corresponding to aliphatic chains appear at 1386 and 1471 cm^{-1} and are due to methyl and methylene groups' deformation vibrations.

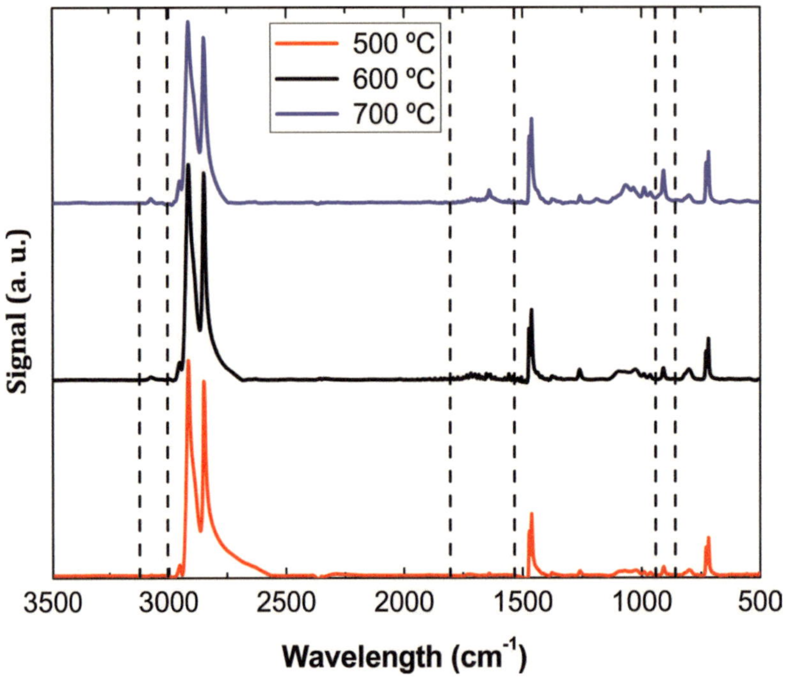

Figure 6. FTIR spectra of the waxes obtained.

The peaks which demonstrate the presence of olefinic bonds are the following: the bands at 1646 and 1725 cm^{-1} are due to the stretching of C=C bonds, which confirms the presence of olefins. Another significant difference between commercial waxes is the more pronounced olefinic nature of pyrolysis waxes, which is explained by the formation of C=C groups by the radical degradation mechanisms. This olefinic nature is confirmed by the bands at 910 and 995 cm^{-1}, which correspond to the $R-CH=CH_2$ group [29]. The olefinic nature is clearly observed in the waxes obtained at 700°C.

3.1.1.3. Heating value

The higher heating value has been measured in a calorimetric bomb. The values obtained for the waxes collected at 500, 600, and 700°C are 35.6, 44.1, and 45.4 MJ/kg, respectively. The values for the waxes obtained at 500 and 600°C are comparable to the original polyolefins and not much lower than the values corresponding to gasoline and natural gas. Thus, the direct combustion of the waxes to obtain energy is an alternative to pyrolysis, in case the use of waxes as raw materials to obtain fuels or chemical products would not be of commercial interest.

3.1.2. Application of waxes

The waxes can be used to prepare new materials and as feed in the FCC units of the refineries.

3.1.2.1. New materials

The waxes obtained in the pyrolysis of polyolefins may substitute those obtained in the Fischer-Tropsch synthesis in order to mix them with asphalt, producing WMA (warm mix asphalt). This mixture can be used to prepare asphalt at lower temperatures minimizing the environmental problems associated with the emission of volatiles and moreover, improving the resistance to fatigue and stiffness due to its peculiar crystalline structure [30].

They can be used to prepare reinforced plastics with natural fibres [31]. They can also be used as binders in the preparation of composites (as WC-TiC-Co carbide) by the means of moulding, improving its mechanical properties [32]. Furthermore, they can be used for metal coating, in this way, decreasing the infrared emissivity of metals (as a consequence of high transparency and low emissivity of waxes) and reducing the aging of metal.

Waxes have a high fusion latent heat, and this property can be useful for incorporating them into construction materials, and by means of encapsulation techniques, in textile materials.

3.1.2.2. Feed for the Fluid Catalytic Cracking (FCC) unit of a refinery

As waxes are not composed of many aromatic compounds and do not contain heteroatomic compounds (S, N, metals), they are an acceptable feed for the FCC unit of a refinery, either in their pure form, dissolved in the usual feed of FCC units, or dissolved in alternative ones.

Arandes et al. [33] studied the cracking of waxes in an FCC simulator unit (riser). The waxes were previously obtained in the pyrolysis carried out in a conical spouted bed reactor like the one described in section 2.1. They also mixed the waxes (20 wt%) with the standard FCC unit feed and vacuum gas oil (VGO) to analyse their effect on the yields and on the composition of the products obtained. The experiments were carried out in the 500–550°C range, with a catalyst/feed weight ratio of 5.5, and with a contact time value between 3 and 12 s. The catalyst used was an equilibrated commercial catalyst based on an HY zeolite.

The degrewe of conversions obtained by using these feeds follows this order: waxes > mixture > VGO. These results were explained as a consequence of the olefinic nature of the waxes (as opposed to the aromatic content of VGO, around 40%), which is more favourable for cracking.

With respect to the yields of the products obtained, at 550°C, the gasoline fraction was the one with the higher yield, around 40 wt% when waxes were fed purely, followed by the waxes and the VGO-mixture feed and by the VGO-pure feed.

3.2. Catalytic pyrolysis

The use of the catalysts in the conical spouted bed reactor changes the distribution of products obtained in pyrolysis and permits the increase of selectivity towards products of high interest as olefins, gasoline, or diesel fractions.

Elordi et al. [34] studied the effect of an HZSM-5 zeolite catalyst placed *in-situ* in the pyrolysis reactor on the product distribution carried out at 450–570°C. They used 30 g of catalyst in the bed in the form of pellets with a particle diameter between 0.6 and 1.2 mm. HDPE was fed at a flow rate of 1 g/min into a bed of 30 g of catalyst. The products were lumped into the following fractions: (C_2–C_4) olefins, light alkanes (C_4–), non-aromatics (C_5–C_{11}), single-ring aromatic (C_{12}–C_{20}) fractions, and waxes (C_{21+}). The evolution of different fraction yields in relation to the quantity of HDPE fed at 500°C are shown in **Figure 7**.

The main differences obtained with respect to pyrolysis without catalyst are the very high yield of olefins and the very low yield of the C_{12}–C_{20} fraction and waxes. These results are a consequences of the cracking reactions promoted by the catalyst that leads to obtaining lighter products. The results also show the low deactivation of the catalyst behaviour as HDPE is fed.

In the case of LDPE and PP, the results have been quite similar.

The use of other catalysts modifies the distribution of products obtained in pyrolysis. Elordi et al. [35] studied the effect of Hß and HY zeolite catalysts on product distribution. The yields of different product-fractions obtained with Hß zeolite catalyst and HY zeolite catalyst, respectively, at 500°C are shown in **Figures 8** and **9**. Using a Hß zeolite catalyst, the non-aromatic C_5–C_{11} fraction is the one with the highest yield, followed by the light olefin fraction (around 47 wt.% and 35 wt%, respectively). Using a HY zeolite catalyst, the non-aromatic C_5–C_{11} fraction (around 45 wt%) is also the main fraction, followed again by light olefins (around 22 wt%). In this case, the yield of C_{11+} fraction is higher than that obtained using the HZSM-5 and the Hß zeolite catalysts, around 10 wt%, and this yield increases with time. Contrary to what occurs with the HZSM-5 zeolite catalyst, with Hß and HY zeolite catalysts, the yield of wax increases with time due to the deactivation that both catalysts undergo.

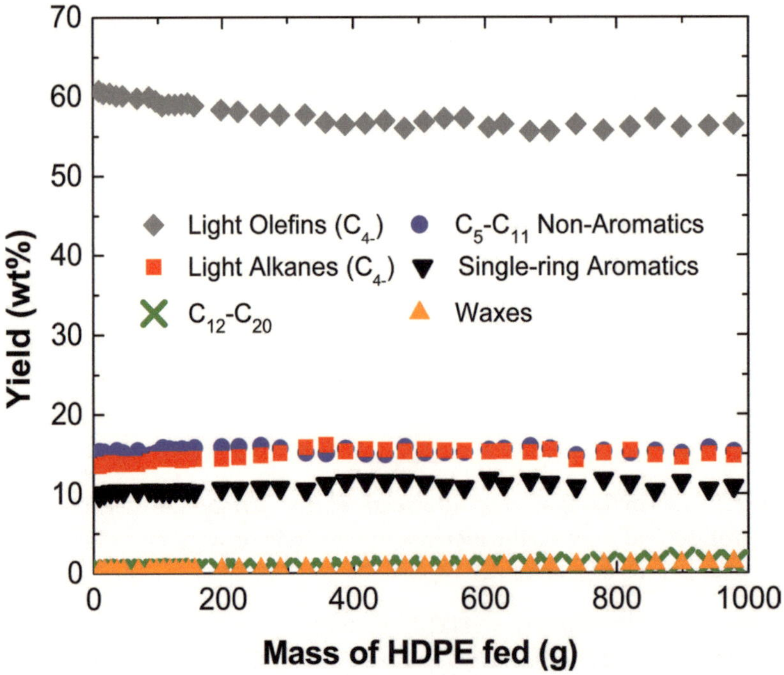

Figure 7. Evolution of product fraction yield with the quantity of HDPE fed at 500°C.

http://dx.doi.org/10.5772/67706

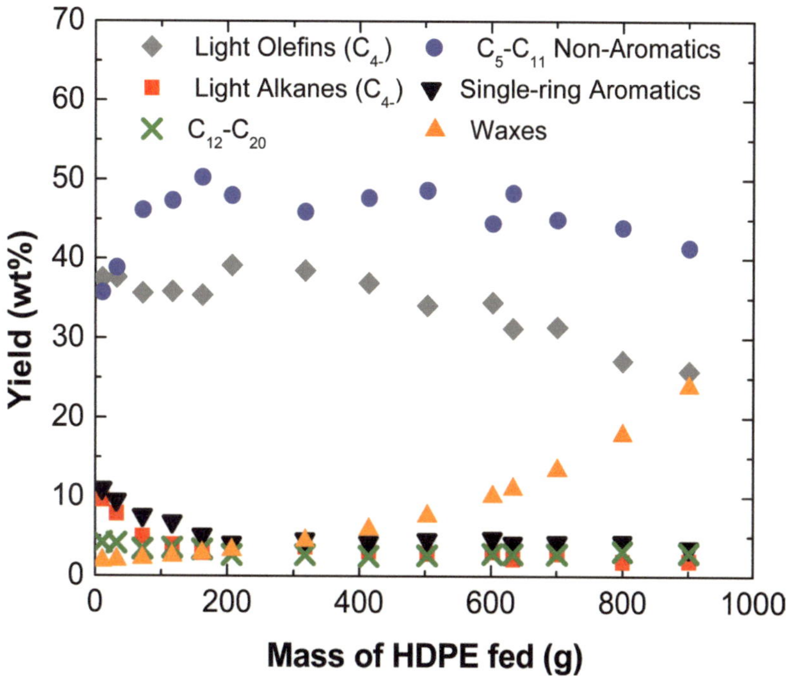

Figure 8. Yields of fractions throughout time using the Hß zeolite catalyst.

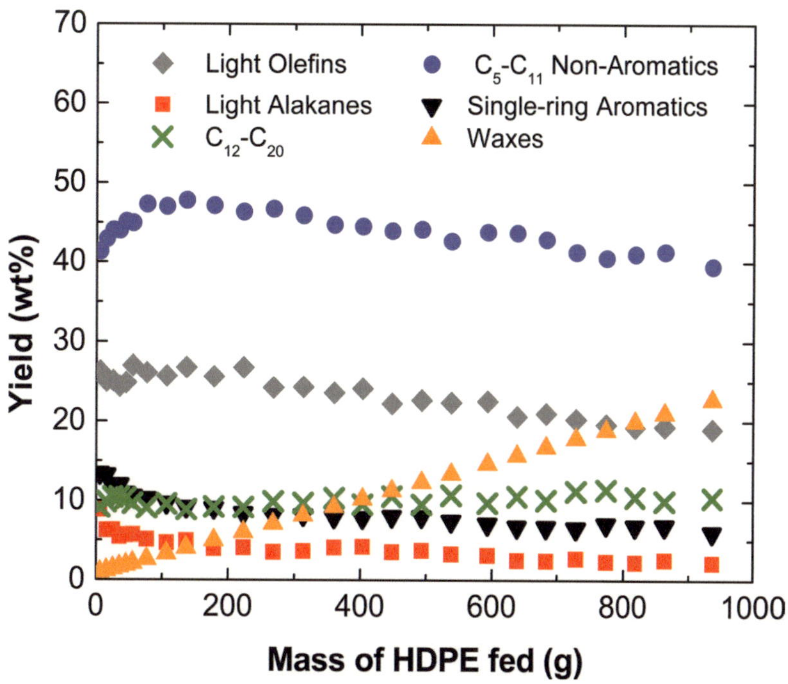

Figure 9. Yields of fractions throughout time using the HY zeolite catalyst.

This different fraction distribution is a consequence of the pore size and the acid strength of each catalyst. Thus, the HZSM-5 zeolite, which has the lowest pore size, 5.5 Å, related to the shape selectivity of this zeolite, gives way to the highest yield of light olefins, whereas the lowest yields are obtained with the HY zeolite catalyst. In the same way, the yields of non-aromatic C_5–C_{11} and C_{11+} fractions are higher when using the catalyst with a larger pore size (HY zeolite). The high acid strength of the HZSM-5 zeolite is related to the higher yield of single-ring aromatics obtained with this zeolite.

Olazar et al. [36] obtained 8 wt% of gaseous fraction, 22 wt% of C_5–C_9 fraction, and 69 wt% of C_{10+} fraction (diesel fraction) at 475°C, using a commercial FCC catalyst based on a HY zeolite active phase (30 g), after a severe steaming was applied to the catalysts. These results are a consequences of the decrease in acidity that the catalyst undergoes in the aforementioned treatment.

3.3. Other applications

Another option to modify the product distribution consists of carrying out the reforming of volatiles formed in the pyrolysis by means of a second reactor. Artetxe et al. [37] carried out the pyrolysis of HDPE in a conical spouted bed reactor at 500°C (using 50 g of sand), and the volatiles formed were passed through a second fixed bed reactor, where an HZSM-5 zeolite catalyst (8 g), heated at 500°C, was placed. The conditions used were appropriate to obtain a high yield of olefins, around 58% wt%, similar to that obtained when placing the catalyst in-situ, although with a lower catalyst space time.

Another application that might be used combined with the pyrolysis of HDPE, gives production hydrogen yield [38]. The volatiles formed in the pyrolysis, using steam as a fluidization agent, are reformed in a fixed bed reactor, placed in line using an Ni commercial catalyst. The pyrolysis was carried out at 500°C and the reforming at 700°C. They observed that steam is inert at 500°C since the products obtained are very similar to those obtained using N_2. Moreover, the use of steam in the first reactor simplifies the pyrolysis-reforming process. The H_2 yield obtained was 81.5% with respect to the stoichiometric one, which ressults in 34 g H_2 per 100 g HDPE fed into the reactor.

Barbarias et al. [39] studied the same process, but using a fluidized bed reactor to carry out the reforming of the volatiles formed in the pyrolysis step. They obtained a H_2 yield of 92.5% with respect to the stoichiometric one, which results in 38 g H_2 per 100 g HDPE fed. This higher yield compared to that obtained in a fixed bed reactor is attributed to the better mixing of the catalyst particles, higher temperature of isothermicity in the bed, and the gasification of the coke formed during the reaction promoted by the fluidized bed reactor.

4. Conclusion

The conical spouted bed reactor is adequate to carry out the pyrolysis process and catalytic pyrolysis of polyolefins. There is not any problem of bed defluidization using plastic materials that melt and cover the sand particles of the bed at a temperature at around 500°C due to the vigorous cyclic movements of the particles.

In the pyrolysis process without the catalyst, the wax is the fraction obtained in a higher yield at low temperatures, being around 67 wt% in weight at 500°C. This wax can be used to prepare new materials like asphalt at lower temperatures, reinforced plastics with natural fibres, composites or for metal coating, and can be incorporated into construction materials and into textile materials due to its high fusion latent heat. On other hand, it has good properties to be fed into an FCC unit of a refinery. On the contrary, at higher temperatures, 700°C, the yield of the wax fraction decreases considerably and the gas fraction, formed principally of olefins (ethene, propene and butene), becoming the main fraction, around 40 wt %, as a consequence of the cracking reactions that are favoured by the high temperature.

With respect to catalytic pyrolysis, it significantly modifies the product distribution. Thus, the wax fraction decreases dramatically as a consequence of the cracking reactions that are promoted by the acidic zeolite-based catalysts (HZSM-5, HY, and Hß). Thus, olefins constitute the main fraction with the HZSM-5 zeolite catalyst, whereas the non-aromatic C_5–C_{11} fraction is the main fraction when the HY and Hß zeolite catalysts are used.

H_2 may be produced by reforming the volatiles obtained in the pyrolysis of polyolefins with steam. 38 g of H_2 were obtained per 100 g of HDPE fed.

All the product fractions obtained in these processes can be used in refinery units to obtain commercial products such as ethene, propene, butenes, gasoline and diesel,either mixing them with conventional streams of a refinery and/or directly applying some treatments to them.

Author details

Miriam Arabiourrutia*, Gorka Elordi, Martin Olazar and Javier Bilbao

*Address all correspondence to: miriam.arabiourrutia@ehu.eus

Department of Chemical Engineering, University of the Basque Country, UPV/EHU, Bilbao, Spain

References

[1] Plastics Europe-Plastics-the Facts 2015. An analysis of European plastic production, demand and waste data, [Internet]. 2015. Available from: www.plasticseurope.org/Document/plastics---the-facts-2015.aspx [accessed on 21 November 2016].

[2] Plastics News 2013. [Internet]. 2013. Available from: www.plasticsnews.com/article/20130909/NEWS/130909948 [accessed on 21 November 2016].

[3] Mastral JF, Berrueco C, Ceamanos J. Theorical prediction of product distribution of the pyrolysis of high density polyethylene. Journal of Analytical and Applied Pyrolysis. 2007;80(2):427-438. doi:10.1016/j.jaap.2006.07.009.

[4] Siddique R, Khatib J, Kaur I. Use of recycled plastic in concrete: a review. Waste Management. 2008;28(10):1835-1852. doi:10.1016/j.wasman.2007.09.011.

[5] Zucchelli L, Rizzi R. Fuel composition which combusts instantaneously, method and plant therefore. US Patent 7,488,358 B2, Pirelli Ambiente S.p.A. 2009.

[6] Vasudevan R, Ramalinga Chandra Sekar A, Sundarakannan B, Velkennedy R. A technique to dispose waste plastics in an ecofriendly way-application in construction of flexible pavements. Construction and Building Materials. 2012;28(1):311-320. doi:10.1016/j.conbuildmat. 2011.08.031.

[7] Wahid SA, Rawi SM, Desa NM. Utilization of plastic bottle waste in sand bricks. Journal of Basic and Applied Scientific Research. 2015;5(1):35-44.

[8] Suganthy P, Chandrasekar D, Sathish KPK. Utilization of pulverized plastic in cement concrete as fine aggregate. International Journal of Research in Engineering and Technology. 2013;2(6):1015-1019.

[9] Cerruti P, Avolio R, Gentile G, Carfagna C, Persico P, Errico ME, Malinconico M, Avella M. Up-cycling end-of-use materials: highly filled thermoplastic composites obtained by loading waste carbon fiber composite into fluidified recycled polystyrene. Polymer Composites. 2014;35(8):1621-1628. doi:10.1002/pc.22815.

[10] Hugo AM, Scelsi L, Hodzic A, Jones FR, Dwyer-Joyce R. Development of recycled polymer composites for structural applications. Plastics, Rubber and Composites. 2011;40(6-7):317-323. doi:10.1179/1743289810Y.0000000008.

[11] Siddiqui MN, Redhwi HH. Pyrolysis of mixed plastics for the recovery of useful products. Fuel Processing Technology. 2009;90:545-552. doi:10.1016/j.fuproc.2009.01.003.

[12] López A, de Marco I, Caballero BM, Laresgoiti MF, Adrados A. Pyrolysis of municipal solid wastes: influence of raw material composition. Waste Management. 2010;30:620-627. doi:10.1016/j.wasman.2009.10.014.

[13] Liu W, Hu C, Yang Y, Liu S, Tong D, Li G, Zhu, L. Influence of ZSM-5 on the pyrolytic intermediates from the co-pyrolysis of pubescens. Energy, Conversion and Management. 2010;51:1025-1032. doi:10.1016/j.enconman.2009.12.005.

[14] De Marco I, Caballero BM, Cabrero MA, Laresgoiti MF, Torres A, Chomón MJ. Recycling of automobile shredder residues by means of pyrolysis. Journal of Analytical and Applied Pyrolysis. 2007;79:403-408. doi:10.1016/j.jaap.2006.12.002.

[15] Guo Q, Yue X, Wang M, Liu Y. Pyrolysis of scrap printed circuit board plastic particles in a fluidized bed. Powder Technology. 2010;198:422-428. doi:10.1016/j.powtec.2009.12.011.

[16] Kang BS, Kim SG, Kim JS. Thermal degradation of poly(methyl methacrylate) polymers: kinetics and recovery of monomers using a fluidized bed reactor. Journal of Analytical and Applied Pyrolysis. 2008;81(1):7-13. doi:10.1016/j.jaap.2007.07.001.

[17] Na D, Zhang Y, Wang Y. Thermogravimetric analysis and kinetic study on pyrolysis of representative medical waste composition. Waste Management. 2008;28:1572-1580. doi:10.1016/j.wasman.2007.05.024.

[18] Kaminsky W, Schmidt H, Simon CM. Recycling of mixed plastics by pyrolysis in a fluidized bed. Macromolecular Symposia. 2000;152:191-199. doi:10.1002/1521-3900(200003)152:1<191:: AID-MASY191>3.0.CO;2-2.

[19] Donaj PJ, Kaminsky W, Buzeto F, Yang W. Pyrolysis of polyolefins for increasing the yield of monomers' recovery. Waste Management. 2012;32(5):840-846. doi:10.1016/j.wasman.2011.10.009.

[20] Aguado R, Prieto R, San José MJ, Alvarez S, Olazar M, Bilbao J. Defluidization modelling of pyrolysis of plastics in a conical spouted bed reactor. Chemical Engineering and Processing. 2005;44(2):231-235. doi:10.1016/j.cep.2004.02.016.

[21] Elordi, G., Thermal and catalytic pyrolysis of polyolefins in a conical spouted bed reactor [thesis], University of the Basque Country, Bilbao; 2010.

[22] Elordi G, Olazar M, Lopez G, Artetxe M, Bilbao J. Product yields and compositions in the continuous pyrolysis of high density polyethylene in a conical spouted bed reactor. Industrial & Engineering Chemistry Research. 2011;50:6650-6659. doi:10.1021/ie200186m.

[23] Olazar M, San José MJ, Izquierdo MA, Ortiz de Salazar A, Bilbao J. Effect of operating conditions on solid velocity in the spout, annulus and fountain of spouted beds. Chemical Engineering Science. 2001;56:3585-3594.

[24] Arabiourrutia M, Elordi G, Lopez G, Borsella E, Bilbao J, Olazar M. Characterization of the waxes obtained by the pyrolysis of polyolefin plastics in a conical spouted bed reactor. Journal of Analytical and Applied Pyrolysis. 2012;94:230-237. doi:10.1016/j.jaap.2011.12.012.

[25] Hájeková E, Bajus M. Recycling of low-density polyethylene and polypropylene via copyrolysis of polyalkene oil/waxes with naphta: product distribution and coke formation. Journal of Analytical and Applied Pyrolysis. 2005;74:270-281. doi:10.1016/j.jaap.2004.11.016.

[26] Predel M, Kaminsky W. Pyrolysis of mixed polyolefins in a fluidized-bed reactor and on a pyro-GC/MS to yield aliphatic waxes. Polymer Degradation and Stability. 2000;70:373-385.

[27] Kirkwood KC, Leng SA, Sims DW. Polymer cracking. US Patent 5,364,995. 1992.

[28] Vogel A. Vogel's Textbook of Practical Organic Chemistry. New York: Longman; 1981.

[29] Chaala A, Darmstadt H, Roy C. Vacuum pyrolysis of electric cable wastes, Journal of Analytical and Applied Pyrolysis.1997;39:79-96.

[30] Petit C, Millien A, Canstrari F, Pannunzio V, Virgili A. Experimental study on shear fatigue behavior and stiffness performance of Warm Mix Asphalt by adding synthetic wax. Construction and Building Materials. 2012;34:537-544. doi:10.1016/j.conbuildmat.2012.02.010.

[31] Arrakhiz FZ, El Achaby M, Benmoussa K, Bouhfid R, Essassi EM, Qaiss A. Evaluation of mechanical and thermal properties of pine cone fibers reinforced compatibilized polypropylene. Materials and Design. 2012;40:528-535. doi:10.1016/j.matdes.2012.04.032.

[32] Qu X, Gao J, Qin M, Lei C. Application of a wax-based binder in PIM of WC-TiC-Co cemented carbides. International Journal of Refractory Metals & Hard Materials. 2005;23:273-277. doi:10.1016/j.ijrmhm.2005.04.006.

[33] Arandes JM, Torre I, Castaño P, Olazar M, Bilbao J. Catalytic cracking of waxes produced by the fast pyrolyse of polyolefins, Energy & Fuels. 2007;21:561-569. doi:10.1021/ef060471s.

[34] Elordi G, Olazar M, Lopez G, Artetxe M, Bilbao J. Continuous polyolefin cracking on an HZSM-5 zeolite catalyst in a conical spouted bed reactor. Industrial & Engineering Chemistry Research. 2011;50:6061-6070. doi:10.1021/ie2002999.

[35] Elordi G, Olazar M, Lopez G, Amutio M, Artetxe M, Aguado R, Bilbao J. Catalytic pyrolysis of HDPE in continuous mode over zeolite catalysts in a conical spouted bed reactor. Journal of Analytical and Applied Pyrolysis. 2009;85:345-351. doi:10.1016/j.jaap.2008.10.015.

[36] Olazar M, Lopez G, Amutio M, Elordi G, Aguado R, Bilbao J. Influence of FCC catalyst steaming on HDPE pyrolysis product distribution. Journal of Analytical and Applied Pyrolysis. 2009;85:359-365. doi:10.1016/j.jaap.2008.10.016.

[37] Artetxe M, Lopez G, Amutio M., Elordi G, Bilbao J, Olazar M. Light olefins from HDPE cracking in a two-step thermal and catalytic process. Chemical Engineering Journal. 2012;207-208:27-34. doi:10.1016/j.cej.2012.06.105.

[38] Erkiaga A, Lopez G, Barbarias I, Artetxe M, Amutio M. HDPE pyrolysis-steam reforming in a tandem spouted bed-fixed bed reactor for H_2 production. Journal of Analytical and Applied Pyrolysis. 2015;116:34-41. doi:10.1016/j.jaap.2015.10.010.

[39] Barbarias I, Lopez G, Alvarez J, Artetxe M, Arregi A, Bilbao J, Olazar M. A sequential process for hydrogen production based on continuous HDPE fast pyrolysis and in–line steam reforming. Chemical Engineering Journal. 2016;296:191-198. doi:10.1016/j.cej.2016.03.091.

Chapter 13

Pyrolysis: Pathway to Coal Clean Technologies

Andrew O. Odeh

Additional information is available at the end of the chapter

http://dx.doi.org/10.5772/67287

Abstract

Pyrolysis remains key to all coal utilisation processes such as combustion, gasification and liquefaction. Understanding the thermochemical changes accompanying these processes through pyrolysis would help in defining the technical performance of the processes. With the recent concern for the environment and renewed interest in research on clean coal technology (CCT), hydrogen from coal through the integrated gasification combined cycle has been considered for the proposed hydrogen economy.

Keywords: char, coal, pollution control, emissions, pyrolysis

1. Introduction

What is pyrolysis: pyrolysis is a thermochemical decomposition of carbonaceous materials such as biomass, plastic, tyre, coal, etc. at elevated temperatures of 200°C and above in the absence of oxygen. It is an irreversible chemical reaction in which there is a simultaneous change of chemical composition and physical phase of the matter. This reaction involves the molecular breakdown of larger molecules (polymer) into smaller molecules in the presence of heat. Pyrolysis is also referred to as thermal cracking, thermolysis, depolymerisation, etc.

What is coal pyrolysis: coal pyrolysis involves subjecting coal to high temperature of 400–450°C, in the absence of oxygen. When oxygen or steam is present, coal will start burning, and the process is no longer known as pyrolysis but rather referred as combustion and gasification. The benefits of coal pyrolysis are enormous and are listed below:

- Converts waste (char) to energy.
- The in-product can be used as fuel in existing industrial boilers and furnaces.

- The end products can also be used for generating electricity.
- It offers renewable energy source.
- Solid waste management.

Coal and coal products will continue to play an increasingly important role in fulfilling the energy needs and economies of nations. This is because of the abundant reserves of coal and its low cost [1, 2]. Coal accounts for roughly 25% of the world's energy supply and 40% of carbon emissions but even with the high percentage of emissions, it is very unlikely that any of these countries that are into coal exploration and production will turn their back on coal very soon [3]. Economic growth requires energy growth [4]. With the recent concern for the environment and renewed interest in research on alternative energy from renewable sources such as fuel cells and wind, hydrogen from coal through the integrated gasification combined cycle has been considered for the proposed hydrogen economy [5, 6]. Gasification has been tipped as the twenty-first century clean coal conversion technology than the other coal utilisation processes such as liquefaction and combustion because it is high energy efficient [7], non-polluting [8] and economical [9]. It also has the merit of going beyond the use of coal for the generation of power [10], metal processing and the production of chemicals [11], as coal could be converted to useful gases and liquids [12]. Coal is a complex carbonaceous material consisting of organic and inorganic matter [13]. During gasification, the organic and inorganic matter undergoes various chemical and physical transformations [14]. In order to maximise the gasification efficiency, there is a need to understand the mechanism of the chemical and physical transformation, as this will assist in the reduction of carbon emissions in the process especially when gasifying low rank coal [15–17]. Several options are used to control the feed rate of coal during gasification: fixed bed, fluidised bed, and entrained flow gasifiers [18]. Fluidised bed gasifiers have the potential advantage that low-grade coals rich in ash and inertinites, such that South African coals, can be processed more efficiently than in conventional pulverised coal boilers [19–21].

Therefore, the design of coal utilisation processes will require a deeper understanding of coal's intrinsic properties and the ways in which it is chemically transformed under process conditions [22, 23]. One of the ways to get this understanding is through pyrolysis which serves as an enroute to all coal utilisation processes [19]. Hence in this communication, the evaluation of six southern hemisphere coals would be used to illustrate the intermediary role played by pyrolysis in the coal utilisation processes.

2. Influence of changes in chemical and physical properties on coal performance

Currently, research efforts on the utilisation of coal and coal products are driven towards clean coal technology (CCT) [20, 24]. Previous studies on CCT for the past 30 years have been on the chemical cleaning of coal and of recent on carbon capture and storage (CCS)

[20, 25]. Research efforts have been limited to laboratory-scale in the determination of molecular and structural parameters such as aromaticity, degree of condensation that defines the technical performance of coal during coal utilisation processes [20, 26–28]. The essence of chemical cleaning in coal is to remove or reduce the mineral content in coal as it has been reported that the mineral content in coal melts when it is subjected to heat treatment during coal conversion processes [20, 29] which results in blocking the carbon active sites [30] thereby reducing the reactivity of the coal and decreasing the emission of pollutants [20, 31].

Coal is a complex carbonaceous polymer made up of organic and inorganic substances [32, 33]. The organic materials are known as macerals, while the inorganic impurities are considered as the minerals [34]. When exposed to heat-treatment; the physical, chemical, thermal, mechanical and electrical properties of coal undergo transformation [20, 35]. One of the key parameters that are used in measuring the chemical stability of this transformation is the aromaticity [20, 36]; it gives a good representation of the maceral to char transformation, which stands as a good indicator of coal maturity due to the realignment of the carbon [20, 37].

The change in carbonaceous structure due to the modification of the organic and inorganic constituents in coal and its subsequent char is stated to be one of the main factors that affect the reactivity of coal/char in coal conversion processes [20, 38, 39]. The chemical transformation involves the change in the organic chemical structure (**Tables 1–3**) while the physical transformation involves a change in the char morphology and porosity (**Table 4, Figures 1–12**).

Coal	SPL	SM	BCH	SSL	NGR	GER
wt% inherent moisture (air dried)	1.5	1.0	2.1	4.2	9.6	15.4
wt% ash (air-dried)	11.2	17.3	16.2	29.1	9.0	12.4
wt% volatile matter (air-dried)	5.3	7.6	26.7	21.4	37.6	45.7
wt% fixed carbon (air-dried)	82	74.1	55.0	45.3	43.8	26.4
wt% carbon (daf)	90.2	90.4	81.6	77.5	75.6	70.5
wt% hydrogen (daf)	2.7	3.5	4.6	4.5	5.2	6.6
wt% nitrogen (daf)	2.2	2.0	2.0	2.2	1.7	0.6
wt% oxygen (daf)	2.7	3.3	10.7	15.4	16.9	18.5
wt% sulphur (daf)	2.3	0.9	1.2	0.4	0.7	3.7
Gross calorific value (MJ/kg)	29.6	28.7	26.8	20.0	24.6	21.2
H/C	0.4	0.5	0.7	0.7	0.8	1.1
f_a	0.91	0.85	0.73	0.72	0.65	0.49

Table 1. Proximate analysis, ultimate analysis, calorific values and calculated H/C and aromaticity values for untreated coal.

Coal	SPL	SM	BCH	SSL	NGR	GER
wt% inherent moisture(air dried)	2.5	2.3	2.7	1.3	1.9	1.7
wt% ash (air-dried)	1.5	1.8	1.2	3.3	2.0	0.8
wt% volatile matter (air-dried)	6.8	9.6	27.2	25.0	43.2	60.3
wt% fixed carbon (air-dried)	89.2	86.3	68.9	70.4	53.0	37.3
wt% carbon (daf)	85.6	89.0	83.4	80.9	75.1	69.2
wt% hydrogen (daf)	2.4	3.3	4.6	4.2	5.2	6.2
wt% nitrogen (daf)	2.0	1.8	2.0	2.3	1.8	0.6
wt% oxygen (daf)	7.7	5.0	9.1	12.3	17.4	20.3
wt% Sulphur (daf)	2.1	0.7	1.0	0.3	0.1	2.7
Gross calorific value (MJ/kg)	32.7	33.3	32.0	30.0	29.3	28.9
H/C	0.3	0.4	0.7	0.6	0.8	1.1
$f_{a\,(CA)}$	0.92	0.86	0.74	0.76	0.65	0.52
$f_{a\,(FTIR)}$	0.98	0.84	0.72	0.74	0.58	0.40
$f_{a\,(C\text{-}NMR)}$	0.98	0.94	0.76	0.80	0.58	0.43
$f_{a\,(XRD)}$	0.89	0.87	0.78	0.74	0.70	0.66

Table 2. Proximate analysis, ultimate analysis, calorific values and calculated H/C and aromaticity values for acid-treated coal.

Coal	450	500	550	600	650	700
	GER					
H/C	0.5	0.4	0.3	0.3	0.2	0.1
$f_{a\,(CA)}$	0.86	0.89	0.95	0.95	0.99	1.00
$f_{a\,(FTIR)}$	0.66	0.69	0.73	0.74	0.76	0.79
$f_{a\,(XRD)}$	0.66	0.67	0.68	0.72	0.74	0.76
	NGR					
H/C	0.5	0.4	0.3	0.3	0.2	0.1
$f_{a\,(CA)}$	0.86	0.90	0.93	0.96	1.00	1.03
$f_{a\,(FTIR)}$	0.75	0.78	0.81	0.84	0.87	0.90
$f_{a\,(XRD)}$	0.67	0.69	0.70	0.74	0.78	0.80
	SSL					
H/C	0.4	0.4	0.3	0.3	0.2	0.1
$f_{a\,(CA)}$	0.87	0.91	0.93	0.96	1.00	1.05
$f_{a\,(FTIR)}$	0.84	0.88	0.90	0.93	0.97	1.00
$f_{a\,(XRD)}$	0.91	0.94	0.96	0.97	0.97	0.97
	BCH					
H/C	0.5	0.4	0.3	0.3	0.2	0.1

Coal	450	500	550	600	650	700
$f_{a\,(CA)}$	0.86	0.89	0.92	0.95	0.98	1.03
$f_{a\,(FTIR)}$	0.83	0.86	0.89	0.92	0.95	1.00
$f_{a\,(XRD)}$	0.93	0.94	0.97	0.98	0.99	0.99
	SM					
H/C	0.4	0.4	0.3	0.3	0.2	0.1
$f_{a\,(CA)}$	0.88	0.89	0.92	0.95	0.99	1.03
$f_{a\,(FTIR)}$	0.94	0.95	0.98	1.00	1.00	1.00
$f_{a\,(XRD)}$	0.96	0.98	0.99	0.99	0.99	0.99
	SPL					
H/C	0.3	0.3	0.3	0.3	0.2	0.1
f_a	0.94	0.95	0.95	0.97	0.98	1.03
$f_{a\,(FTIR)}$	0.97	0.98	1.00	1.00	1.00	1.00
$f_{a\,(XRD)}$	0.96	0.97	0.98	0.99	0.99	0.99

Table 3. Calculated H/C and aromaticity values for heat-treated coal.

Coal	450	500	550	600	650	700
	GER					
O/C	0.132	0.103	0.092	0.073	0.064	0.056
BET surface area (m^2/g)	169.96	193.97	230.41	241.82	262.61	268.56
	NGR					
O/C	0.130	0.110	0.083	0.075	0.067	0.061
BET surface area (m^2/g)	155.78	182.61	183.19	234.10	238.14	239.74
	SSL					
O/C	0.081	0.076	0.063	0.052	0.051	0.042
BET surface area (m^2/g)	136.60	153.47	199.72	200.38	214.46	224.19
	BCH					
O/C	0.064	0.057	0.044	0.039	0.037	0.029
BET surface area (m^2/g)	130.17	158.68	183.89	206.40	215.40	224.95
	SM					
O/C	0.039	0.042	0.033	0.033	0.037	0.032
BET surface area (m^2/g)	137.94	148.17	170.35	186.54	194.60	196.99
	SPL					
O/C	0.039	0.048	0.063	0.039	0.037	0.036
BET surface area (m^2/g)	113.93	135.18	136.74	150.98	162.47	164.40

Table 4. Calculated atomic O/C and BET surface area values from SEM and ASAP 2020 for heat-treated coal.

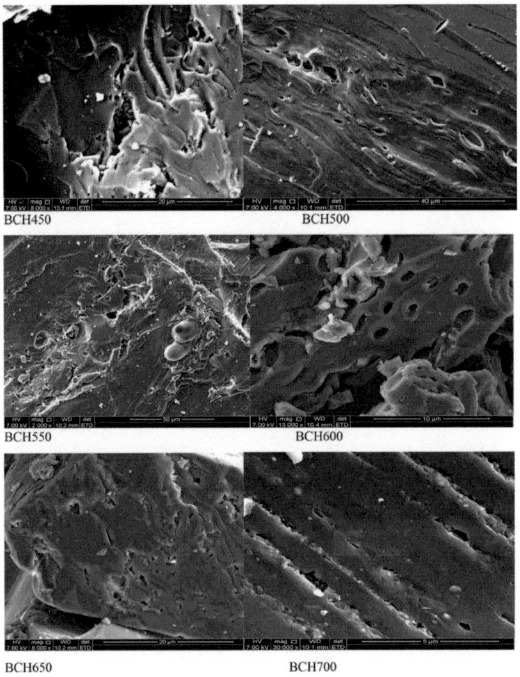

Figure 1. SEM micrographs of the transition of BCH coal to char.

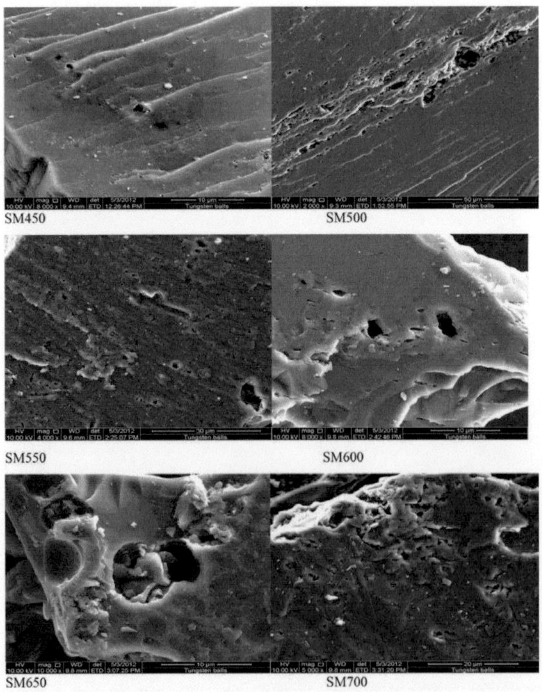

Figure 2. SEM micrographs of the transition of SM coal to char.

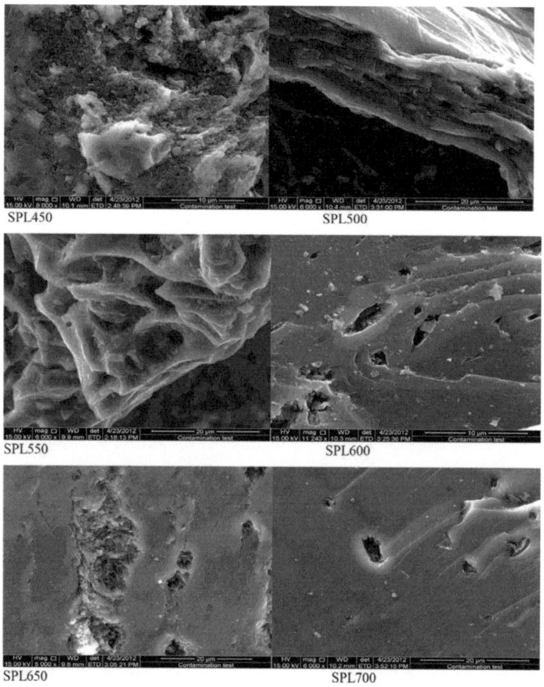

Figure 3. SEM micrographs of the transition of SPL coal to char.

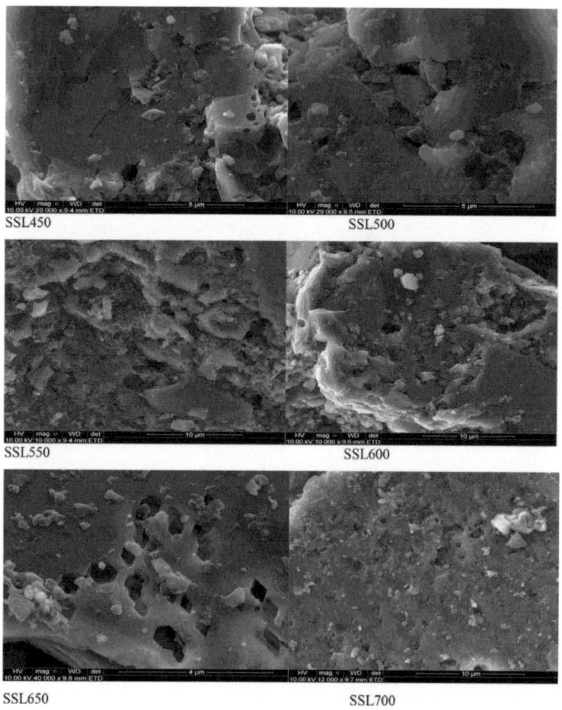

Figure 4. SEM micrographs of the transition of SSL coal to char.

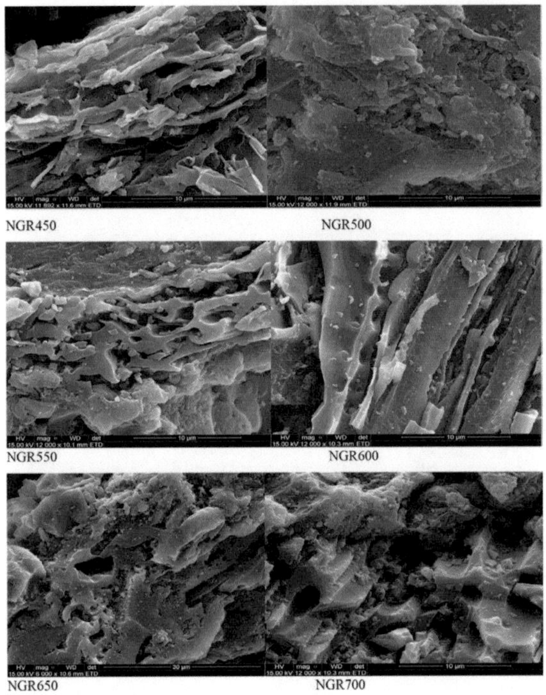

Figure 5. SEM micrographs of the transition of NGR coal to char.

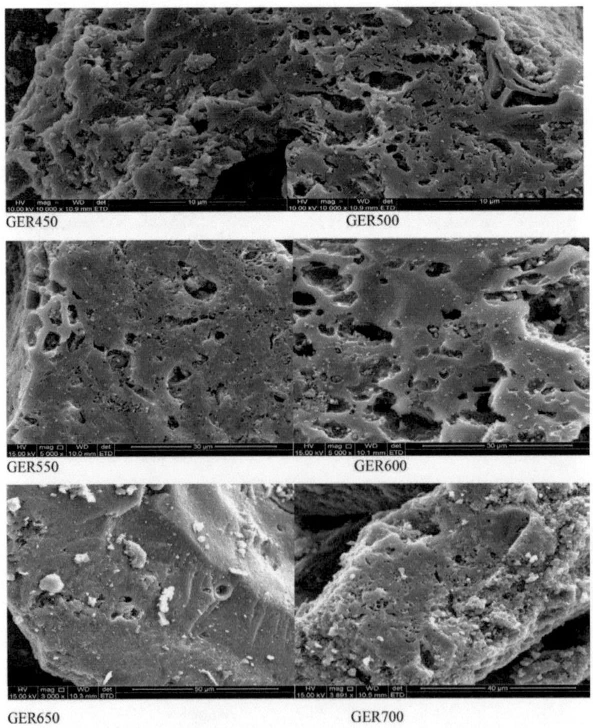

Figure 6. SEM micrographs of the transition of GER coal to char.

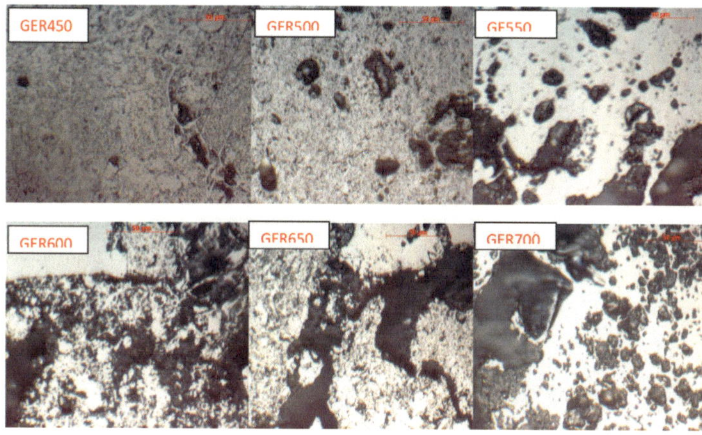

Figure 7. Petrographic pictures of the transition of coal to char for GER suites.

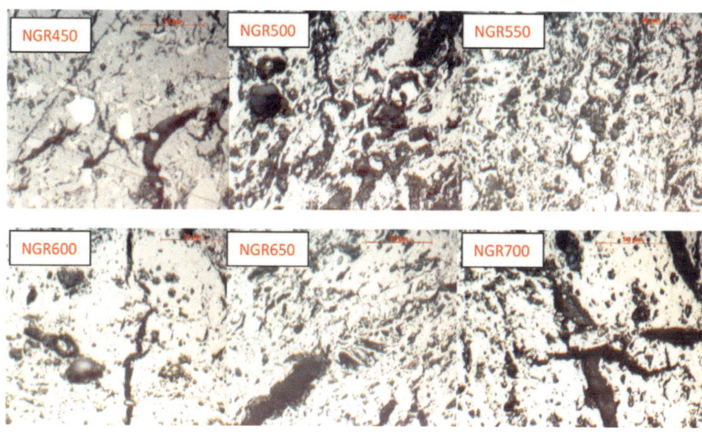

Figure 8. Petrographic pictures of the transition of coal to char for NGR suites.

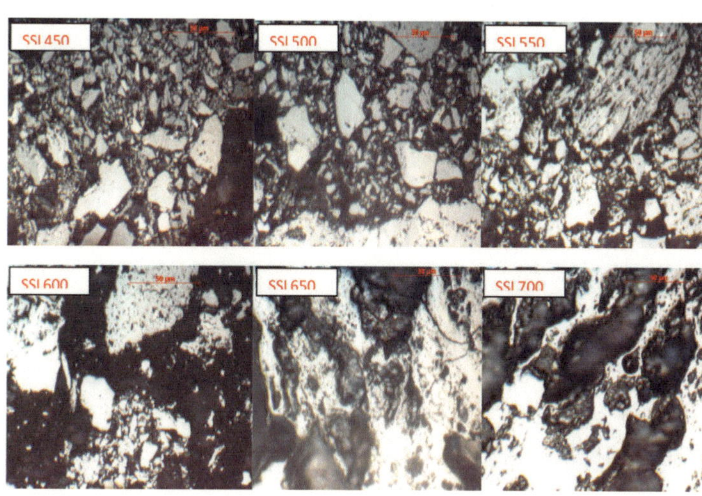

Figure 9. Petrographic pictures of the transition of coal to char for SSL suites.

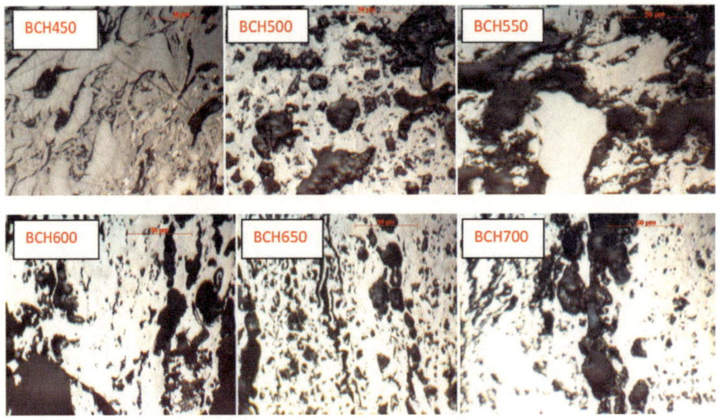

Figure 10. Petrographic pictures of the transition of coal to char for BCH suites.

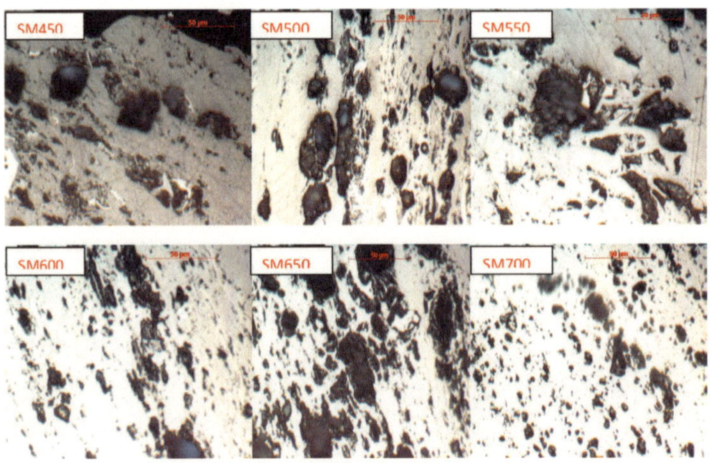

Figure 11. Petrographic pictures of the transition of coal to char for SM suites.

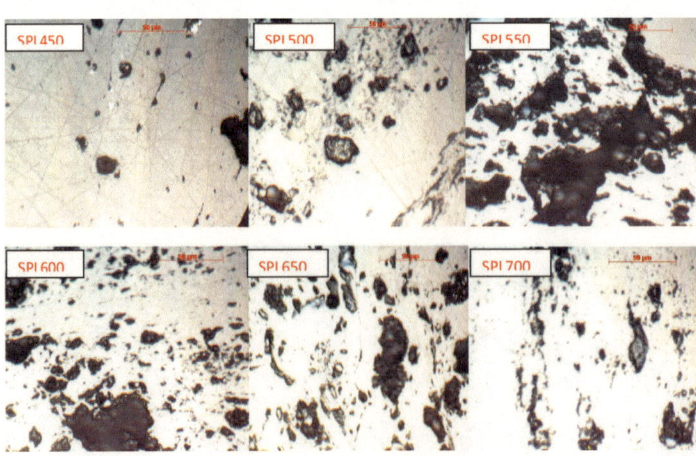

Figure 12. Petrographic pictures of the transition of coal to char for SPL suites.

Author details

Andrew O. Odeh

Address all correspondence to: odehandy@yahoo.com

Department of Chemical Engineering, University of Benin, Benin City, Nigeria

References

[1] Ahmed, I.I. and Gupta, A.K. 2011. Particle size, porosity and temperature effects on char conversion. Applied Energy 88: 4667-4677.

[2] Choudhury, N., Biswas, S., Sarkar, P., Kumar, M., Ghosal, S., Mitra, T., Mukherjee, A. and Choudhury, A. 2008. Influence of rank and macerals on the burnout behaviour of pulverised Indian coal. International Journal of Coal Geology 74: 145-153.

[3] Cousins, A., Paterson, N., Dugwell, D.R., Kandiyonti, R. 2006. An investigation of the reactivity of chars formed in fluidized bed gasifiers: the effect of reaction conditions and particle size on coal char reactivity. Energy & Fuels 20: 2489-2497.

[4] Hurt, R., Sun, J.K. and Lunden, M. 1998. A kinetic model of carbon burnout in pulverized coal combustion. Combustion and Flame 113: 181-197.

[5] Ibarra, J.V., Munoz, D. and Moliner, R. 1996. FTIR study of the evolution of coal structure during the coalification process. Organic Geochemistry 24: 725-735.

[6] Jasienko, S., Matuszewska, A. and John, A. 1995. Properties and structure of hard coals from the borehole Niedobczyce IG-1 in the Rybnik Coal District, Upper Silesian Coal Basin, their petrographic and group constituents. 2. Variations in petrographic composition of the coals along the depth of borehole and alterations in structure of the coals characterized by vitrinites spectroscopic analyses (X-ray, IR). Fuel Processing Technology 41: 221-232.

[7] Li, C.Z. 2007. Some recent advances in the understanding of the pyrolysis and gasification behaviour of Victorian brown coal. Fuel 86: 1664-1683.

[8] Li, Y., Cao, X., Zhu, D., Chappel, M.A., Miller, L.F. and Mao, J. 2012. Characterisation of coals and their laboratory-prepared black carbon using advanced solid-state ^{13}C nuclear magnetic resonance spectroscopy. Fuel Processing Technology 96: 56-64.

[9] Li, Z., Fredericks, P.M., Rintoul, L. and Ward, C.R. 2007. Application of attenuated total reflectance micro-Fourier transform infrared (ATR–FTIR) spectroscopy to the study of coal macerals: examples from the Bowen Basin, Australia. International Journal of Coal Geology 70: 87-94.

[10] Lu, L., Sahajwalla, V. and Harris, D. 2000. Characteristics of chars prepared from various pulverized coals at different temperatures using drop-tube furnace. Energy & Fuels 14: 869-876.

[11] Lu, L., Sahajwalla, D. Kong, C. and Harris, D. 2001. Quantitative X-ray diffraction analysis and its application to various coals. Carbon 39: 1821-1833.

[12] Lunden, M.M., Yang, N.Y.C., Headley, J. and Shaddix, C.R. 1998. Mineral-char interactions during char combustion of a high volatile coal. Proceedings of Combustion Institute 27 2: 1695-1702.

[13] Maity, S. and Choudhury, A. 2008. Influence of nitric acid treatment in different media on X-ray structural parameters of coal. Energy & Fuels 22: 4087-4091.

[14] Matsuoka, K., Suzuki, Y., Eylands, K., Benson, S. and Tomita, A. 2006. CCSEM study of ash forming reactions during lignite gasification. Fuel 85: 2371-2376.

[15] Matsuoka, K., Akahane, T., Aso, H., Sharma, A. and Tomita, A. 2008. The size of polyaromatic layer of coal char estimated from elemental analysis data. Fuel 87: 539-545.

[16] McBeath, A.V., Smernik, R.J., Schneider, M.P.W. and Schmidt, M.W.I. 2011. Determination of the aromaticity and degree of aromatic condensation of a thermosequence of wood charcoal using NMR. Organic Geochemistry 42: 1194-1202.

[17] Neavel, R.C. 1981. "Origin, Petrography and Classification", Chemistry of Coal Utilization. 2nd Supplement Volume. Wiley: New York.

[18] Oboirien, B.O., Engelbrecht, A.D., North, B.C., du Cann, V.M., Verryn, S. and Falcon, R. 2011. Study on structure and gasification characteristics of selected South African bituminous coal in fluidised bed gasification. Fuel Processing Technology 92: 735-742.

[19] Odeh, A. 2015. Qualitative and quantitative ATR-FTIR analysis and its application to coal char of different ranks. Journal of Fuel Chemistry and Technology 43: 129-137.

[20] Odeh, A. 2015. Comparative study of the aromaticity of the coal structure during the char formation process under both conventional and advanced analytical techniques. Energy & Fuels 29: 2676-2684.

[21] Odeh, A. 2015. Exploring the potential of petrographics in understanding coal pyrolysis. Energy 87: 555-565.

[22] Orrego-Ruiz, R.A., Cabanzo, R. and Mejia-Ospino, E. 2011. Study of Colombian coals using photoacoustic Fourier transform infrared spectroscopy. International Journal of Coal Geology 85: 307-310.

[23] Painter, P., Sobkowiak, M., Heidary, S. and Coleman, M. 1994. Current status of FT-IR in the analysis of coal structure. Fuel 39: 49-53.

[24] Perry, S.T., Hambly, E.M., Fletcher, T.H., Solum, M.S., and Pugmire R.J. 2000. Solid-state C NMR characterization of matched tars and chars from rapid coal characterization. Proceedings of the Combustion Institute 28: 2313-2319.

[25] Pusz, S., Duber, S. and Kwiecinska, B. 2002. The study of textural and structural transformations of carbonized anthracites. Fuel Processing Technology 77: 173-180.

[26] Rodrigues, S., Suarez-Ruiz, I., Marques, M., Camean, I. and Flores, D. 2011. Microstructural evolution of high temperature treated anthracites of different rank. International Journal of Coal Geology 87: 204-211.

[27] Rouzand, J.N. and Oberlin, A. 1989. Structure, microtexture, and optical properties of anthracene and saccharose-based carbons. Carbon 27: 517-529.

[28] Sekine, Y., Ishikawa, K., Kikuchi, E., Matsukata, M. and Akimoto, A. 2006. Reactivity and structural change of coal during steam gasification. Fuel 85: 122-126.

[29] Senneca, O., Salatino, P. and Masi, P. 1998. Microstructural changes and loss of gasification reactivity of chars upon heat treatment. Fuel 77: 1483-1493.

[30] Senneca, O., Salatino, P. and Menghini, D. 2007. The influence of thermal annealing on oxygen uptake and combustion rates of a bituminous coal char. Proceedings of the Combustion Institute 31: 1889-1895.

[31] Strydom, C.A., Bunt, J.R., Schobert, H.H. and Raghoo, M. 2011. Changes to the organic functional groups of an inertinite rich medium rank bituminous coal during acid treatment process. Fuel Processing Technology 92: 764-770.

[32] Suarez-Ruiz, I. and Garcia, A.B. 2007. Optical parameters as a tool to study the microstructural evolution of carbonized anthracites during high-temperature treatment. Energy & Fuels 21: 2935-2941.

[33] Solum, M.S., Pugmire, R.J., Jagtoyen, M. and Derbyshire, F. 1995. Evolution of carbon structure in chemically activated wood. Carbon 33: 1247-1254.

[34] Supaluknari S., Larkins, F.P., Redlich, P. and Jackson, W.R. 1989. Determination of aromaticities and other structural features of Australian coals using solid state 13C NMR and FTIR spectroscopies. Fuel Processing Technology 23: 47-61.

[35] Tsai, C.-T. and Scaroni, A.W. 1986. The structural changes of bituminous coal particles during the initial stages of pulverized–coal combustion. Fuel 66: 200-206.

[36] van Niekerk, D., Pugmire, R.J., Solum, M.S. and Mathews, J.P. 2008. Structural characterisation of vitrinite-rich and inertinite-rich Permain-aged South African bituminous coals. International Journal of Coal Geology 76: 290-300.

[37] Wang, J., Du, J., Chang, L. and Xie, K. 2010. Study on the structure and pyrolysis characteristics of Chinese western coals. Fuel Processing Technology 91: 430-433.

[38] Zhang, H., Pu, W., Ha, S., Li, Y., and Sun, M. 2009. The influence of included mineral on the intrinsic reactivity of chars prepared at 900°C in a drop tube furnace and muffle furnace. Fuel 88: 2303-2310.

[39] Zhu, W., Song, W. and Lin, W. 2008. Effect of particle size on pyrolysis and char reactivity for two types of coal and demineralised coal. Energy & Fuels 22: 2482-2487.

Chapter 14

Pyrolysis of Low-Rank Coal: From Research to Practice

Fan Nie, Tao Meng and Qiumin Zhang

Additional information is available at the end of the chapter

http://dx.doi.org/10.5772/67498

Abstract

Low-rank coal (LRC), as a conventional fossil fuel, has wealth of reserves and a wide range of distribution around the world, and pyrolysis is thought to be an easy way for clean and efficient conversion of LRC. In this chapter, the characteristics and world's reservation of LRC are introduced. Then, the chemical reactions and product formation process during pyrolysis of LRC are described. Meanwhile, how the factors, such as temperature, minerals in coal, heating rate, particle size and atmosphere, influence the pyrolysis process are discussed. Finally, three LRC pyrolysis-based polygeneration systems are illustrated for recent developments on LRC industrial practice.

Keywords: low-rank coal, pyrolysis, polygeneration process

1. Introduction: An overview of low-rank coal

1.1. Low-rank coal and its characteristics

Coal is a kind of combustible black or brownish-black sedimentary rock. The degree of change undergone by a coal as it matures from peat to anthracite—known as coalification—has an important bearing on its physical and chemical properties and is referred to as the "rank" of the coal [1]. Determined by temperature, burial pressure and the length of time during formation, coal is diverse mainly on its content of moisture, volatile matters and fixed carbon. According to "ISO 11760-2005: Classification of coals," low-rank coal (LRC) is with bed moisture <75% and mean random vitrinite reflectance ($\overline{R}_r$) <0.5%, which includes lignite (brown coal) and sub-bituminous coal. Additional subcategories of low-rank coals are listed in **Table 1**.

Subcategory	Description
Low-rank C (Lignite C)	$\overline{R}_r < 0.4\%$ and bed moisture >35% and <75%, ash-free basis
Low-rank B (Lignite B)	$\overline{R}_r < 0.4\%$ and bed moisture , ash-free basis
Low-rank A (Sub-bituminous coal)	$0.4 \leq \overline{R}_r < 0.5\%$

Table 1. Subcategories of low-rank coal.

Typical LRC is a soft, friable material with a dull, earthy appearance. It is characterized by high moisture, low carbon content and low calorific value, but relatively high volatiles. As to element constitutes of organic matters, LRC has relatively high hydrogen and oxygen content. **Table 2** lists the results of proximate and ultimate analysis of some typical LRC [2–5].

The organic structure of coal is complex and compositionally heterogenetic from different coal deposits. Due to the lower degree of coalification, the organic structure in LRC generally has smaller aromatic cores, but more and longer branched chains linking to them [6–8]. Another critical characteristic of LRC is abundant oxygen-containing functional group as a part of organic matrix [9], and it consists of carbonyl, hydroxyl (including phenolic), carboxyl, methoxyl and other inactive oxygen-containing functional groups(oxygen heterocycles). It was found that oxygen in the form of hydroxyl accounted for the major fraction of the total oxygen, followed by the carbonyl oxygen [10–12], and alkali and alkaline earth metallic (AAEM) species are able to exist as ion-exchangeable cations associated with the carboxyl groups forming part of the organic coal substance. Although the accurate organic structure of LRC has not been proposed yet, compared with higher rank of coal, basic characteristics of LRC organic structure can be summarized as below:

a. LRC has relatively less aromaticity value, less number of carbon atoms per aromatic lamellae and higher ratio of amorphous carbon.

b. The aromatic core is not highly condensed but with a proportion of biphenyl and phenylnaphthalene skeleton structure. Each aromatic core is linked by bridged bond, such as polymethylene and ether bridged bonds.

c. Oxygen is a predominate heteroatom in LRC. Abundant oxygen-containing functional group can directly link to the aromatic core or combine different aromatic cores.

Thus, LRC has more "incompact" macromolecule structure than other higher rank of coal. Those abundant "C—C" and "C—O" bridged linkages could play an important role on depolymerization of LRC macromolecules. Due to the characteristics of organic structure, LRC tends to have better reactivity than higher rank of coal, and it could be good material for chemical conversion, such as pyrolysis and gasification.

Categories	Source	Proximate analysis (wt%)				Ultimate analysis (wt%, dry ash-free basis)				
		M_{ar}	A_d	V_d	FC_d^*	C	H	N	S	O*
Lignite	Western Australia	64.9	3.1	52.6	44.3	66.4	5.1	0.48	0.21	25.5
	Morwell, Australia	15.3	5.4	35.3	59.3	74.85	4.96	1.11	0.80	18.28
	Indonesia	32.70	7.56	48.83	43.60	58.09	3.39	0.46	0.17	13.90
	Schleenhain, Germany	15.5 (dried)	11.6	52.4	36.0	70.6	5.1	-	4.3	20.0 (N+O)
	Xiaolongtan, China	14.04	10.42	45.14	44.44	53.86	3.65	1.41	1.27	39.81
	Pingzhuang, China	10.91	29.08	30.56	40.36	46.24	2.78	0.68	1.20	49.1
Sub-bituminous coal	Indonesia	12.41	9.58	42.06	48.36	72.13	6.67	1.4	0.22	19.58
	Diyarbakır-Hazro,Turkey	1.32 (ad)	43.80	25.71	30.49	70.40	5.65	0.68	7.12	16.15
	India	4.57 (dried)	13.49	38.67	47.84	60.12	6.84	1.47	0.46	31.11
	Shenfu,China	5.24 (ad)	6.95	29.65	63.40	79.38	4.78	1.71	0.36	13.77
	Hongliulin,China	10.60	9.45	32.65	57.90	80.45	4.80	1.09	0.40	13.26
	Zhundong,China	8.79 (ad)	5.68	32.96	61.36	75.23	3.18	0.70	0.53	20.36

*By difference.

Table 2. Proximate and ultimate analysis of some LRC.

1.2. Distribution and utilization around the world

At the end of 2015, the total coal reservation around the world has proved to be 891,531 million tonnes, in which LRC (sub-bituminous and lignite) is 488,332 million tonnes, accounting for 54.77% of the total. As to the distribution, above 70% of LRC reserves in United States (26.37%), Russian Federation (22.10%), China (10.71%), Germany (8.29%) and Australia (8.05%) [13].

Although LRC's recoverable reserves are a lot, their high moisture content, greater tendency to combust spontaneously, high degree of weathering and the dusting characteristics restrict widespread use of such coals [14]. But for a few countries, LRC can be important in the respective national energy mix. Possible reasons are mainly on the considerations of native fossil resource endowment, security of supply, price and industry maintenance [15].

Traditionally, the major way for LRC utilization is direct burning, and its chemical energy can be converted into heat, power or electricity. Problems, such as emissions of CO_2, SO_x, NO_x, heavy metal and microdust, may easily be caused during such traditional utilizations of LRC. With the improvement of environmental awareness, cleaner and more efficient ways for LRC utilization are desirable. Among them, technologies based on pyrolysis are promising ways for LRC being converted into not only clean fuels but also chemicals. Since the industrial development of coal pyrolysis at early nineteenth century, people have been seeking ways for high yield and quality of pyrolytic products, and lots of views on pyrolysis mechanism or impact factors are proposed. According to these researches, new technologies and equipment are carried out as well. Recently, with the increasing pursue of cleanliness and efficiency during process, pyrolysis of LRC tends to combine with other technology forming series of polygenerative processes. And in this chapter, some new views and new technologies or ideas during recent years on LRC pyrolysis have been reviewed.

2. Pyrolysis of low-rank coal and its products

2.1. Pyrolysis of low-rank coal and its products

LRC pyrolysis is a complex process with both physical changes and chemical reactions at elevated temperatures in the absence of oxygen or in an inert atmosphere. And it is thought to be a promising and convenient method for direct generation of clean fuels and high-valuable chemicals from the abundant hydrocarbons in LRC. Pyrolysis reactions are also an important medium process for LRC conversion technologies, such as coking, combustion, gasification and direct liquefaction. Referred to different standards, LRC pyrolysis can be classified as follows [16]:

- According to the final temperature, there are low-temperature pyrolysis (450–650°C), medium-temperature pyrolysis (600–900°C), high-temperature pyrolysis (900–1200°C) and ultrahigh-temperature pyrolysis (>1200°C).

- According to heating rate, there are slow pyrolysis (1°C/s), medium-speed pyrolysis (5–100°C/s), fast pyrolysis (500–10⁶°C/s) and flash pyrolysis (>10⁶°C/s).
- According to forms, there are normal (or traditional) pyrolysis, plasma pyrolysis, hydropyrolysis and catalytic pyrolysis.

During pyrolysis, the organic structures in LRC are cracked by heat and amount of fragments devolatilized from the particles. Cooled by cold medium, the condensable part of volatiles forms as tar; the incondensable part forms as gas. Simultaneously, there also remain some nonvolatiles, which are solid products (coke or char).

The yields and properties of pyrolytic products are determined not only by the properties of the raw LRC but also by the operating condition of pyrolysis. As to the normal pyrolysis of LRC, temperature can be an important factor of what needs to be considered for desirable products. For example, low-temperature pyrolysis is usually for tar production, while medium-temperature pyrolysis and high-temperature pyrolysis are for char and coke production, respectively. For plasma pyrolysis, coal is heated rapidly to ultrahigh temperatures; if conditions are favorable, the emitted volatiles will consist mainly of acetylene [17]. However, for better recovery of the abundant volatiles in LRC, low-temperature pyrolysis and medium-temperature pyrolysis are preferred.

Due to the low degree of coalification, the products generated from LRC pyrolysis are also differed from the ones generated from higher rank of coal. The abundant oxygen-containing functional groups bring more moisture and oxygen-containing compounds (especially of phenols) into tars [18]. But after physical separation, tars are able to be good raw materials for production of transport fuels. On the contrary, little oxygen remains in char or coke, which elevates their calorific value and reduces the tendency to combust spontaneously. In addition, most of the pollutants (such as sulfur, volatile of organics) and heavy metals (such as mercury) are removed so that char can be much cleaner as solid fuels than the raw LRC [16]. Moreover, the developed pore structure is also a typical character of chars derived from LRC pyrolysis. With further activation methods, chars can be served as adsorbing materials or catalyst supports [19–21]. Gas products, derived from LRC pyrolysis, are full of H_2, CO and CH_4, which are able to be fuels or chemical synthesis after purification. However, on account of compositionally heterogeneity of LRC in different deposits, suitable ways for the development of downstream products need to be considered and tested individually.

2.2. Reactions during pyrolysis and the formation of products

2.2.1. Reactions during LRC pyrolysis

Chemical reactions during LRC pyrolysis are thought to be a series of complex sequential and parallel free radical reactions [22]. It involves the generations of radical fragments from the thermal decomposition of organic macromolecules and the generations of volatiles and residues (coke or char) from reactions between radical fragments [23].

In general, reactions during pyrolysis come from the thermal instability of structures. The thermostability of organic structure is much determined by the bond energy between atoms.

Generally, the thermostability of hydrocarbons is decreasing as follows: condensed aromatics > aromatics > cyclic hydrocarbons > olefins > alkanes. And this is also suitable for LRC pyrolysis [24]. Moreover, because of the conjugated structure, aromatic heteroatom (such as sulfur, nitrogen and oxygen) structures are much more thermostable than aliphatic ones. And it is also reported that the fragmentation progress of LRC is more significant than higher rank of coal [25, 26]. The instable structures in the macromolecules of LRC homolytic cleave into radicals, and small molecules are generated with stabilization of these radicals, while the stable structures associated with condensed products remain in char or coke. Due to the complexity of organic structure, there are no universal or accurate models for the description of reactions during LRC pyrolysis. But with mounts of studies, the types of pyrolysis reactions are as follows:

- Primary reactions: "Primary" indicates the reactions directly take place in the macromolecules of LRC. During pyrolysis, cracking reactions are the dominating primary reactions. According to the organic structures in the macromolecules of LRC, the four probable types of cracking reaction are listed below [24]:

(i) Cracking of bridged bond: In the organic structure of LRC, there are many bridged bonds, such as $-CH_2-$, $-O-$, $-CH_2-CH_2-$, $-CH_2-O-$, $-S-$, $-S-S-$ etc., linking to the aromatic cores. The bond energy of these "bridges" is relatively low, and they can be more easily damaged by heat or other radicals to generate new radicals. This is also the key for depolymerization of macromolecules in LRC [18]. And the thermal degradation of sulfurated and oxygenated cross-links was more substantial than that of alkyl linkages [27].

(ii) Cracking of aliphatic side chain: The stability of aliphatic side chain decreases with the increasing number of carbon and aromatic rings. Long side chains tend to crack to generate radicals or gaseous hydrocarbons of small molecules.

(iii) Cracking of aliphatics: The low molecular weight compounds with mainly aliphatic structure are melted and simultaneously cracked to generate volatiles.

(iv) Cracking of heteroatom functional groups: Sulfur, nitrogen and especially oxygen are the major heteroatoms in the organic structure of LRC. Aliphatic heteroatom structures are able to decompose at relative low temperature, while the temperature for decomposition of aromatic heteroatom structures needs to be 500°C or higher.

- Secondary reactions : During the process of devolatilization, high temperature in the region of both coal particles and surroundings will facilitate for secondary reactions of volatiles (primary volatiles). Major secondary reactions are listed below:

(i) Cracking reactions: The hydrocarbons derived from primary reactions are able to further crack at high-temperature circumstances. Consequently, more compounds with smaller molecular weight are generated.

(ii) Dehydrogenation reactions: Cycloalkanes can be converted into cyclenes or further into aromatics by dehydrogenation reactions Hydroaromatics can be added aromatic rings. Simultaneously, hydrogen gas is generated.

(iii) Hydrogenation reactions: Some chemical active heteroatom-containing compounds may be attacked by hydrogen radicals. As a result, the heteroatoms are removed and form as gas molecules (such as H_2S, NH_3 and H_2O).

(iv) Condensation reactions: Aromatics in volatiles may condense into polycyclic aromatic hydrocarbons (PAHs).

- Condensation reactions: With the increasing number of aromatic rings, the volatility of aromatics decreases. The nonvolatile parts will condense and release hydrogen gas, and it is more obvious at temperature above 700°C.

Besides the reactions listed above, there are also reactions, such as hydrogen transfer, rearrangements and cross-linking reactions, for final stabilization of free radicals.

2.2.2. *Pyrolysis process and formation of products*

Due to the diversity on thermostability of organic structure in LRC, the pyrolysis process is mainly determined by temperature. With the raising temperature, the extents of thermal cracking of organic macromolecules in LRC increase, and the constituents of pyrolytic products vary simultaneously.

Generally, there are few changes on the major hydrocarbon structures in LRC below 300°C. But noncovalent bonds in LRC dissociate and devolatilize. Saxena [28] summarized that pyrolysis reactions started with the cleavage of weak bonds. Due to the minimum amount of energy requirement for overcome the C—C bond energy, major pyrolysis reactions would not commence until the temperature was around 400°C. Guan [29] studied the yield and properties of the products derived from lignite (Huolinhe, China) pyrolysis at temperature below 400°C. It was observed that the condensate volatile was visibly colorless and transparent water with very a little value of chemical oxygen demand (CODcr), volatile phenol and ammonia nitrogen at pyrolysis temperature below 200°C. But when temperature is above 250°C, organic matters are obviously found in the condensates and the contents increase with the increasing pyrolysis temperature. The condensates derived above 350°C show as a mixture with oil and water layer. Analyzed by surface structure of solid residues below 200°C, no evident changes of phenolic hydroxyl and carboxyl groups on the surface were observed. When temperature is above 250°C, despite the major hydrocarbon structures are changed, some unstable functional group and side chain, such as carboxyl, aliphatic side chain and ether structure, are starting to crack and decompose. As to the organics in the condensates, alkanes, olefins, aromatics and phenols are obviously detected. Meanwhile, due to the low thermostability of carboxyl and carboxylate, it will decompose into CO_2 around 250°C.

During the process of LRC pyrolysis, the major organic volatile-producing stage is 350–600°C. When temperature raises to above 350°C, the linkages of polymethylene or ether oxygen, side chain or heteroatom-containing functional group are obviously damaged by heat, and some hydroaromatic rings start to dehydrogenate. As a result, high-reactive fragments and radicals are generated. Then, radicals may be stabilized by hydrogen transfer, rearrangements, cross-linking reactions, condensation reactions or capturing hydrogen by colliding with other fragments or radicals. And the primary products that directly generated from the macromolecules will be secondarily reacted during the devolatilization. Zhan et al. [30] investigated the mechanism of initial thermal decomposition associated with pyrolysis of a kind of sub-bituminous coal by molecular dynamic simulation. The calculation results showed that the primary decomposition reactions of Hatcher sub-bituminous model [27] began with intramolecular changes such as the cleavage of unstable C—C and C—O bonds. Castro-Marcanoet al. [31] utilized ReaxFF reactive force field to perform pyrolysis simulations on a large-scale molecular model of coal [32]. It was found that pyrolysis mainly initiated by release of hydroxyl groups and dehydrogenation of hydroaromatic structures followed by breakage of heteroatom-containing cross-links. And further analysis of ReaxFF simulations show that aryl and alkyl C—S bonds are weaker than aryl and alkyl C—C bonds. Hence, the cleavage of the C—S bonds resulted in more extensive fragmentation leading to larger quantities of aliphatic and aromatic structures. Li et al. [33] discussed the mechanism of coal-based model compound (anisole, phenyl ethyl ether and p-methyl anisole) by detecting the reactants, radicals and products using vacuum ultraviolet single-photon ionization time-of-flight mass spectrometry. It was found that PhO—C homolytic bond scission was the first step for the radical reaction, while β–H was an important factor for the nonradical reactions.

Tromp [34] and Miura [23] believe that, as shown in **Figure 1a**, the primary devolatilization reactions are very rapid reactions, which consist of radical formation reactions, polymerization-condensation reactions, radical recombination reactions, hydrogen addition reactions, etc., and subsequent secondary gas-phase reactions are decomposition reactions of the volatile products produced through the primary reactions. And the char particles come from the polymerization and condensation reactions. But with recent researches, especially on the radicals in pyrolytic products and flash pyrolysis of LRC, it is not able to be observed the devolatilization of a whole coal particle and the absent generation of char. Wu et al. [35] measured the free radicals in some LRC pyrolytic chars at different temperatures (350–600°C) by electron paramagnetic resonance (EPR). It was found radicals also existed in chars. Liu [36] believes that a part of coal structure will be recombined simultaneously with devolatilization. During the recombination, covalent bonds still crack to generate smaller molecules and condensed as chars (as shown in **Figure 1b**). As to the tars, He et al. [37] examined the reactivity of fast pyrolysis tars and observed that tars contained high concentrations of radicals and were highly reactive at temperatures higher than 400°C to generate more radicals and form coke. Thus, the content of heavy components increased which would reduce the quality of tars.

During the process of pyrolysis and devolatilization, both physical and chemical properties of pyrolytic residues also vary with the temperature. As to LRC pyrolytic solid residues, typical changes are on the oxygen content and pore's structures. As to noncaking coal (such as lignite), they almost inherit the main skeleton structures of the raw coal during pyrolysis. Wu [38] studied the properties' evolution of Hesigewula lignite (Inner Mogolia, China) pyrolytic solid residues via temperature (75–550°C). As shown in **Figure 2a**, the content of carbon in solid residue is increasing with the increasing temperature, while oxygen content is dramatically decreasing. The decreasing oxygen content mainly comes from the decomposition of oxygen-containing functional group and crystal water of minerals. Generally, the thermostability of oxygen-containing functional group is decreasing as oxygen heterocyclic

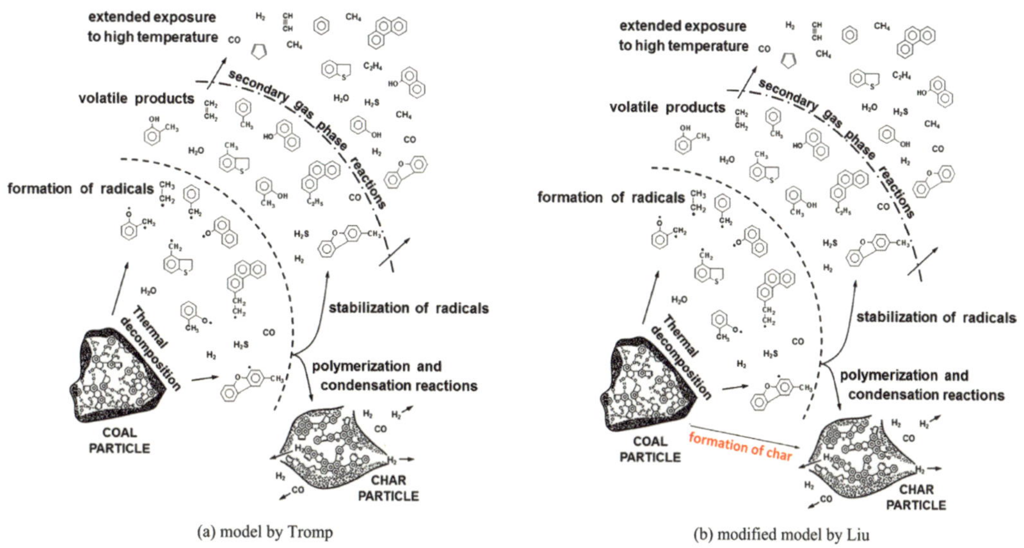

Figure 1. The mechanism model of coal pyrolysis. (a) Model by Tromp; (b) modified model by Liu.

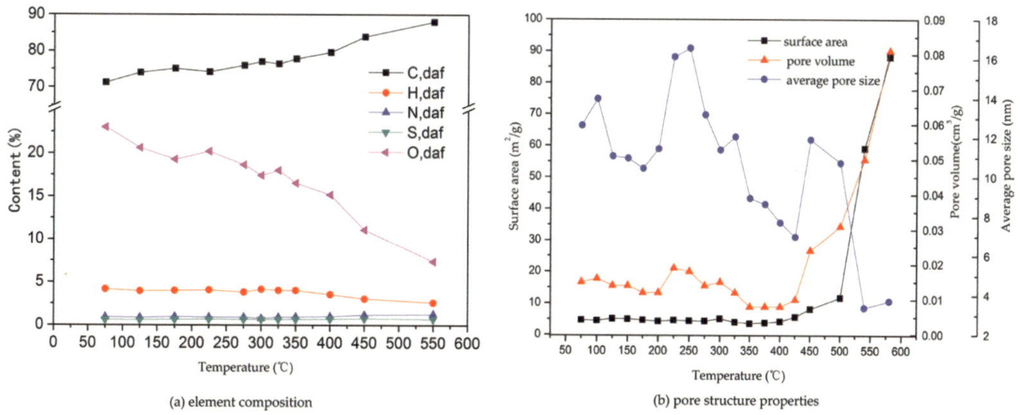

Figure 2. The evolution of element composition and pore structure properties during pyrolytic solid residue. (a) Element composition; (b) pore structure properties.

ring > hydroxyl > carbonyl > carboxyl > methoxyl. The thermal decomposition of oxygen-containing functional groups generates oxygen-containing gas, such as H_2O, CO, CO_2 and COS. And with the condensation process, H_2O will be liquefied accompany with tars. Both pyrolysis and devolatilization will strengthen the pore structures in pyrolytic solid residues. As shown in **Figure 2b**, the surface area and pore volume increase obviously after 400°C. Higher temperature will further facilitate dehydrogenation and condensation of aromatic cores. But as to coal with caking property, coal particles will soften and melt forming colloids together with gas-liquid-solid three-phase during the pyrolysis process. With devolatilization, colloids are swelling and finally solidifying as char or coke [39].

In general, pyrolysis of LRC could be described as a process of thermostablization accompany with releasing thermos-unstable substances. The pyrolytic products and their source are summarized in **Table 3**.

2.3. The influence factors during LRC pyrolysis

As mentioned above, LRC pyrolysis and its formation of pyrolytic products are mainly concerned with chemical reactions and volatile transfer process. Although the factors that influence LRC pyrolysis are various and complex, these factors, besides the composition of raw coal itself, ultimately influence the chemical reactions and volatile transfer process. And here, some important factors and their effects on LRC pyrolysis are discussed as follows.

2.3.1. Temperature

As mentioned above, thermal cleavage of covalent bonds in LRC is a single step and mainly determined by the temperature. However, the reactions of the volatiles involve multiple steps and can be influenced by many factors especially for the gas phase temperature, which is generally higher than that of the coal [40]. Generally, without effective cracking of organics at temperature higher than 400°C, desirable product yield (especially for liquid products) could not be achieved. However, higher temperature can not only promote the depolymerization of LRC macromolecules but also the secondary reactions of volatiles. Due to the competition of primary and secondary reactions on temperature, there exists a maximum yield of liquid product at a certain temperature (usually 450–550°C) according to the results of most experiments. But if the condition of volatile transfer is improved, higher yield of liquid products is also able to be achieved. For example, Zhang et al. [41] proposed and tested an indirectly

Products	Source	Process
Tar	Dissociative organics, side chains, organic units connecting by weak bonds	Distillation + pyrolysis
CO_2	Carboxyl and carbonates	Decomposition of carboxyl and carbonates minerals
CO (<500°C)	Carbonyl, ether	Decarbonylation
CO (>500°C)	Oxygen heterocyclic rings	Ring-opening reactions
H_2O	Moisture, hydroxyl, crystal water of minerals	Evaporation, Dehydroxylation, decomposition of minerals
CH_4, C_nH_m (light hydrocarbons)	Aliphatic or aliphatic side chain	Dealkylation and thermocracking
H_2	Hydroaromatics	Dehydrogenation and condensation
Char or coke	Solid residue and PAHs	Pyrolytic remains and condensation of PAHs

Table 3. The products and their major source during LRC pyrolysis.

heated fixed-bed reactor with internals to regulate the pyrolysis gas flow direction inside the reactor. As shown in **Figure 3**, the central gas collection pipe leads to the different flow patterns of the pyrolytic volatile inside the reactor, which does not only avoid deep secondary cracking from the annular layer, but also condense and capture the heavy species in volatiles by relative low-temperature core layer. As a result, the tar yield still increases from 600 to 1000°C in reactor B, while the tar yield decreases with the same temperature region in reactor A. At the tested highest temperature of 1000°C, the tar yield in reactor B was 2.23 times of that in reactor A.

2.3.2. Minerals in coal

Minerals in LRC are various. Some minerals could decompose or work catalytically, and some are inactive during pyrolysis. Many studies have been done by selective demineralization of coal with acids to investigate the effect of individual elements in coal on pyrolysis yield and distribution of products. The brown coal in the Latrobe Valley, Victoria, Australia, has a very low ash yield and contains highly dispersed alkali and alkaline earth metallic (AAEM) species, either as carboxylates forming part of its organic matter or as NaCl dissolved in its moisture [42]. Li [43] concluded that the AAEM cations in the coal substrate were able to cause decrease in the yields of tar and/or large aromatic rings. The ion-exchangeable AAEM species in coal also seem to change the aromatic/aliphatic composition of the resulting tars. Liu et al. [44] investigated the effect of mineral matter in coal on reactivity and kinetic characteristics of LRC pyrolysis, which found that inherent mineral had no evident effect on the reactivity and kinetics of pyrolysis. But CaO, K_2CO_3 and Al_2O_3 could work catalytically during pyrolysis of LRC, and their effects were closely related to temperature region and coal types. Zeng et al. [45] studied the influence of minerals on pyrolysis yield and content of phenols. Under the same operating condition, acid-washed (demineralization) coal has higher tar yields, but lower yields of char and gas at 500°C. It was inferred that the minerals in coal had catalytic effects on the generation of CO and ethylene, and the decomposition of tar. Meanwhile, the decomposition of intermediates or the producing of final phenols during coal pyrolysis may be suppressed. However, due to the differences in composition and properties for different

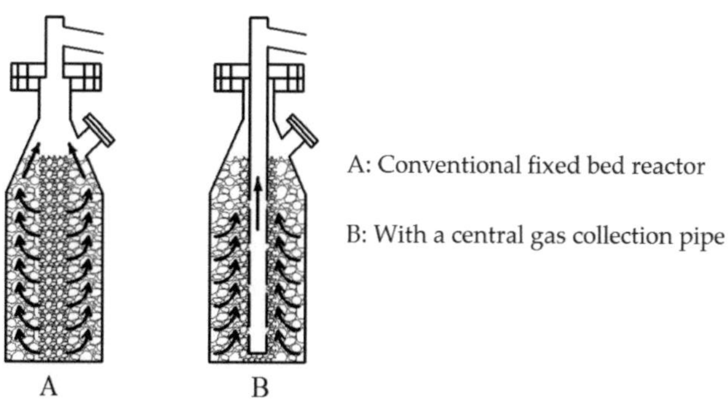

Figure 3. Schematic of fixed reactors with and without a central gas collection pipe. (A) Conventional fixed bed reactor; (B) with a central gas collection pipe.

coals, there were discrepancies in the results obtained by different researchers for different coal samples using different methods.

2.3.3. *Heating rate and particle size of coal*

Heat transfer is an important factor which directly influences the extent of LRC's pyrolysis reactions. Except microwave-assisted heating, the heat is transferring from the outside to the inside of the particle. It is known that, if the temperature around a pyrolyzing coal particle exceeds a threshold value (for thermal cracking reactions), external heat transfer controls the extent of pyrolysis [46]. Nevertheless, heating rate and particle size of coal can greatly influence the intensity and efficiency of heat transfer.

Heating rate mainly influences the course of temperature and time. Generally, the difference of heating rate in half order of magnitude could have tiny differences on pyrolysis yields and distribution of pyrolytic products. Thus, the comparisons of effects brought by heating rates are usually between slow, medium-speed, fast and flash pyrolysis of LRC. Higher heating rate means the coal particles are to be heated to the temperature in shorter time by the higher intensity of heat. Then the organic structures in LRC are damaged more quickly and generate mounts of volatiles in particles. Observed from the differential thermogravimetric (DTG) curves, the weight loss region during pyrolysis of LRC is shifting to the high-temperature region with increasing heating rate, and the products' evolutions are overlapped [47]. Some believe that higher heating rate increases the pressure of volatiles in particles, which provides stronger driving force for devolatilization from particles to outer space and shorten the residence time during heat transfer of fragments in particles. Therefore, the secondary reaction could be weakened and finally obtain more liquid and less gas products. But some believe that higher heating rate does not only mean higher temperature of coal particle but also the circumstances of devolatilization (in char and in outer space). Due to the opposite direction of heat and mass transfer, volatiles are heated again, which benefits for their secondary reactions [40]. The both views do make sense, while the key point is the residence time of fragments in high-temperature region. However, without a suitable residence time for secondary cracking, some active and high mass weight fragments will bring in tars which will probably lead further condensations and sacrifice the quality of liquid products [48].

Heating rate brings the intensity of heating, while particle size determines the effect of heating. In fact, the heat conductivity of LRC is poor. The particle size can greatly influence the yield and properties of pyrolytic products. The rate at which volatile matter is produced during pyrolysis is demonstrated to be controlled not at all by mass transfer, but by heat conduction to a moving reaction front inside a coal particle, provided its diameter exceeds ~3 mm [49]. When the particle size increases, it will not only increase the time of heat transfer to the centre of the particles but also bring high tendency of secondary reactions due to the temperature difference from the surface to the centre of particles. Thus, small particles are preferable for fast heating, but the deposit of small particles in the reaction bed would possibly increase the pressure drop and hinder the devolatilization. A proper reactor is needed to be considered for pyrolysis of small particle size of LRC, while as to production of lump char, slow rate of heating is desirable of pyrolysis of lump LRC.

2.3.4. Atmosphere

Generally, pyrolysis of LRC is in an inert atmosphere, such as N_2, He and Ar. But in order to increase the yields or improve the quality of pyrolytic products, lots of studies have been done on pyrolysis in reducing atmosphere (active atmosphere), such as H_2, CO and CH_4. Liu et al. [50] investigated the influence of atmosphere on pyrolysis of a sub-bituminous coal under N_2, H_2, CH_4 and H_2—CO (2:1) atmosphere. It was found that the tar yield at different atmosphere was increasing as $H_2 > CH_4 > H_2$—CO > N_2, and more saturates were generated in active atmosphere than in N_2. Wang et al. [51] found that higher tar yield was achieved in CO_2 than H_2, CH_4 or N_2. And tars from CH_4 atmosphere had relatively lower phenol content but higher cresol and C_2 alkyl-substituted phenols compared with those from other atmosphere. Meanwhile, reactive atmosphere is beneficial for heteroatom removal during pyrolysis. It was found that CO promoted the pyritic sulfur to form as COS; CH_4 and CO_2 inhibited the evolution of sulfur-containing gases at temperatures below 600°C, but, respectively, promoted the formation of COS and H_2S at higher temperature; and H_2 was more likely to promote the sulfur transferring to the gaseous product, especially as H_2S [52]. As to the mechanism of effects brought by active gas, many agree that active gas produces new radicals to stabilize the volatile fragments from condensations, which increases the yield and qualities of liquid products [53]. And improving the partial pressure or with the help of catalyst can facilitate the reactivity and efficiency of reactive gas. In order to improve the activity of CH_4, Hu et al. [54–57] propose a series of studies on integrated process of coal pyrolysis with CO_2/steam reforming of methane, which improves greatly of liquid yields compared with pyrolysis simply in reactive atmosphere.

2.4. Further research and interest on LRC pyrolysis

In general, the process of product formation during LRC pyrolysis is complex. It includes both chemical reactions and heat/mass transfer process. Although thousands of studies have been done on pyrolysis, there are still many problems need to be further considered:

- Pyrolysis reactions: Due to the influence of secondary reactions and complexity of coal macromolecules, it is also difficult for observing the primary reactions during pyrolysis. Recent strategies for LRC pyrolysis mechanism and process are to study the relatively simple coal-based compounds or use in situ analysis methods.
- Dynamics, heat/mass transfer: As discussed above, it is a comprehensive process with lots of factors for the final yields and properties of pyrolytic productions. However, pyrolysis dynamics of LRC is developed generally according to the thermogravimetric results. It only represents the dynamics of devolatilization, but some reactions remain in the solid (such as formation of char). It may be not beneficial for the design of pyrolysis reactor. Thus, more accurate dynamic models need to be established [36]. The heat and mass transfer process also dramatically influence the pyrolysis process. Better understanding of the two processes will also greatly help for better designing the reactors or their internals.
- Catalyst: Pyrolysis of LRC is a direct liquefaction process. Volatiles generated from pyrolysis are firstly cooled to obtain liquid products. But the liquid products are complex, high moisture and relative poor quality. Traditionally, the liquid products need to be heated for

further conversion. If the volatiles could be in situ conversed during pyrolysis, it would be great efficient for the whole process. Thus, proper catalysts are the key points for in situ catalytic conversions of volatiles. Together with active atmosphere, integrated process of coal pyrolysis with catalytic conversion could be an important method for utilization of LRC.

3. Pyrolysis and its polygeneration technology

In order to meet the needs of continuous operation for industrial applications, various technologies on pyrolysis of LRC are proposed and tested. According to the motion state of coal particles, pyrolysis reactor can be designed as moving bed, fluidized bed or entrained bed. As to heating method, there are internal heating (coal contact directly with the heating source), external heating or hybrid heating. And for heat carrier, it may be solid, gas or the both.

Since nineteenth century, pyrolysis of LRC has been industrially applied for fuels or chemicals production. During 1910–1950s, the low-temperature pyrolysis of LRC was quickly and well developed in Germany, typically producing liquid fuels by lignite pyrolysis in three-stage Lurgi. Whereafter, with the development of petroleum chemical industry, large-scale industrial production temporarily was interrupted. But due to strategic needs, a series of fundamental studies and industrial tests have been carried out. Typical technologies of LRC pyrolysis developed during this period are COED (multistage fluidized bed, USA), Toscoal (horizontal rotary kiln, ceramic ball as heat carrier, USA), Garret (entrained flow bed, char as heat carrier, USA), L-R (moving bed, char as heat carrier, Germany), CSIRO (double fluidized bed, char as heat carrier, Australia), DG (moving bed, char as heat carrier, China), etc. From 1990s, the increasing price of crude promoted a new round of researches and development of LRC pyrolysis around the world. But large scale of industrial tests were mainly reported in China, such as continuous vertical retorts (SH series, SJ series and ZNZL series), rotary kiln technology (by China Coal Research Institute), Guofu retorts (by Beijing Guodianfutong Science and Development Co., Ltd. China), BT technology (by Institute of Process Engineering, Chinese Academy of Science, PR China), etc. Other typical technologies are LFC (Bluegrass Coal Development Company, USA) and CCI (Covert Coal Company, USA).

Recently, with the increasing concern on environment and efficiency, cleaner and more efficient technologies of LRC pyrolysis are required. Due to the best use of chemical resources in LRC, polygeneration system based on pyrolysis is a good way to meet the requirements. According to properties of the raw coal and market demands, lots of conceptions have been carried out [58, 59]. The major concept is for clean fuels and chemicals or materials of high added value. However, minority of them is applied or industrial tested. Here, three application examples of LRC-pyrolysis-based polygeneration technologies are briefly introduced.

3.1. Heat-electricity-gas-tar polygeneration system

Heat-electricity-gas-tar polygeneration system is designed by Zhejiang University, China. As shown in **Figure 4**, LRC is heated to around 600°C by hot ash in a circulating fluidized bed

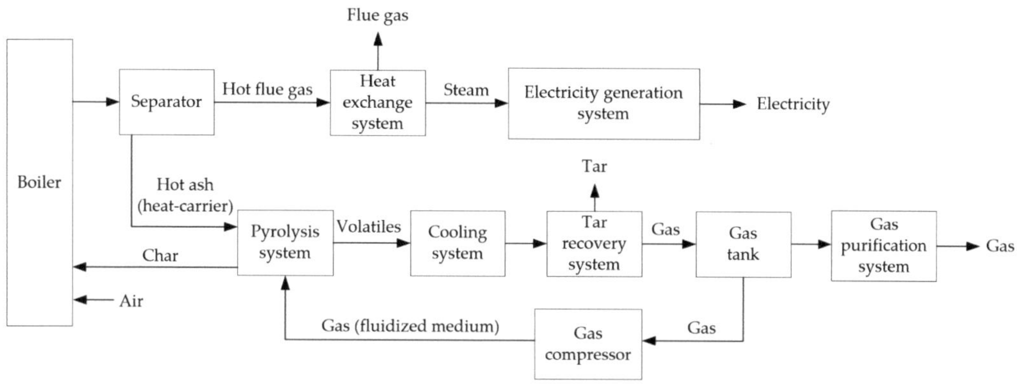

Figure 4. Circulating fluidized bed polygeneration system.

reactor. The produced volatiles sequentially go through cooling system and tar recovery system to obtain tar and gas. A part of gas is conveyed by gas compressor to the pyrolysis system as fluidized medium. And the rest part of gas is used for methanol synthesis after purification by purification system. The produced char is moved into the boiler and burns. And the high-temperature ash and hot flue gas are separated in separator. A part of hot ash goes into pyrolysis system as solid heat carrier. And the energy of hot flue gas is recovered by producing steam in heat exchange system. The produced steam is used for electricity generation. Recently, the system has developed 12 and 25 MW circulating fluidized bed polygeneration system and constructed a 75 t/h testing apparatus [60].

3.2. Clean solid-liquid fuels polygeneration system

Based on DG process, Dalian University of Technology (China), Shenmu Fuyou Energy Technology Co., Ltd. (China) and China National Chemical Engineering Co., Ltd. design and construct a 60×10^4 t/a LRC pyrolysis and12 $\times 10^4$ t/a medium/low-temperature tar processing polygeneration system [61]. As shown in **Figure 5**, the raw coal (sub-bituminous coal, particle size <6 mm) is firstly dried by the hot flue gas and mixed with hot char feeded into the moving bed pyrolysis reactor (around 500°C). A part of produced char is moved into char heating and lifting system and partial combusted to be heat carrier(750–800°C). And the hot flue gas is cooled to around 500°C for the desiccation of raw coal. Another part of char is going to shaping system to generate char balls, which are clean solid fuels for civil use. The produced volatiles sequentially go through purification system and cooling and recovery system to obtain tar, gas and water. Gas fraction is used to generate H_2 by pressure swing adsorption (PSA) process. And H_2 is utilized for hydrogenation and refining of tar by a total fraction hydrogenation technology. Finally, clean liquid fuels can be generated.

3.3. Coal to calcium carbide/acetylene polygeneration system

Coal to acetylene polygeneration system is proposed by Shenwu Environmental Technology Co., Ltd. (China). Traditional calcium carbide production is using char or coke as carbon

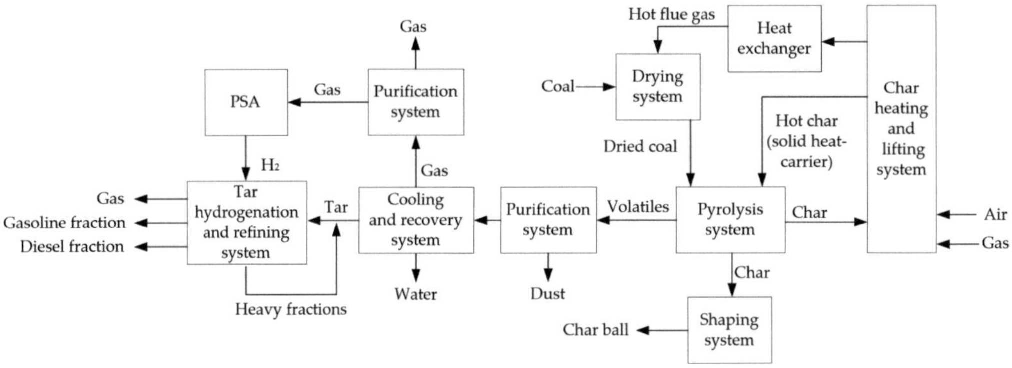

Figure 5. Clean solid-liquid fuels polygeneration system.

source [62]. In this system (**Figure 6**), lower price of LRC is firstly mixed with quick lime and proportionally shaped as calcium-coal ball. Then, calcium-coal balls are conveyed into a rotary hearth furnace for pyrolysis. The generated volatiles are going to tar and gas recovery system to obtain tar and gas. The produced calcium-char balls are conveyed in thermal state to calcium carbide production system. And the exhaust produced by calcium carbide furnace is used for calcining the limestone after purification. Moreover, the calcium carbide can be generated acetylene with water. Recently, based on the system, an 80 × 10^4 t/a polyethylene project is under construction in Baotou, China. Besides polyethylene, liquefied natural gas, tar and polybutylene are also by-produced. But due to the high content of CO in exhaust produced by calcium carbide furnace, it can be used for glycol synthesis with H_2 from gas produced by pyrolysis in the future.

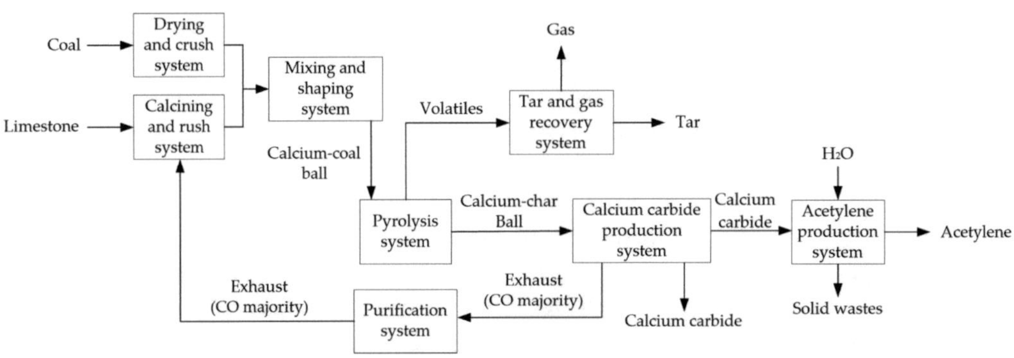

Figure 6. Coal to calcium carbide/acetylene polygeneration system.

Acknowledgements

Supports from National Energy Administration of China (NY 20130302513-1) are greatly acknowledged.

Author details

Fan Nie[1], Tao Meng[2] and Qiumin Zhang[1,3]*

*Address all correspondence to: zhangqm@dlut.edu.cn

1 School of Chemical Engineering, Dalian University of Technology, Dalian, China

2 Beijing Guodian Longyuan Environmental Engineering Co., Ltd., Haidian District, Beijing, China

3 State Key Laboratory of Fine Chemicals, Dalian University of Technology, Dalian, P. R. China

References

[1] World Coal Association. The coal resources: a comprehensive overview of coal. London, world coal. org/resources/wca-publications, 2009.

[2] Zhang D., Yani S. Sulphur transformation during pyrolysis of an Australian lignite. Proceedings of the Combustion Institute. 2011;**33**(2):1747-1753. doi:10.1016/j.proci.2010.07.074

[3] Chen L., Bhattacharya S. Sulfur emission from Victorian brown coal under pyrolysis, oxy-fuel combustion and gasification conditions. Environmental Science &Technology, 2013:**47**(3):1729-1734.

[4] Wang N., Yu J., Tahmasebi A., et al. Experimental study on microwave pyrolysis of an Indonesian low-rank coal. Energy & Fuels. 2013;**28**(1):254-263. doi:10.1021/ef401424p

[5] Zhong S., Baitalow F., Nikrityuk P., et al. The effect of particle size on the strength parameters of German brown coal and its chars. Fuel. 2014;**125**:200-205. doi:10.1016/j.fuel.2014.02.022

[6] Kwan J T., Yen T F. Aromaticity determination of coal, oil shale, and their derivatives by x-ray diffraction. American Chemical Society (ACS) Division of Energy and Fuels (United States), 1976;**21**(7).

[7] Manoj B., Kunjomana A G. Study of stacking structure of amorphous carbon by X-ray diffraction technique. International Journal of Electrochemical Science, 2012;7(4):3127-3134.

[8] Lv J H., Wei X Y., Yu Q., et al. Insight into the structural features of macromolecular aromatic species in Huolinguole lignite through ruthenium ion-catalyzed oxidation. Fuel, 2014;**128**:231-239.

[9] Meyers R., (ed).Coal Structure. Elsevier, U.K.,1982.

[10] Ogunsola O I. Distribution of oxygen-containing functional groups in some Nigerian coals. Fuel, 1992;**71**(7):775-777. doi:10.1016/0016-2361(92)90128-B

[11] Angle C., Berkowitz N. Distribution of low rank coals oxygen forms in Alberta. Fuel, 1991;**70**(7):891-896.

[12] Murata S., Hosokawa M., Kidena K., et al. Analysis of oxygen-functional groups in brown coals. Fuel Processing Technology, 2000;**67**(3):231-243.

[13] BP plc. B.P. Statistical review of world energy. U.K.; 2016.

[14] Lee S., et al. Efficient use of low rank coal: current status of low rank coal utilization. In: Zhao B., editor. Cleaner Combustion and Sustainable World. Berlin: Springer; 2013. pp. 893-895. doi:10.1007/978-3-642-30445-3_120

[15] Mills S. Global Perspective on the Use of Low Quality Coals. IEA Clean Coal Centre. U.K.; 2011.

[16] Gao J. Coal Pyrolysis Coking and Coal Tar Process. 1st ed. Beijing: Chemical Industry Press; 2010.

[17] Nicholson R., Littlewood K. Plasma pyrolysis of coal. Nature, 1972;**236**(5347):397-400.

[18] Hodek W., Kirschstein J., Heek K H V. Reactions of oxygen containing structures in coal pyrolysis. Fuel, 1991;**70**(3):424-428.

[19] Ma B., Zhang Q. The Utilization of Char. Beijing: Metallurgical Industry Press; 2014. 378 p.

[20] Zhang H., Shi H., Chen J., et al. Elemental mercury removal from syngas at high-temperature using activated char pyrolyzed from biomass and lignite. Korean Journal of Chemical Engineering, 2016;**33**(11):3134-3140.

[21] Feng Y., Dou J., Tahmasebi A., et al. Regeneration of Fe–Zn–Cu sorbents supported on activated lignite char for the desulfurization of coke oven gas. Energy & Fuels, 2015;**29**(11):7124-7134.

[22] Petrakis L., Grandy D W. Free Radical in Coal and Synthetic Fuels. Amsterdam: Elsevier; 1983:5-24.

[23] Miura K. Mild conversion of coal for producing valuable chemicals. Fuel Process Technology, 2000;**62**:119-135.

[24] He X. Coal Chemistry. 2nd ed. Beijing: Metallurgical Industry Press; 2010. 287 p.

[25] Cui T., Zhou Z., Dai Z., et al. Primary fragmentation characteristics of coal particles during rapid pyrolysis. Energy & Fuels. 2015;**29**(10):6231-6241. doi:10.1021/acs.energyfuels.5b01289

[26] Cui T., Fan W., Dai Z., et al. Variation of the coal chemical structure and determination of the char molecular size at the early stage of rapid pyrolysis. Applied Energy. 2016;**179**:650-659. doi:10.1016/j.apenergy.2016.06.143

[27] Hatcher P G. Chemical structural models for coalified wood (vitrinite) in low rank coal. Organic Geochemistry, 1990;**16**:959-68.

[28] Saxena S C. Devolatilization and combustion characteristics of coal particles. Progress in Energy and Combustion Science. 1990;**16**(2):55-94. doi:10.1016/0360-1285(90)90025-X

[29] Guan J. Characteristics of Lignite Thermal Upgrading Process and Water Quality [thesis]. Dalian: Dalian University of Technology; 2013.

[30] Zhan J., Wu R., Liu X., et al. Preliminary understanding of initial reaction process for subbituminous coal pyrolysis with molecular dynamics simulation. Fuel, 2014;**134**:283-292. doi:10.1016/j.fuel.2014.06.005

[31] Castro-Marcano F., Russo M F., van Duin A C T., et al. Pyrolysis of a large-scale molecular model for Illinois no. 6 coal using the ReaxFF reactive force field. Journal of Analytical and Applied Pyrolysis. 2014;**109**:79-89. doi:10.1016/j.jaap.2014.07.011

[32] Castro-Marcano F., Lobodinb V., Rodgers R., McKenna A., Marshall A., Mathews J. A molecular model for Illinois no. 6 Argonne Premium coal: Moving toward capturing the continuum structure. Fuel, 2012;**95**:35-49 doi:10.1016/j.fuel.2011.12.026

[33] Li G., Li L., Tang Z., Hu H. et al. Experimental and theoretical study on the pyrolysis mechanism of three coal-based model compounds. Energy &Fuel, 2014;**28**(2):980-986. doi:10.1021/ef402273t

[34] Tromp P J J. Coal pyrolysis [thesis]. Amsterdam: University of Amsterdam; 1987.

[35] Wu A., Pan T., Shi X. et al. Study on free radicals in low rank coal pyrolysis process. Coal Conversion. 2012;**25**(2):1-5.

[36] Liu Z. Advancement in coal chemistry: structure andreactivity. ScientiaSinicaChimica. 2014;**44**(9):1431-1438. doi:10.1360/N032014-00159

[37] He W., Liu Z., Liu Q., et al. Behaviors of radical fragments in tar generated from pyrolysis of 4 coals. Fuel, 2014;**134**:375-380. doi:10.1016/j.fuel.2014.05.064

[38] Wu M. Effect on the Re-absorption Ability of Low-rank Coal under Air Atmosphere after Heat Treatment [thesis]. Dalian: Dalian University of Technology; 2015.

[39] Yu J., Lucas J A., Wall T F. Formation of the structure of chars during devolatilization of pulverized coal and its thermoproperties: A review. Progress in Energy and Combustion Science, 2007;**33**(2):135-170. doi:10.1016/j.pecs.2006.07.003

[40] Liu Z., Guo X., Shi L., He W., Wu J., Liu Q., Liu J. Reaction of volatiles – A crucial step in pyrolysis of coals. Fuel, 2015;**154**:361-369. doi:10.1016/j.fuel.2015.04.006

[41] Zhang C., Wu R., Xu G. Coal pyrolysis for high-quality tar in a fixed-bed pyrolyzer enhanced with internals. Energy Fuels, 2014;**28**(1):236-244. doi:10.1021/ef401546n

[42] Brockway D J., Higgins R S. Brown coal sampling, analysis and composition. In: Durie R A., editor. The Science of Victorian Brown Coal. Oxford: Elsevier; 1991. pp. 247-278.

[43] Li C. Some recent advances in the understanding of the pyrolysis and gasification behaviour of Victorian brown coal. Fuel, 2007;**86**:1664-1683. doi:10.1016/j.fuel.2007.01.008

[44] Liu Q., Hu H., Zhou Q., Zhu S., Chen G. Effect of inorganic matter on reactivity and kinetics of coal pyrolysis. Fuel, 2004;**83**(4):713-718. doi:10.1016/j.fuel.2003.08.017

[45] Zeng F., He D., Guan J., Zhang Q. Effect of minerals in coal on the product yield of coal pyrolysis and phenols in the low-temperature tar. Advance Materials Research, 2012;**512**:1129-1136.

[46] Chern J., Hayhurst A. A simple theoretical analysis of the pyrolysis of an isothermal particle of coal. Combustion and Flame, 2010;**157**(5):925-933. doi:10.1016/j.combustflame.2009.12.003

[47] Tian B., Qiao Y., Tian Y., Liu Q. Investigation on the effect of particle size and heating rate on pyrolysis characteristics of a bituminous coal by TG–FTIR. Journal of Analytical and Applied Pyrolysis, 2016;**121**:376-386. doi:10.1016/j.jaap.2016.08.020

[48] Liu Z. Origin of common problems in fast coal pyrolysis technologies for tar: the countercurrent flow of heat and volatiles. CIESC Journal. 2016;**67**(1):1-5. doi:10.11949/j.issn.0438-1157.20151718

[49] Chern J., Hayhurst A. A model for the devolatilization of a coal particle sufficiently large to be controlled by heat transfer. Combustion and Flame, 2006;**146**(3):553-571. doi:10.1016/j.combustflame.2006.04.011

[50] Liu Y., He X., Yang F., Zhang Y., Ren X., Zhou A. Impacts of pyrolysis temperature and atmosphere on product distribution of Shenfu coal pyrolysis. Journal of China Coal Society, 2015;**40**(S2):497-504. doi:10. 13225/j.cnki.Jccs. 2015. 1005

[51] Wang P., Jin L., Liu J., Zhu S., Hu H. Analysis of coal tar derived from pyrolysis at different atmospheres. Fuel, 2013;**104**:14-21. doi:10.1016/j.fuel.2010.06.041

[52] Zhou Q., Hu H., Liu Q., Zhu S., Zhao R. Effect of atmosphere on evolution of sulfur-containing gases during coal pyrolysis. Energy Fuels, 2005;**19**(3):892-897. doi:10.1021/ef049773p

[53] Shi X., Wang Q., Ma W., Wu D. Study on reactivity of pyrolysis of Pingshuo coal in different reactive gas. Coal Conversion, 2014;**37**(3):5-9.

[54] Dong C., Jin L., Zhou Y., Zou L., Hu H. Integrated process of coal pyrolysis with steam reforming of improving the tar yield. Energy Fuels, 2014;**28**(12):7377-7384. doi:10.1021/ef501796a

[55] He X., Hu H., Jin L., Wei H. Integrated process of coal pyrolysis and CO_2 reforming of methane with and without using dielectric barrier discharge plasma. Energy Sources, Part A: Recovery, Utilization, and Environmental Effects, 2016;**38**(5):613-620. doi:10.1080/15567036.2011.650277

[56] Liu J., Hu H., Jin L., Wang P., Zhu S. Integrated coal pyrolysis with CO_2 reforming of methane over Ni/MgO catalyst for improving tar yield. Fuel Processing Technology, 2010;**91**(4):419-423. doi:10.1016/j.fuproc.2009.05.003

[57] Wang P., Jin L., Liu J., Zhu S., Hu H. Isotope analysis for understanding the tar formation in the integrated process of coal pyrolysis with CO_2 reforming of methane. Energy Fuels, 2010;**24**:4402-4407. doi:10.1021/ef100637k

[58] Guan J., He D., Zhang Q. The technology of improving lignite quality through pyrolysis and the concept of poly-generation. Coal Chemical Industry, 2011;**6**:1-4.

[59] Zhou X. Poly-generation technology of low-rank coal. Clean Coal Technology, 2013;**19**(6):47-51.

[60] Fang M., Cen J., Shi Z., Wang Q., Luo Z. Experimental study on 75 t/h circulating fluidized bed poly-generation system. Proceedings of the CSEE. 2010;**30**(29):9-15.

[61] Shang J., Ma B., Zhang Q., Shen H. Poly-generation technologies for grading conversion of low-rank coal. Beijing: China Coal Industry Publishing House; 2013. 335 p.

[62] Shenwu Environmental Technology Co., LTD. Available from: http://www.swet.net.cn/index.html [Accessed: 2017.1.5].